The Biology of the Avian Respiratory System

John N. Maina
Editor

The Biology of the Avian Respiratory System

Evolution, Development, Structure and Function

Springer

Editor
John N. Maina
Department of Zoology
University of Johannesburg
Johannesburg, South Africa

ISBN 978-3-319-44152-8 ISBN 978-3-319-44153-5 (eBook)
DOI 10.1007/978-3-319-44153-5

Library of Congress Control Number: 2017940480

Printed on acid-free paper

This Springer imprint is published by Springer Nature
The registered company is Springer International Publishing AG
The registered company address is: Gewerbestrasse 11, 6330 Cham, Switzerland

This book is dedicated to my family, friends, and past and present students.

Preface

Among the extant terrestrial (air-breathing) vertebrates, comprising ~10,000 extant species (http://www.birds.cornell.edu/clementschecklist/updates-corrections-august-2014/) that occupy practically every habitat on earth and pursuing various lifestyles, birds are the most successful taxon. Their evolution from reptiles, accomplishment of flight and attainment of exceptionally efficient respiratory system are the foremost factors that help explain the biology of modern avian taxa. The avian respiratory system, the lung-air sac system, is structurally the most complex (Maina 2005, 2015, 2016) and functionally the most efficient (Scheid 1980; Fedde 1998). In the recent past, its development, morphology and physiology have received intense interest. Particularly, application of modern techniques such as computational modelling, X-ray computer tomography and molecular biology have allowed more robust findings to be made and therefore its biology to be better understood. Plentiful information continues to accrue on various aspects of its biology. However, such accounts are widely scattered amongst different publications. They may not all be accessible to individual investigators. The last comprehensive reviews of the biology of the avian respiratory system were made about three decades ago by King and McLeland (1989) and previous to that by Seller (1987). In science, regular reviews of literature are of the essence. By delineating the controversial areas and gaps in knowledge, they help minimize isolation between investigators and therefore avert costly duplication of effort. The current book was probably overdue. It is hoped that it will stimulate further investigations on the biology of birds in general and the structure, function and development of the respiratory system in particular. The field of avian respiratory biology is too broad to be covered in a single book. Contingent on the availability of contributors, the topics were carefully selected to provide an adequately illustrative sample of field. The common theme in the reviews was to evaluate the earlier investigations and synthesize, critically assess and reconcile such data with the latest findings. The book presents cutting-edge understanding on evolution of birds and certain important aspects of their biology.

In Chap. 1, Min Wang and Zhonghe Zhou offer a synthesis on the evolution of birds, the most speciose clade of modern terrestrial vertebrates. They point out that the biological success of birds can be attributed to evolutionary innovations that comprise feathers, powered flight and a unique respiratory system. Data on the most recent discoveries that have been made on exceptionally well-preserved feathered non-avian dinosaurs and primitive birds are presented and the most compelling evidence that supports the hypothesis that birds are descendants of theropod dinosaurs adduced.

In Chap. 2, Dominique G. Homberger states that the lingual and laryngeal apparatus of birds serve as a gateway to the respiratory and alimentary systems in terrestrial vertebrates with the two organs being involved in alimentation and vocalization. Like the skull and jaws, their structures and functions differ profoundly between birds and mammals. The author underscores that in birds the movements of the lingual and laryngeal apparatus are interdependent and work together with the movements of the jaw apparatus in complicated and inadequately understood ways. Instead of reviewing the diversity of the lingual morphology of birds per se, the author focuses on the functional-morphological interdependences and interactions of the lingual and laryngeal apparatus with each other and with the skull and jaw apparatus. The following aspects are surveyed: the salient features of the mammalian head which sets it apart from the avian one; the general morphological features of the lingual and laryngeal apparatus within the context of the skull and jaw apparatus; the fundamental functional-morphological differences that occur among the jaw and the lingual and laryngeal apparatus of birds and the models that describe the movements of the various parts of the two apparatus based on biomechanical analyses; and the integration of the functional models with behaviours such as thermoregulation, respiration, feeding, drinking and vocalization. Detailed morphological and functional analyses can be tested and expanded by using 3D visualization and animation. The chapter places the available data in an evolutionary framework.

In Chap. 3, while highlighting the great diversity of pulmonary morphologies observed in different vertebrate lineages, C. G. Farmer reviews the structure of the lung as a gas exchanger and refutes the view that increase of aerobic capacities explains the diversity of the vertebrate respiratory organs. The author contends that the basis of the evolution of some of the most interesting changes of the vertebrate lung is unidentified. One of the evident changes in the evolution of vertebrate gas exchangers, especially those of birds and mammals, is that of a highly branched conducting airway system. Bird lungs have a particularly complex arrangement of airways through which air flows unidirectionally during the respiratory phases. In mammals, which also have an intricate but regularly branched airway system, air flow occurs tidally. It is emphasized that the discovery of unidirectional airflow in crocodilian and lizard lungs shows that several entrenched hypotheses for the selective driversity of this ventilatory mechanism cannot be correct and that neither endothermy nor activity determined the evolution of unidirectional air flow in the lung.

In Chap. 4, Graham R. Scott and Neal J. Dawson highlight that birds that fly at high altitudes have to maintain intense exercise in oxygen-thin environments. The attributes that allow high fliers to maintain the high rates of metabolism required for flight at elevation are delineated. The various characteristics of the O_2 transport pathway that differentiate birds from other vertebrates are outlined. They include increased pulmonary gas-exchange efficiency, maintenance of O_2 delivery and oxygenation in the brain during hypoxia, increased O_2 diffusion capacity in peripheral tissues and a high aerobic capacity. The contributors emphasize that these features are not high-altitude adaptation since they occur in lowland birds but are nonetheless important for hypoxia tolerance and exercise capacity. The distinguishing adaptations that have evolved for high-altitude flight comprise an increased hypoxic ventilatory response and/or an efficient breathing pattern, large lungs, haemoglobin with a high O_2 affinity, specializations which enhance the uptake, circulation and efficient utilization of O_2 during high-altitude hypoxia.

In Chap. 5, Rute S. Moura and Jorge Correia-Pinto underscore that the avian pulmonary system develops from a succession of complex events that comprise coordinated growth and differentiation of particular cellular components. After lung specification of the anterior foregut endoderm, branching morphogenesis ensues producing a complex system of airways. The process is determined by epithelial–mesenchymal interactions that are closely regulated by a network of conserved signaling pathways that drive cellular processes and control temporal-spatial expression of various molecular factors that are vital to the development of the lung. Furthermore, remodeling of the extracellular matrix provides the proper environment for the transfer of diffusible regulatory factors that control the cellular processes. The molecular mechanisms that direct the development of the avian lung are systematically reviewed. Fibroblast Growth Factor, Wnt, Sonic Hedgehog, Transforming Growth Factor β, Bone Morphogenetic Protein, Vascular Endothelial Growth Factor and regulatory mechanisms such as microRNAs regulate cell proliferation, cell differentiation and patterning of the embryonic chick lung.

The origin of the vertebrate lung from the endoderm is explained in Chap. 6 by Andrew N. Makanya. It occurs in the region of the primitive foregut where the epithelium gives rise to the airways and the respiratory units. The mesenchyme forms the connective tissue, muscles and blood vessels. The primary bronchus forms in the bird lung and gives rise to secondary bronchi that in turn give rise to parabronchi. The mechanisms that produce a thin blood–gas barrier correspond to those of exocrine secretion, but they occur in a programmed, time-limited manner. The required thickness is attained by cutting or decapitation of cells. By the processes of secarecytosis and peremerecytosis, a high columnar epithelium undergoes considerable size reduction and cellular polarization is lost. Vascularization occurs concomitantly with that of the development of the airways. Blood vessels form by the process of vasculogenesis and the network is outspread and remodelled by that of intussusceptive angiogenesis.

By comparing the avian and mammalian systems, in Chap. 7, John B. West explains the interesting properties of the avian pulmonary circulation, particularly those about the blood capillaries. The lungs are rigid and the respiratory system is separated into ventilatory and gas-exchanging (parenchymatous) parts: the air sacs form the former while the lungs form the latter. In the avian lung, ventilation is unidirectional, while in the compliant mammalian lung it is tidal. Compared to those of the mammalian lung, the pulmonary blood capillaries of the avian lung operate under a totally different setting. Unlike those in the mammalian lung that are spread along the interalveolar walls, the avian pulmonary blood capillaries interdigitate closely with the air capillaries. The mechanical support of the blood capillaries in the avian lung is therefore profoundly different. The blood–gas barrier of the avian lung is remarkably thin and even. Unlike in the mammalian lung where a type-I collagen chain is needed to reinforce the blood capillaries, in the avian lung where such a cable is lacking, a much thinner blood–gas barrier has developed. The avian pulmonary capillaries are also extraordinarily rigid compared to those in mammalian lungs which undergo recruitment and distension when blood pressure increases.

In Chap. 8, John N. Maina explains the structure and function of the avian respiratory system, the lung-air sac system. With the lung intercalated between a cranial and a caudal group of air sacs, the lung is ventilated continuously in craniocaudal direction by synchronized activity of the air sacs. Delicate, transparent and compliant, the air sacs are not directly involved in gas exchange. The avian lungs which are deeply attached to the ribs and the vertebrae on the dorsolateral aspects and to the horizontal septum on the ventral one are practically rigid. It has allowed the exchange tissue (parenchyma) to be intensely subdivided into very small terminal respiratory units, the air capillaries. Comprising of a three-tier system of airways that comprise a primary bronchus, secondary bronchi and tertiary bronchi (parabronchi), the airways form a continuous loop. The orthogonal orientation between the flow of air in the parabronchial lumen and that of venous blood in the exchange tissue produces a cross-current system; the relationship between the directions of the flow of air in the air capillaries and that of blood in the blood capillaries constitute an auxiliary countercurrent system; and the sequential interaction between the blood capillaries and the air capillaries in the exchange tissue of a parabronchus forms a multicapillary serial arterialization system. In addition to these features, large capillary blood volume, large surface respiratory surface area and exceptionally thin blood–gas (tissue) barrier explain the exceptional efficiency of the avian respiratory system in gas exchange.

Regarding the transition of animal life from water to land and subsequent diversification of life, the importance of evolution of a cleidoic egg in amniotes is highlighted in Chap. 9 by John N. Maina. In developing avian egg, the exchange of O_2 and CO_2 and loss of metabolically produced water vapour occurs entirely by passive diffusion across the eggshell and fertile incubated egg chorioallantoic membrane (CAM). The shell presents compromise functional engineering. It has to be thin to allow optimal exchange with the external environment but not too thin to be assailed by pathogens and injurious substances and prone to breaking. Furthermore, it must not be too strong (thick) for the chick not to break out at hatching. The thickness of the eggshell and the numbers, shapes and sizes of the pores determine the hatchability of the egg and probably the incubation period of the egg. While the number of pores may be fixed at the formation of the eggshell, the surface area and vascularization of the CAM increase during incubation. The functional design of the shell and that of the CAM are described.

In summary, this book explains the evolution of birds and certain physiological, developmental and structural aspects that make birds what they are.

Johannesburg, South Africa John N. Maina

References

Fedde MR. Relationship of structure and function of the avian respiratory system to disease susceptibility. Poult Sci. 1998;77:1130–1138.

King AS, McLelland J, editors. Form and function in birds, vol. 4. London: Academic Press; 1989.

Maina JN. The lung air sac system of birds: development, structure, and function. Heidelberg: Springer; 2005.

Maina JN. The design of the avian respiratory system: development, morphology and function. J Ornithol. 2015;156:41–63.

Maina JN. Pivotal debates and controversies on the structure and function of the avian respiratory system: setting the record straight. 2016. doi:10.1111/brv.12292

Scheid P. Avian respiratory system and gas exchange. In: Sutton JR, Coates G, Remmers JE, editors. Hypoxia: the adaptations. Burlington, ON: BC Decker Inc., 1990; p. 4–7.

Seller TJ, editor. Bird respiration, vols. I and II. Boca Raton, FL: CRC Press; 1987.

Acknowledgement

I thank all the contributors to this book for preparing excellent treatises.

Contents

Editor and Contributors

About the Editor

John N. Maina served as a professor of veterinary anatomy for 17 years at the University of Nairobi (Kenya) and of human anatomy for 13 years at the University of the Witwatersrand (South Africa). He currently works as a research professor at the University of Johannesburg. He has been investigating the structure of the avian respiratory system for nearly four decades and published 150 papers in peer-reviewed journals, 12 book chapters and five books on the comparative functional design of the gas exchangers. He has collaborated with others both nationally and internationally, has held appointments in various academic institutions, and has been invited to give lectures at different meetings/conferences and academic institutions. He has lectured in courses pertaining to the structure and function of the animal and the human bodies.

Contributors

J. Correia-Pinto Life and Health Sciences Research Institute (ICVS), School of Health Sciences, University of Minho, Braga, Portugal

ICVS/3B's – PT Government Associate Laboratory, Braga/Guimarães, Portugal

Biology Department, School of Sciences, University of Minho, Braga, Portugal

Department of Pediatric Surgery, Hospital de Braga, Braga, Portugal

Neal J. Dawson Department of Biology, McMaster University, Hamilton, ON, Canada

C.G. Farmer Department of Biology, University of Utah, Salt Lake City, UT, USA

Dominique G. Homberger Department of Biological Sciences, Louisiana State University, Baton Rouge, LA, USA

John N. Maina Department of Zoology, University of Johannesburg, Johannesburg, South Africa

Andrew N. Makanya Department of Veterinary Anatomy and Physiology, University of Nairobi, Nairobi, Kenya

Route S. Moura Life and Health Sciences Research Institute (ICVS), School of Health Sciences, University of Minho, Braga, Portugal

ICVS/3B's – PT Government Associate Laboratory, Braga/Guimarães, Portugal

Biology Department, School of Sciences, University of Minho, Braga, Portugal

Graham R. Scott Department of Biology, McMaster University, Hamilton, ON, Canada

Min Wang Key Laboratory of Vertebrate Evolution and Human Origins of Chinese Academy of Sciences, Institute of Vertebrate Paleontology and Paleoanthropology, Chinese Academy of Sciences, Beijing, China

John B. West Department of Medicine, University of California San Diego, La Jolla, CA, USA

Zhonghe Zhou Key Laboratory of Vertebrate Evolution and Human Origins of Chinese Academy of Sciences, Institute of Vertebrate Paleontology and Paleoanthropology, Chinese Academy of Sciences, Beijing, China

The Evolution of Birds with Implications from New Fossil Evidences 1

Min Wang and Zhonghe Zhou

Abstract

Birds have evolved on the planet for over 150 million years and become the most speciose clade of modern vertebrates. Their biological success has been ascribed to important evolutionary novelties including feathers, powered flight, and respiratory system, some of which have a deep evolutionary history even before the origin of birds. The last two decades have witnessed a wealth of exceptionally preserved feathered non-avian dinosaurs and primitive birds, which provide the most compelling evidence supporting the hypothesis that birds are descended from theropod dinosaurs. A handful of Mesozoic bird fossils have demonstrated how birds achieved their enormous biodiversity after diverging from their theropod relatives. On basis of recent fossil discoveries, we review how these new findings add to our understanding of the early avian evolution.

Keywords

Birds • Cretaceous • Enantiornithes • Flight • Fossil • Jehol Biota • Ornithuromorpha • Theropod dinosaurs

M. Wang (✉) • Z. Zhou
Key Laboratory of Vertebrate Evolution and Human Origins of Chinese Academy of Sciences, Institute of Vertebrate Paleontology and Paleoanthropology, Chinese Academy of Sciences, Beijing 100044, China
e-mail: wangmin@ivpp.ac.cn; zhouzhonghe@ivpp.ac.cn

Contents

1.1 Introduction

The origin of birds has been the subject of debates since the advent of evolutionary theory (Ostrom 1974; Martin 1983; Cracraft 1986; Feduccia 2001; Zhou 2004a; Xu et al. 2014).

J.N. Maina (ed.), *The Biology of the Avian Respiratory System*, DOI 10.1007/978-3-319-44153-5_1

Over the past two decades, numerous exquisitely preserved feathered non-avian dinosaurs and early birds from the Late Jurassic and Early Cretaceous deposits have significantly enriched our understanding about this issue (Xu et al. 2014; Brusatte et al. 2015). It is now widely accepted that birds are derived non-avian dinosaurs and deeply nested within the theropod phylogeny, and their closest relatives are small-sized, feathered dromaeosaurids and troodontids (Gauthier and Gall 2001; Norell and Makovicky 2004; Xu et al. 2011; Turner et al. 2012) (Fig. 1.1). As one

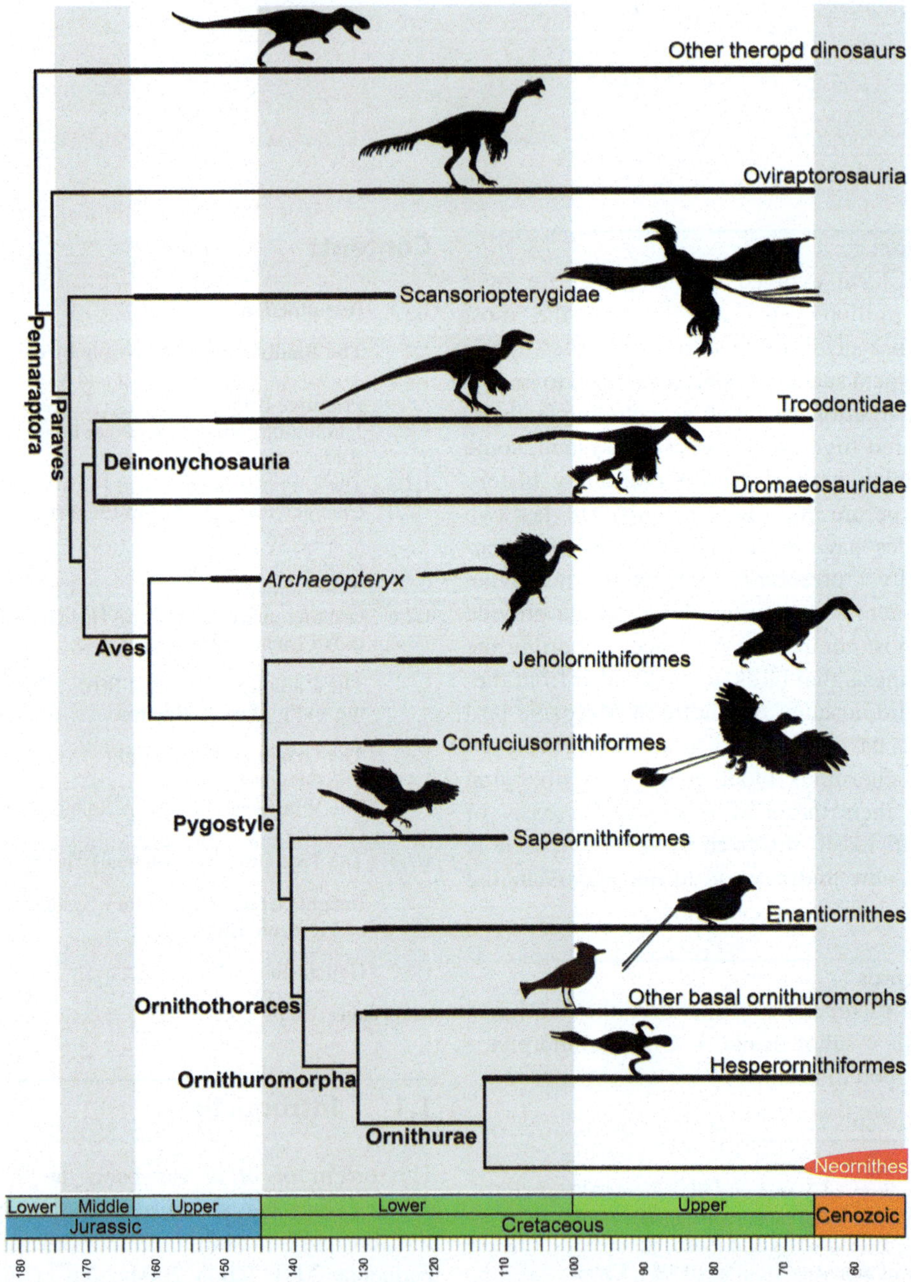

Fig. 1.1 Simplified time-scaled Mesozoic avian phylogeny. The cladogram is based on recent studies (Turner et al. 2012; Wang et al. 2015b, 2016b) and shows that birds are deeply nested within theropod phylogeny. The *thicker lines* represent the dating of upper and lower boundaries of terminal taxon-bearing deposits, modified from Wang and Lloyd (2016)

of the most diverse living vertebrates, birds encompass over 10,000 species and exhibit a huge range of morphological, physiological, and ecological diversity (Gill 2007). In short, birds have undergone two phases of large-scale radiations: diversification of the stem-group birds in the Cretaceous and the subsequent radiation of the crown-group birds in the Paleogene (Chiappe 1995; Novacek 1999; Feduccia 2003; Mayr 2009). Nearly all the stem-group birds perished at the end-Cretaceous mass extinction alongside their dinosaur relatives (Longrich et al. 2011); only a few derived clades persisted beyond that catastrophe and embraced an extremely rapid radiation in wake of Cretaceous, giving rise to modern birds (Mayr 2009; Prum et al. 2015). The fossil records of Mesozoic birds used to be rare, and thus it is interesting to know how the modern birds evolved from the reptile-like *Archaeopteryx*. Fortunately, thousands of complete and articulated bird skeletons have been unearthed recently from the Cretaceous terrestrial sediments, particularly the Jehol Biota of China, providing unprecedented information regarding early avian evolution. The Early Cretaceous Jehol Biota is the second oldest bird-bearing deposits (130.7–120 Ma; He et al. 2006), only surpassed by the Upper Jurassic Solnhofen limestones that preserved *Archaeopteryx* in Germany (Jin et al. 2008). To date, over 40 avian species have been recognized from the Jehol Biota, accounting more than half of the known global diversity of Mesozoic birds (Zhou and Wang 2010); the numerous specimens, some of which preserve feathers and soft tissues, have bridged the considerable large morphological and biological gaps between the "Urvogel" *Archaeopteryx* and modern birds (Zhang et al. 2006; Zhou and Zhang 2006a; Xu et al. 2010; Zheng et al. 2011, 2013; O'Connor and Zhou 2015; Wang et al. 2016a). We present a short review about the early evolution of birds with comments about the origin of flight. Hereafter, the term "bird" refers to the most inclusive clade containing *Passer domesticus* Linnaeus 1758 but not *Dromaeosaurus albertensis* Matthew and Brown 1922 or *Troodon formosus* Leidy 1956. We use "dinosaur" to refer simply as "non-avian dinosaur" and "theropod" for "theropod dinosaur." To avoid the controversy about the homologies of the manual digits of bird and other non-avian tetanurans ("I–II–III" versus "II–III–IV"; see review in Xu and Mackem 2013), we use the alular, major, and minor digits to refer the manual digits of birds, respectively.

1.2 The Biodiversity of Mesozoic Birds

1.2.1 *Archaeopteryx*: The Oldest and Most Primitive Bird

Ever since its discovery in 1861, *Archaeopteryx* has attracted scientific attention for its central role in regarding the origin of birds (Wellnhofer 2010). As the "missing link," *Archaeopteryx* possesses the mosaic combination of dinosaurian and avian characters, leading Ostrom to revitalize the hypothesis that birds evolved from dinosaurs first proposed by Huxley over a hundred years ago (Huxley 1868; Ostrom 1969). There are ten skeletons and one single feather of *Archaeopteryx*, all collected from the Upper Jurassic plattenkalk, the Solnhofen lithographic limestone of Germany (Elzanowski 2002; Mayr et al. 2005; Foth et al. 2014; Lee et al. 2014). *Archaeopteryx* is widely regarded as the most primitive bird (Zhou 2004a; Lee and Worthy 2011; Foth et al. 2014), although some studies alternatively suggested that it may be a primitive troodontid or dromaeosaurid (Xu et al. 2011). The taxonomy of the known specimens of *Archaeopteryx* is far from conclusive (Wellnhofer 2008). Despite all considered as members of the clade Archaeopterygidae, the known specimens have been regarded as different genera or species, or different ontogenetic series of the single taxon *Archaeopteryx lithographica* (Elzanowski 2002).

The anatomy of *Archaeopteryx* has been studied intensively, demonstrating that the first bird is more like a dinosaur rather than a bird in skeletal morphologies (Fig. 1.2a). The toothed skull shows primitive features in having many bony elements such as the postorbital and

Fig. 1.2 The long-tailed basal birds. (**a**) Thermopolis specimen of *Archaeopteryx* (courtesy of Gerald Mayr); (**b**) lower jaw of *Archaeopteryx* showing the interdental plates (*arrowheads*), the Munich specimen (courtesy of Oliver Rauhut); (**c**) *Jeholornis prima* with preservation of seeds in abdomen (*circle*); (**d**) lower jaw of *Jeholornis prima*; (**e**) forelimb and the lateral trabecula of the sternum of *Jeholornis prima* (the *arrowhead* indicates a foramen in the lateral trabecula that is likely perforated by the clavicular air sac)

squamosal (Elzanowski and Wellnhofer 1996), which otherwise have either been incorporated into the braincase or lost in living birds (Baumel and Witmer 1993). In contrast to modern birds, the antorbital fenestra is larger than the external naris. The maxilla is massive and forms the major part of the facial margin. As in many theropods, the maxilla bears a large dorsal process that houses the maxillary and premaxillary fenestrae, whereas the dorsal process is reduced, and both fenestrae are absent in more derived birds (Xu et al. 2011). As in most theropods and some Triassic archosaurs, there exist bony plates between the teeth, called interdental plates (Fig. 1.2b), a plesiomorphic character lost in other birds as well as crocodilians (Elzanowski and Wellnhofer 1996). Apart from aforementioned primitive features, the palatine of *Archaeopteryx* exhibits distinctive avian morphology. To be specifically, the palatine is triradiate and loses the jugal process, whereas the bone is tetraradiate in most theropods with the jugal process well developed (Elzanowski and Wellnhofer 1996). As in modern birds, the maxillary process of the palatine is shorter than the pterygoid wing, but the opposite is true in theropods (Elzanowski and Wellnhofer 1996). The choanal process is hook shaped and encloses the choana (Elzanowski 2002).

The postcranial skeleton lacks most features that characterize a bird. As in most theropods, the coossification of the postcranial skeletons is poorly developed. For example, the elements that constitute to the fusion of compound bones such as the carpometacarpus, tibiotarsus, and tarsometatarsus remain separated (Mayr et al. 2007), and a long bony tail is retained, including 21–23 elongated free caudal vertebrae (Zhou and Zhang 2003a). The shoulder girdle is minimally modified compared with other dinosaurs, but not to the degree adaptive to powered flight. For instance, as in some dromaeosaurids, the coracoid is subrectangular and lacks a procoracoid process. In modern birds, the procoracoid process and the shoulder ends of the furcula and scapula form the triosseal canal, through which the tendon of the M. supracoracoideus passes. Because of the triosseal canal, the M. supracoracoideus can elevate rather than depress the wing to complete the upstroke movement (Baumel and Witmer 1993). As the major anchor for the flight muscles in living birds, an ossified sternum is absent in *Archaeopteryx* (Zheng et al. 2014a). The deltopectoral crest of the humerus is small, which serves as the attachment for the flight muscles—*M. pectoralis* and cranial head of the *M. deltoideus* major (Baumel and Witmer 1993). The hand retains three elongated, clawed digits with a formula of 2-3-4, recalling the condition in most theropods (Zhang et al. 2002; Mayr et al. 2007). Neither the synsacrum nor the pelvic girdle is fused, and thus the pelvis lacks the rigidity of living birds. The pubis is ventrally rather than caudally directed. As in most theropods, the ischium bears an obturator process distally (Elzanowski 2002). A reversed hallux is lacking (Mayr et al. 2005). The body plan, although minimized, was generally inherited from their dinosaurian ancestors in terms that the forelimb is shorter than the hind limb and the ulna is shorter than the humerus, whereas the opposite is true in other early birds and most living volant birds. Despite the lack of skeletal modifications akin to flight, the wings of *Archaeopteryx* are remarkably similar to that of modern birds in the following aspects: the primary and secondary remiges are asymmetrical, pennaceous; the rachis of the flight feather is robust and curved; the dorsal coverts overlay the primary remiges and measure about half the length of the latter (Longrich et al. 2012; Foth et al. 2014), suggesting that it might be capable of some primitive flapping flight.

1.2.2 Jeholornithiformes: The Other Long-Tailed Birds

Jeholornithiformes are the only known avian clade that has a long bony tail, except for *Archaeopteryx*. Recent phylogenetic studies resolve Jeholornithiformes only more derived than *Archaeopteryx* (Zhang et al. 2008a; Zhou et al. 2008; Wang and Lloyd 2016; Wang et al. 2016b) (Fig. 1.1). The known specimens of Jeholornithiformes are all collected from the Lower Cretaceous Jehol Group of northeastern

China (Zhou and Zhang 2002a, 2003a; O'Connor et al. 2011a; Lefèvre et al. 2014). The holotype of *Jeholornis prima* is turkey-sized and much larger than *Archaeopteryx* (Fig. 1.2c). The skeletal morphologies of *Jeholornis* show many derived features related to flight with respect to *Archaeopteryx*. The coracoid is elongated. A procoracoid process is reported from *Jeholornis curvipes*, suggesting the presence of a triosseal canal (Lefèvre et al. 2014); however, this feature is absent in other jeholornithids. The glenoid faces dorsolaterally (Zhou and Zhang 2003a), allowing large excursion of the humerus in the dorsoventral plane (Gatesy and Middleton 2007). By contrast, the glenoid faces laterally in *Archaeopteryx* and other theropods (Phil 2006; Heers and Dial 2012). An ossified sternum is present, but a keel is lacking (Zhou and Zhang 2002a). A pair of lateral trabeculae is developed and bears a round fenestra (Fig. 1.2e), likely caused by the pneumatization of the clavicular air sac (Zhou and Zhang 2003a). A rigid pelvis is retained by forming a synsacrum with six fused sacral vertebrae. The hand is modified in that a carpometacarpus is fused, the major metacarpal is robust, and the minor metacarpal is strongly bowed, making the hand more suitable for the attachment of the primary remiges. The hallux is partially reversed caudally, documenting the transitional stage toward advanced perching ability (Zhou and Zhang 2002a). Despite all these derived features, *Jeholornis* surprisingly has a long bony tail composed of 27 caudal vertebrae (Fig. 1.2c), even exceeding the number in *Archaeopteryx* (Zhou and Zhang 2003a; O'Connor et al. 2011a). The feeding habit of *Archaeopteryx* remains speculative, but direct fossil evidence points herbivorous diet for Jeholornithiformes (Zhou and Zhang 2002a) (Fig. 1.2c), which partially explains why Jeholornithiformes have a robust jaw and fewer teeth (Fig. 1.2d).

1.2.3 The Origin of Short-Tailed Birds

The most easily recognizable feature that distinguishes birds from dinosaurs is the absence of the long bony tail (Gatesy and Middleton 1997). Nevertheless, little is known about how the abbreviation of the tail took place without relevant fossils documenting that transition. A short tail first evolved in Confuciusornithiformes and Sapeornithiformes, most members of which are endemic to the Jehol Biota (Chiappe et al. 1999; Zhou and Zhang 2002b). In those stem-group birds, the tail is considerably abbreviated by reducing the number of caudal vertebrae to six or seven and by the fusion of the several caudal vertebrae to a compound element, called the pygostyle (Zhou and Zhang 2006a). However, confuciusornithids and sapeornithids co-lived with jeholornithids and are only phylogenetically more derived than the latter clade (Fig. 1.1). Therefore, such rapid change of tail morphology awaits new taxa (phylogenetically intermediate between jeholornithids and sapeornithids) to clarify.

Sapeornithids are the largest known Early Cretaceous birds (Fig. 1.3a, b), with the forelimb length about 45 cm (Zhou and Zhang 2002b). Moreover, these birds stand out from other Mesozoic taxa in having a large wing proportion (the forelimb/hind limb length ratio ~1.5, but that ratio is slightly greater than 1.0 in other Cretaceous birds; Zhou and Zhang 2006b). Teeth are considerably reduced compared with *Archaeopteryx*, with their low jaws edentulous (Zhou and Zhang 2003b). As in *Archaeopteryx*, the coracoid is short and rectangular, and an ossified sternum is absent (Zheng et al. 2014a). The furcula is Y shaped with a short hypocleidium, a feature absent in *Archaeopteryx*, Jeholornithiformes, and Confuciusornithiformes. As in *Jeholornis*, the glenoid faces dorsolaterally. The humerus bears a large deltopectoral crest for flight muscle attachment. *Sapeornis* shares a unique feature with *Confuciusornis* in having an elongated elliptical fenestra in the deltopectoral crest (Zhou and Zhang 2003b), which may have helped to reduce the body weight. Unlike in *Archaeopteryx* and *Jeholornis*, the manual digits are reduced in that the minor digit has only two slender phalanges without the claw (Zhou and Zhang 2003b; Gao et al. 2012). The degree of coossification in the hind limb is more advanced than in *Archaeopteryx* and *Jeholornis* in having a true tibiotarsus

Fig. 1.3 The earliest short-tailed birds. (**a**) *Sapeornis chaoyangensis*; (**b**) *Sapeornis chaoyangensis* with preservation of a crop (*sqaure*); (**c**) close-up view of the seeds in the crop; (**d**) *Confuciusornis sanctus*; (**e**) proximal end of the right humerus of *Confuciusornis sanctus* showing the large deltopectoral crest with a round fenestra

and a proximally fused tarsometatarsus. Despite being such primitive, direct fossil evidences shows that *Sapeornis* had already evolved derived digestive system in having a crop (Fig. 1.3a, b), a feature unique to modern seed-eating birds (Zheng et al. 2011). The crop is an expanded section of the esophagus and serves as food storage, where food is softened and regulated (Gill 2007).

Confuciusornithiformes document the earliest record of toothless, beaked birds (Hou et al. 1995) (Fig. 1.3d). The confuciusornithids clade persisted for at least 10 million years (130.7–120 Ma), with its basalmost member *Eoconfuciusornis zhengi* collected from the Huajiying Formation—the second oldest bird-bearing deposits with an age of 130.7 Ma (Zhang et al. 2008b). Like *Archaeopteryx* (Elzanowski and Wellnhofer 1996), *Jeholornis* (Zhou and Zhang 2002a; Lefèvre et al. 2014), and *Sapeornis* (Zhou and Zhang 2003b; Pu et al. 2013), *Confuciusornis* retain a primitive skull with overall lack of bone fusion and reduction, and the cranial morphology of *Confuciusornis* is more primitive on the basis of the following characters. The postorbital is T shaped with a long descending (jugal) process to contact the jugal, and thus the orbit is isolated from the temporal fossa (Chiappe et al. 1999). The contact between the postorbital and the squamosal completely separates the supra- and infratemporal fenestrae (Hou et al. 1999). Taken together, *Confuciusornis* is the first known avian clade that unequivocally retains the primitive diapsid skull (Chiappe et al. 1999). The lower jaw is robust and dorsoventrally deep; the rostral end of the mandible symphysis is notched. The dorsal margin of the dentary is nearly straight, and its ventral margin slants caudoventrally and reaches its lowest point one third of the distance from the dentary tip. Caudally, the dentary bifurcates into long dorsal and ventral processes.

The shoulder girdle is primitive. The coracoid and scapula are fused (Chiappe et al. 1999), a feature unknown in most other Cretaceous birds (Wang et al. 2016b). The furcula is robust and boomerang in shape. The sternum is nearly rectangular with a pair of lateral trabeculae, but a keel is absent. The humerus uniquely has a triangular and large deltopectoral crest. The dorsodistal margin of that crest strongly projects dorsally, and its maximal width exceeds that of the humeral shaft. The deltopectoral crest is perforated by a subrounded fenestra (Fig. 1.3e), which is proportionally smaller than in *Sapeornis*. The hand exhibits many primitive features: it is longer than the humerus; the alular digit is elongated, and its proximal phalanx extends beyond the distal end of the major metacarpus; the minor digit has three phalanges and a strongly recurved claw. The major digit ends with a tiny claw, much smaller than others. However, that claw is at best slightly smaller than that of the alular digit in most other Mesozoic birds and theropods (Chiappe et al. 1999; Hwang et al. 2002; Xu et al. 2011). As in *Sapeornis*, metatarsals II–IV are fused to the distal tarsals proximally, but remain separated throughout their entire length. A metatarsal V is present but reduced, as in *Archaeopteryx*, *Jeholornis*, and *Sapeornis*. The foot is anisodactyl with a reversed hallux; however, as other non-ornithothoracine birds, the feet lack phalangeal proportion specialized for either terrestrial or arboreal locomotion (Hopson 2001; Zhou and Farlow 2001).

1.2.4 *Chongmingia zhengi*: An Enigmatic Basal Bird

The recently described taxon *Chongmingia zhengi* from the Jiufotang Formation (120 million years) of the Jehol Group further enriches our knowledge about biodiversity of basal birds (Wang et al. 2016b). *Chongmingia* preserves many primitive features. The shoulder girdle bones—the coracoid and scapula—are fused into a scapulocoracoid as in *Confuciusornis*; the co-occurrence of this feature in phylogenetically remote *Chongmingia* and *Confuciusornis* is likely the result of convergence. The coracoid lacks a procoracoid process, and thus a triosseal canal is absent. The furcula is robust and boomerang in shape. The alular metacarpal is long relative to the major metacarpal; the proximal phalanx of the alular digit terminates at the same level as the major metacarpal. As in other basal birds, a metatarsal V is present. In Wang et al. (2016b), the systematic position of *Chongmingia* was investigated using two separated data matrixes. *Chongmingia* was recovered as the most basal bird other than *Archaeopteryx* when the matrix consisting primarily of Mesozoic birds was used, whereas *Chongmingia* emerged as the direct out-group to Ornithothoraces when added to the coelurosaurian matrix. Unfortunately, the holotype and the only known specimen of *Chongmingia* is incomplete, lacking the skull and tail. Therefore, this taxon can only be considered as non-ornithothoracine bird at this point.

1.2.5 Enantiornithes: The Dominating Mesozoic Avian Group

Ornithothoraces—the clade that comprises Enantiornithes and Ornithuromorpha (Fig. 1.1)—are the first avian group that have a worldwide distribution during the Cretaceous (Chiappe and Walker 2002; Wang et al. 2015a), whereas all the known non-ornithothoracine birds are temporally and geologically restricted: *Archaeopteryx* from the Late Jurassic of Germany (Wellnhofer 2010) and all the rest from the Early Cretaceous of China and Korea (Zhou and Zhang 2006a). The subclass Enantiornithes was established by Walker in 1981 when he studied a few specimens from the Late Cretaceous Lecho Formation of El Brete, Argentina (Walker 1981). Among those fossil materials, Walker noticed unusual construction of the shoulder girdle. Specifically, the scapular articular facet of the coracoid is convex, and its opposing surface in the scapula is concave; that is, the normal configuration between those two shoulder bones is reversed with respect to modern birds, which earned its name "opposite bird" ("enanti," Greek for "opposite"). Ever since then, skeletons of Enantiornithes have been collected from nearly all the continents except the Antarctic, ranging in age from the Early Cretaceous to the K/Pg boundary (Close et al. 2009; Longrich et al. 2011; de Souza Carvalho et al. 2015; Wang et al. 2015a) (Fig. 1.4). Therefore, Enantiornithes likely represent the first avian group that underwent a large scale of diversification.

Enantiornithes are distinguishable from other birds on basis of the following synapomorphies: the caudal end of the dentary is unforked but sloping caudoventrally; the lateral surface of the thoracic vertebrae is excavated by round fossa; the coracoid bears a convex lateral margin; a medial groove runs over the neck of the coracoid that is perforated by a supracoracoidal nerve foramen; the sternum carries two pairs of caudal trabeculae; the furcula is Y shaped with a hypocleidium; the keel is restricted to the caudal half of the sternum, and its cranial end is forked; the proximal margin of the humerus is concave at its central portion; the radius bears a longitudinal groove over its interosseous surface; the minor metacarpal extends further distally than the major metacarpal; the posterior trochanter of the femur is well developed; and metatarsal IV is thinner than metatarsals II and III (Chiappe and Walker 2002; Wang 2014).

The known enantiornithines, except for the Late Cretaceous *Gobipteryx* (Chiappe et al. 2001), are toothed (O'Connor and Chiappe 2011). Despite lacking the overall coossification, the cranial morphology of Enantiornithes exhibits some derived features. For instance, the postorbital fails to contact the jugal (Fig. 1.4). More importantly, Enantiornithes display varied skull shapes, which could hint niche specialization. For example, the *Longipterygidae* have a rostrum of more than 60% the total skull length (Fig. 1.5a), and teeth are restricted to the premaxilla and rostral-most part of the dentary (Zhang et al. 2001; Hou et al. 2004; O'Connor et al. 2011b) (Fig. 1.5b, c), suggesting that those birds may have specialized to mud-probing lifestyle (O'Connor et al. 2009). *Pengornis* have numerous small, blunt, brachydont teeth (Fig. 1.5d) and were interpreted to feed on soft-shelled arthropods (O'Connor et al. 2013). Apart from those, studies relied on morphological

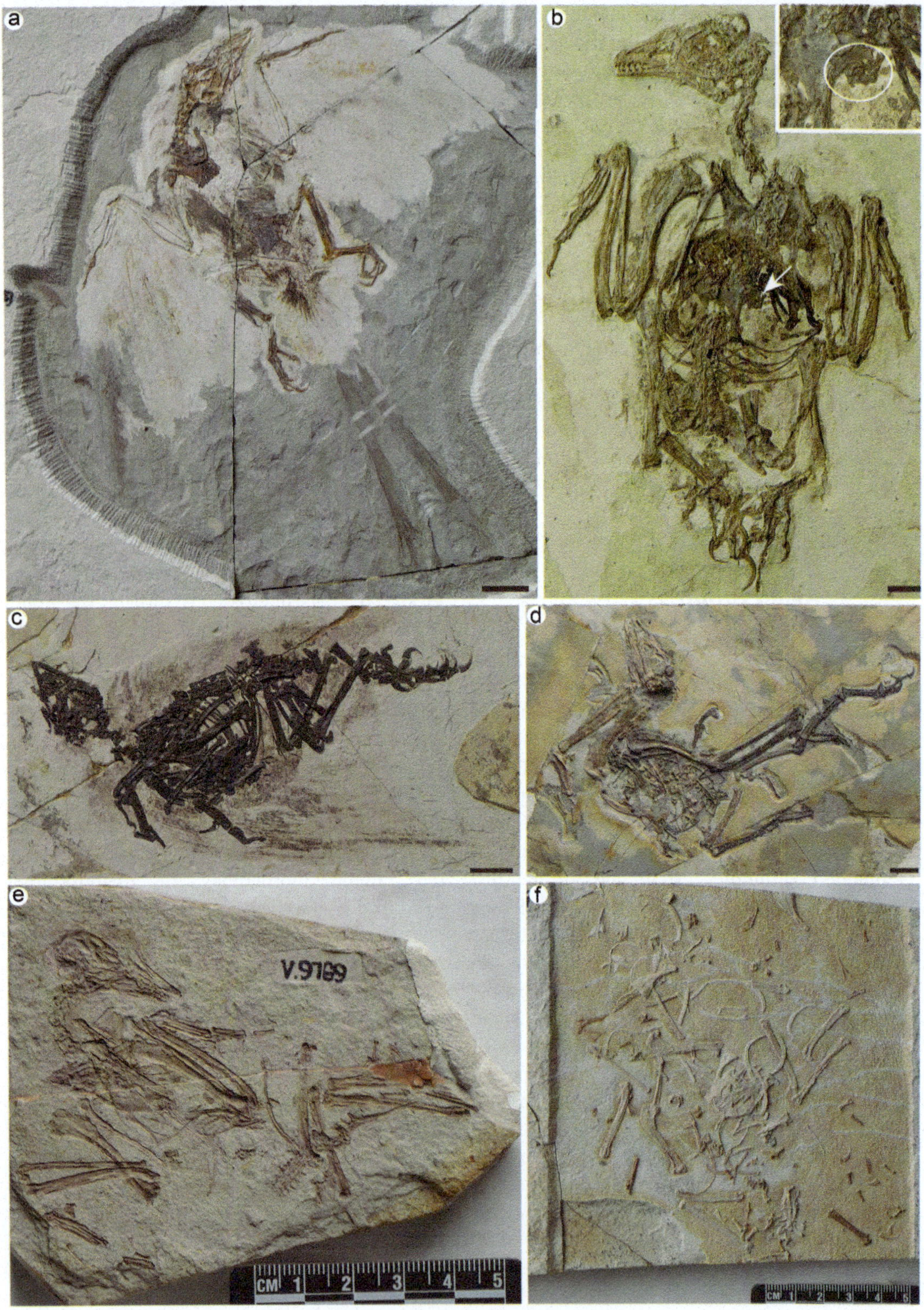

Fig. 1.4 Selected enantiornithine birds from the Early Cretaceous Jehol Biota. (**a**) The oldest enantiornithine *Eopengornis martini* (from the Huajiying Formation, approximately 130.7 million years ago); (**b**) *Bohaiornis*

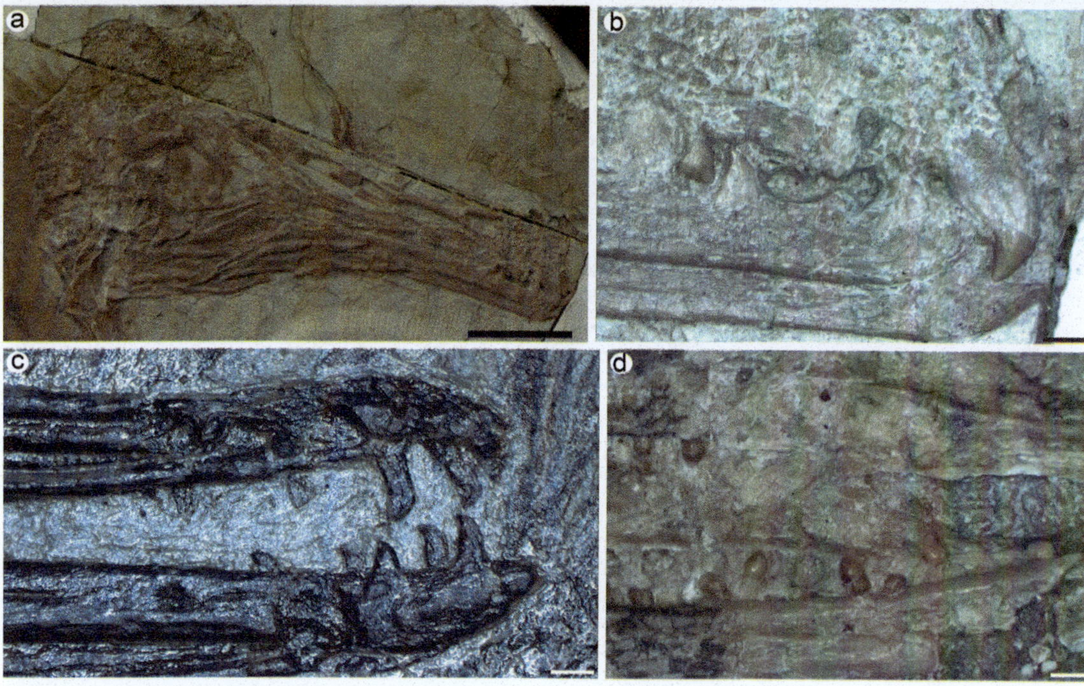

Fig. 1.5 Selected enantiornithine birds exhibiting feeding-specialized rostrum and tooth morphology. (**a**) *Longipteryx chaoyangensis* with an elongated rostrum of more than 60% the skull length; (**b**) teeth of *Longipteryx chaoyangensis* at high magnification; (**c**) rostrum of *Longirostravis hani*, showing that the teeth are restricted to the rostral end; (**d**) jaw of *Pengornis houi*, showing the numerous small, blunt, brachydont teeth. *Scale bars*: 10 mm (**a**), 1 mm (**b–d**)

clues, such as the shape of the rostrum and teeth, as a basis for dietary inferences; direct fossil evidences show trophic partitioning within this diverse clade. The holotype of *Eoalulavis hoyasi* from the Lower Cretaceous of Spain retains crustacean exoskeletons in its abdominal region (Sanz et al. 1996). A referred specimen of *Bohaiornis guoi* preserves a few small rounded stones in the stomach region (Fig. 1.4b); these stones were interpreted as ranges analogous to those swallowed and regurgitated by living raptors in order to purge the alimentary tract (Li et al. 2014). More interesting, Wang et al. (2016a) reported an enantiornithine bird preserving a gastric pellet that includes fish bones (Fig. 1.6), which points to a fish diet, a previously unknown feeding habit for enantiornithines. Moreover, the occurrence of gastric pellet suggests that the alimentary tract of the new enantiornithine resembled those of extant birds in having efficient antiperistalsis and a two-chambered stomach with a muscular gizzard capable of compacting indigestible matter into a cohesive pellet (Fig. 1.6).

The feet of Enantiornithes show features related to arboreal specialization, including a reversed hallux, strongly recurved claws, and elongated penultimate phalanges (Sereno et al. 2002; Zhou and Zhang 2006a; Wang et al. 2014) (Fig. 1.5). Enantiornithes evolved a wide range

Fig. 1.4 (continued) guoi (*arrow* indicates the stones, and enlarged view are in the *inset box*; see text for more explanations); (**c**) a robust, scansorial specialized enantiornithine *Fortunguavis xiaotaizicus*; (**d**) the largest Early Cretaceous enantiornithine *Pengornis houi*; (**e**) *Cathayornis yandica*, the first enantiornithine bird recognized from the Jehol Biota; (**f**) *Pterygornis dapingfangensis*. *Scale bars*: 10 mm (**a–f**)

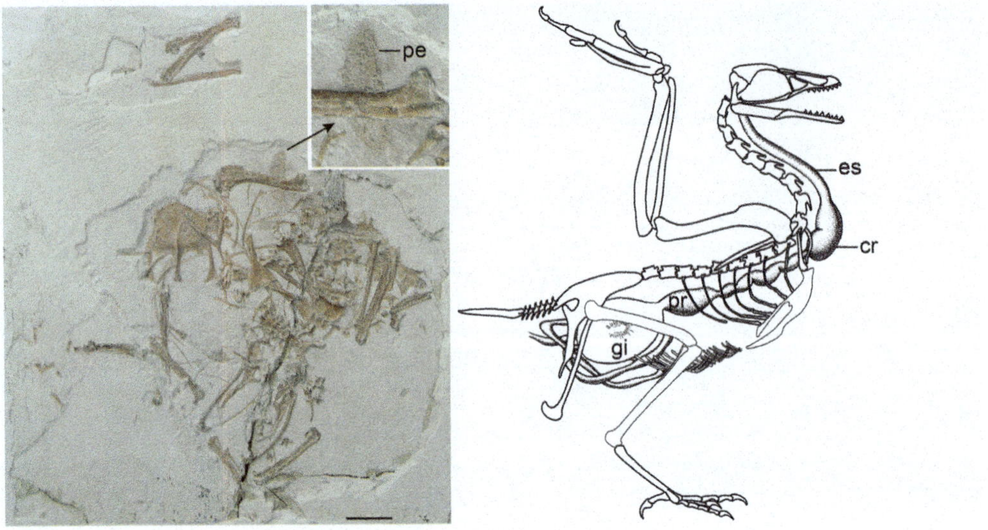

Fig. 1.6 A fish-eating enantiornithine bird from the Early Cretaceous Jehol Biota and a sketch reconstruction of its digestive system. A gastric pellet that includes fish bones is preserved (*arrow*), indicating the enantiornithine has a two-chambered stomach with a muscular gizzard capable of compacting indigestible matter into a cohesive pellet [modified from Wang et al. (2016a)]. Abbreviations: cr, crop; es, esophagus; gi, gizzard; pe, pellet; pr, proventriculus. *Scale bar*: 10 mm

of limb proportions and were likely to have already encompassed the range of flight styles of living birds (Dyke and Nudds 2009). The refined flight capability of Enantiornithes is corroborated by the discovery of alula in some taxa, e.g., *Eoalulavis hoyasi* (Sanz et al. 1996) and *Protopteryx fengningensis* (Zhang and Zhou 2000). The alula, or bastard wing, has a few feathers attached to the first manual digit and located at the leading edge of the wings (Alvarez et al. 2001). As a vortex generator, the alula increases the lift force and enhances maneuverability in flight at a large angle of attack and lower speeds without stalling (Lee et al. 2015).

1.2.6 Ornithuromorpha: A Clade That Gives Rise to All Living Birds

Ornithuromorpha is the clade uniting all living birds and their close Mesozoic relatives but not the Enantiornithes (Chiappe 1996; Wang et al. 2015b). Currently, the oldest record of Ornithuromorpha, *Archaeornithura meemannae*, dates back to 130.7 million years ago (Wang et al. 2015b). Nearly all the Mesozoic birds went extinction at the end of the Cretaceous, and only a few derived clades of Ornithuromorpha survived into the Paleogene and gave rise to all living birds (Chiappe 1995; Brusatte et al. 2015). Prior to recent discoveries, the global records of basal ornithuromorphs are poor and represented by fragmentary materials, including *Ambiortus dementjevi* from the Early Cretaceous of Mongolia (Kurochkin 1985), *Vorona berivotrensis* from the Late Cretaceous of Madagascar (Forster et al. 2002), the flightless *Patagopteryx deferrariisi* from the Late Cretaceous of Argentina (Chiappe 2002), *Hollanda luceria* from the Late Cretaceous of Mongolia (Bell et al. 2010), *Ichthyornis dispar* from the Late Cretaceous of Kansas (Marsh 1880; Clarke 2004), and foot-propelled diving Hesperornithiformes with many fragmentary and taxonomical uncertain specimens from the marine deposits (113–66 Ma) of North America and Eurasia (Martin 1984; Galton and Martin 2002; Bell and Chiappe 2016). Over the last two decades, a wealth of nearly complete and fully articulated skeletons from the Jehol Biota came to light, allowing an unprecedented opportunity to decipher the early history of Ornithuromorpha

(Clarke et al. 2006; Zhou and Zhang 2006a; O'Connor and Zhou 2015; Wang et al. 2016c). Several clades and taxa are well recognized, including the specialized waders Hongshanornithidae—that consists five genera and lasts for over 10 million years (Wang et al. 2015b), the toothless and most basal taxon *Archaeorhynchus spathula* (Zhou and Zhang 2006b; Zhou et al. 2013), and the fish-eating *Yanornis martini* and *Piscivoravis lii* (Zhou and Zhang 2001; Zheng et al. 2014b; Zhou et al. 2014a) (Fig. 1.7).

Despite the large temporal gap (>70 million years), the Early Cretaceous ornithuromorphs have already evolved many derived features characterizing the crown birds, which in turn implies that the origination date of the whole clade could be earlier than currently recognized (Wang and Lloyd 2016). For example, the postorbital is absent, and the squamosal may have been incorporated into the braincase in Early Cretaceous ornithuromorphs; these modifications improve the cranial kinesis by facilitating the craniocaudal movement of the jugal via eliminating the force exerted by the postorbital contacts, e.g., postorbital-jugal, and squamosal-quadratojugal (Zusi 1993). In Early Cretaceous ornithuromorphs, the shoulder bones are morphological indistinguishable from that of modern birds: the coracoid is strut-like and bears a procoracoid process (You et al. 2006; Wang et al. 2016c), and thus a triosseal canal is in place (Fig. 1.8a); the scapula is curved; and the furcula is U-shaped and slender. The sternum is proportionally elongated compared with other basal birds. More importantly, a large keel is developed and extends nearly along the entire length of the sternum, increasing surface area for flight muscle attachment (Clarke et al. 2006; Zhou et al. 2013, 2014a, b) (Fig. 1.8b). The forelimb shows progressively structural and functional modifications as they evolve into wing-like structure: the head of the humerus is globose with a large deltopectoral crest; the major and minor metacarpals are fused not only proximally but also distally, and the proximal phalanx of the major digit is expanded craniocaudally (Clarke et al. 2006; Zhou et al. 2014a). The synsacrum has more incorporated vertebrae than in other early birds. The pygostyle is short and ploughshare shaped, reminiscent that of modern birds (Clarke et al. 2006). The pubis is further caudally directed to flatten the body. The tarsometatarsus is fully fused (Wang and Zhou 2016). An incipient hypotarsus is developed in some Early Cretaceous ornithuromorphs (Clarke et al. 2006).

The known fossil record shows that the tooth reduction happened independently on multiple lineages of Cretaceous ornithuromorphs (Zhou et al. 2010). More specifically, an edentulous beak can only be confirmed in one taxon *Archaeopteryx spathula* (Zhou and Zhang 2006b) (Fig. 1.8c), and the hongshanornithids have but fewer teeth (Zhou and Zhang 2005); by contrast, numerous teeth are present throughout the jaw in *Jianchangornis* (Zhou et al. 2009), *Yanornis* (Zheng et al. 2014b), and *Ichthyornis* (Gregory 1952). The various patterns of tooth count, along with other evidence such as the stomach contents, e.g., fish bone remains and gizzard stones (Zhou 2004b; Chiappe et al. 2014; Zheng et al. 2014b; Wang et al. 2016c) (Fig. 1.7), show that trophic specialization took place early in the history of Ornithuromorpha (Zhou and Zhang 2006a). In contrast to Enantiornithes, basal ornithuromorphs were most likely terrestrial, given that the pedal phalanx decreases its length distally and the claw is weakly curved (Zhou et al. 2013; Wang et al. 2016c) (Fig. 1.8d). The available evidence suggests derived digestive system and trophic partitioning (e.g., herbivorous, piscivorous, durophagous) in enantiornithines and ornithuromorphs (Zheng et al. 2014b; Wang et al. 2016a). This must in turn have contributed to the enormous diversification of both clades since the Early Cretaceous.

1.3 The Radiation of Crown Birds After the K-Pg Mass Extinction

Neornithes, the subgroup of Ornithuromorpha that includes all living birds, is the most diverse clade of modern vertebrates that are adapted to nearly all ecological niches (Gill 2007). The phylogeny of Neornithes has long being hotly

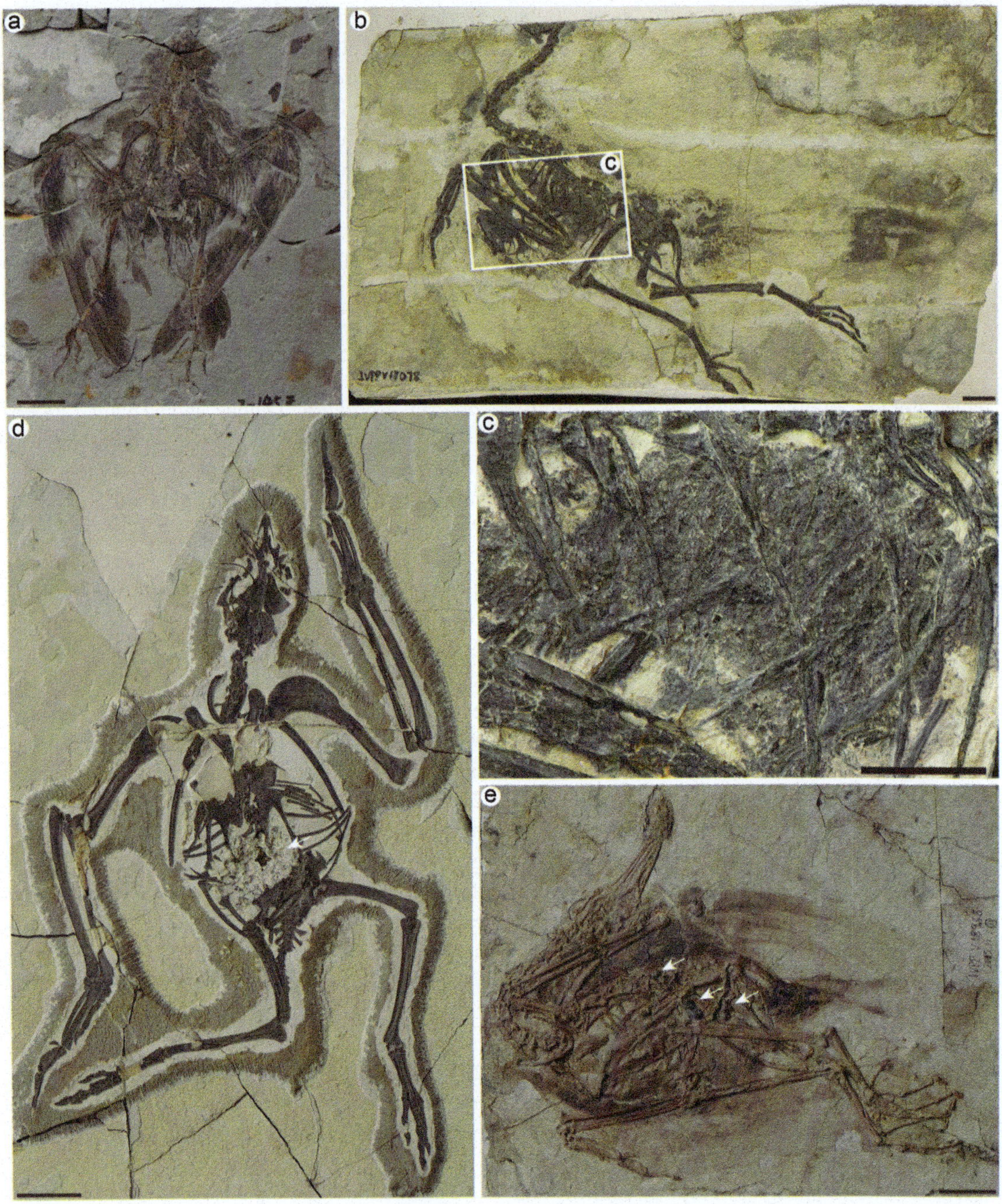

Fig. 1.7 Selected basal ornithuromorph birds from the Jehol Biota. (**a**) The oldest ornithuromorph *Archaeornithura meemannae*; (**b**) a fish-eating ornithuromorph *Piscivorous lii*, preserving fish bones in its stomach region (**c**); (**d**) the most primitive ornithuromorph *Archaeorhynchus spathula*; (**e**) the most derived known ornithuromorph bird from the Jehol Biota *Iteravis huchzermeyeri*. *Arrows* in (**d**, **e**) indicate the gizzard stones (gastroliths). *Scale bars*: 10 mm (**a–e**)

debated, especially the relationships within high clades (Mayr 2011). A robust phylogeny would refine our understanding about how modern birds have achieved their extraordinary diversity in morphology, biology, and ecology. Recently, consensus regarding the interrelationships

Fig. 1.8 Selected basal ornithuromorphs exhibiting derived morphological features. (**a**) Shoulder girdle bones of *Yixianornis grabaui*, showing the well-developed procoracoid process; (**b**) sternum and manus of *Piscivorous lii*, showing that the keel is long and deep and the proximal phalanx of the major digit is expanded; (**c**) skull of *Archaeorhynchus spathula*, showing the toothless jaw; (**d**) left foot of *Archaeorhynchus spathula*, showing that the phalanges decrease length distally, indicative of terrestrial adaptation. All specimens are from the Jehol Biota. Abbreviations: ad, alular digit; co, coracoid; fu, furcula; II–IV, pedal digit II, III, IV; ke, keel; mII–IV, metatarsal II, III, IV; md, major digit; md-1, the proximal phalanx of the major digit; mi, minor digit; pc, procoracoid process; sc, scapula; st, sternum. *Scale bars*: 10 mm (**a–d**)

among the major clades has been approached through genomic phylogenetic studies using increasing molecular data (Ericson et al. 2006; Hackett et al. 2008; Jarvis et al. 2014; Claramunt and Cracraft 2015; Prum et al. 2015). In the most recent phylogenomic study (Prum et al. 2015), over 390,000 bases of genomic sequence data from 198 species that represent the major avian lineages were analyzed using both Bayesian and maximum likelihood methods. The new phylogeny is completely resolved, supports some phylogenetic relationships recognized previously, and also identifies many new clades. In accordance with previous studies based on palate morphology (Pycraft 1900), living birds are divided into two major groups, i.e., Palaeognathae and Neognathae, and the latter group consists the *Galloanserae* and the more derived clade Neoaves (Prum et al. 2015). Within Neoaves, Aequorlitornithes, Gruiformes, Columbaves, and Strisores form successive sister groups to Inopinaves—a diverse landbird clade that includes passerines, falcons, seriemas, parrots, and hoatzin (Prum et al. 2015).

In context of the new phylogeny, many interesting questions about crown birds' evolution can be addressed. For example, how flightless evolves within *Palaeognathae*. *Palaeognathae* comprise the flightless ratites (ostrich, rheas, kiwis, cassowaries, and emu) and the volant tinamous. As in some previous studies, tinamous are deeply nested within the *Palaeognathae*, demonstrating that the ratites are polyphyletic (Harshman et al. 2008; Prum et al. 2015). Therefore, the loss of flight happens either multiple times among these ratites lineages or once in the most recent common ancestors of paleognaths, and the volant ability is regained in tinamous. Given that the loss of flight happens multiple times within other groups (Roff 1994), the first scenario is considered more plausible (Harshman et al. 2008). Within landbirds, vultures, hawks, and owls form out-groups to the remaining taxa, suggesting that the common ancestor of landbirds is a predator (Jarvis et al. 2014; Prum et al. 2015). By using 19 phylogenetically and geologically defined fossil taxa, the time-calibrated avian phylogeny confirmed the hypothesis that the crown birds underwent an explosive radiation in the Paleogene, and representatives of nearly all major lineages appeared within 10 million years after the end-Cretaceous mass extinction, congruent with paleontological record (Prum et al. 2015). Based on biogeographic study, Claramunt and Cracraft (2015) propose that the most common ancestors of the crown birds lived in Western Gondwana during the Cretaceous and expanded to other continents through two routes, the North American-Old World and Antarctica-Australia/New Zealand.

1.4 The Origin of Avian Flight

The origin of avian flight is widely regarded as the key innovation that contributes to the biological success of birds (Gatesy and Dial 1996; Gill 2007; Chatterjee and Templin 2012). The growing fossil evidences have pieced together nearly the whole puzzle of the dinosaur-bird transition. Nevertheless, how protobirds became airborne and performed the first wing stroke remains obscure (Zhou 2004a; Heers and Dial 2012). A better understanding of the evolution of flight requires integrative studies about that transition, including musculoskeletal and neurosensory systems, physiology, aerodynamic principles, etc. Here a short review of recent hypotheses about the origin of avian flight is presented, and for more details readers may turn to relevant reference.

1.4.1 The Ground-Up Theory

There are three popular hypotheses about the origin of avian flight, the ground-up, wing-assisted incline running (WAIR), and trees-down hypotheses. The ground-up theory was initially proposed by Williston (1879) and later elaborated by Nopcsa (1907, 1923) and Ostrom (1974, 1979). The main assumption of it emphasizes that the immediate theropod ancestors of birds were obligated terrestrial bipeds, which used their winged forelimbs for balance during running. A few scenarios derived from the ground-up theory have been generated, which primarily differ in how the cursorial theropods used their wings that directly evolved to the motion of a wingbeat. Ostrom's "insect net" hypothesis depicted *Archaeopteryx* using their wings as a net to catch insects (Ostrom 1974). As the movement of the wings is reinforced as *Archaeopteryx* chases the escaping insects, the precursor of wing stroke is likely to appear. Despite that vivid picture one could image, the "insect net" hypothesis is untenable, because the motion used to catch insect is fundamentally different from the motion used in a flight stroke (Chatterjee and Templin 2012; Heers and Dial 2012; Chatterjee 2015).

Similar to the "insect net" theory, it is hypothesized that the theropod ancestors of birds would use their forelimb to kill prey and that motion finally gave rise to the wing flapping (Gauthier and Padian 1985; Padian and Chiappe 1998). Again, that scenario fails to explain how the dorsoventral movement of wing stroke evolved from the anteroposterior predatory movement, needless to say whether such lifestyle is true for these animals (Bock 1986; Elzanowski 2002). Burgers and Chiappe (1999) proposed that *Archaeopteryx* or its cursorial maniraptoriform ancestors could achieve velocity necessary to take off from ground by running with considerable thrust generated via wing flapping. However, it remains questionable whether *Archaeopteryx* and its maniraptoriform ancestors were capable of performing a modern style flapping (Poore et al. 1997; Chatterjee and Templin 2003). The shoulder bones of *Archaeopteryx* and its closest theropod relatives lack derived features for flapping. For example, the coracoid is short and lacks a procoracoid process—a necessary structure for the pass of the primary elevator muscle supracoracoideus (Mayr et al. 2007; Chatterjee and Templin 2012); a keel is unossified or nonexistent, and thus the major flight muscles pectoralis (responsible for the wing's downstroke) and supracoracoideus (responsible for the wing's upstroke) lack a solid ventral attachment; the glenoid, facing laterally or ventrolaterally, only allows gentle movement of the wing in the dorsoventral plane; the tarsal bones are not fused to the tibia or metatarsus, and a rigid pelvis is lacking, which severely impairs the running ability, and enough leg thrust for taking off from the ground is unlikely available.

The biggest challenge for the ground-up hypothesis is that whether its central argument that the immediate ancestors of birds are obligated cursorial is tenable. For much of the time when the ground-up theory prevailed, all the known theropods are interpreted as cursorial (Padian and Chiappe 1998). However, over the last two decades, a handful of well-preserved theropods from China and elsewhere throw doubt on that stereotype (Bock 1986; Pennycuick 1986; Chatterjee and Templin 2012). The closest relatives of birds are dromaeosaurids and troodontids (Xu et al. 2014), most members of which are small, feathered, and have displayed many arboreal or climbing adaptations (Xu et al. 2000, 2003, 2011; Hu et al. 2009; Burnham et al. 2011). For instance, the penultimate phalanges of the manual digit is elongated in many theropods and particularly those phylogenetically close to birds, including *Deinonychus* (Zhang et al. 2002), *Xiaotingia* (Xu et al. 2011), *Microraptor* (Burnham et al. 2011), and *Anchiornis* (Hu et al. 2009), a feature generally regarded as indicative of arboreal habits (Fröbisch and Reisz 2009). In *Microraptor*, the manual claws correspond well to the tree-climbing morphology, and they are recurved with an angle of 180°, even exceeding the curvature seen in modern climbing birds, strongly indicating that *Microraptor* used their forelimbs for climbing (Burnham et al. 2011). Such notion is reinforced by the discovery of a specimen of *Microraptor* that has the remains of an enantiornithine bird in its abdomen (O'Connor et al. 2011c), and enantiornithines are good fliers (Dyke and Nudds 2009).

The bizarre and enigmatic maniraptoran theropod clade Scansoriopterygidae represents another line of evidence of arboreal adaptation of theropods. Scansoriopterygids uniquely have elongated third manual digit, and the penultimate phalanx is longer than its preceding one in all digits (Zhang et al. 2002, 2008a). Specifically, in *Epidendrosaurus ningchengensis* the penultimate phalanx is approximately 170% as long as the proximal one in the second manual digit (Zhang et al. 2002). Moreover, the hind limb also shows features related to arboreal adaptation: the distal ends of metatarsals II–IV terminate at the same level, a feature widespread among perching birds such as crow and sparrow (Zhang and Tian 2006), and the penultimate phalanx is elongated. All these observations indicate arboreal life for scansoriopterygids (Zhang et al. 2008a). The phylogenetic position of Scansoriopterygidae is yet resolved, and different scenarios are proposed: the immediate sister group to Aves (Zhang et al. 2008a), out-group to Avialae (Turner et al. 2012; Xu et al. 2015), or nested within Aves (Xu et al. 2011; Foth et al. 2014). Despite the phylogenetic inconsistence, a close relationship between scansoriopterygids and birds is widely accepted. Taken together, proponents of ground-up theory have to consider all these paleontological evidences and revise their arguments under the context that the closest theropod ancestors of birds are arboreal rather than terrestrial.

1.4.2 The Wing-Assisted Incline Running Theory (WAIR)

The WAIR hypothesis was proposed by Dial (2003) from his experimental investigation on living bird, chukar partridges. It argues that the posthatching birds vigorously beat their incipient wings to climb steep or even vertical substrate. However, when the wings are trimmed, the climbing performance is severally impaired. By calculating the instantaneous direction and magnitude of whole-body acceleration during the WAIR, Dial (2003) suggested that a significant portion of the wingbeat cycle exerted aerodynamic or inertial forces that pushed the center of the mass of the bird toward the substrate and thus improved the traction of the leg (Bundle and Dial 2003). Dial's behavior experiment is intriguing and provides new insights about the origin of avian flight, which have primarily been investigated in terms of paleontology and aerodynamic principle. During the WAIR, the movement of the wings has an anteroposterior component, which possibly represents an intermediate stage—acquired by protobirds and their ancestors—between the typical forelimb movement of most tetrapods and the wing stroke of birds (Dial 2003). However, the WAIR is based on study of living birds; although immature and with aerodynamic incapable wings, they do have highly derived musculoskeleton that is capable of performing modern style of wing stroke, which is not the case in either *Archaeopteryx* or its theropod ancestors (Elzanowski 2002; Chatterjee and Templin 2012).

1.4.3 The Tree-Down (or Arboreal) Theory

The tree-down (or arboreal) theory was first proposed by Marsh (1880) and has been advanced by many recent studies (Chatterjee and Templin 2003, 2004a, b; Xu et al. 2003; Chatterjee 2015). The tree-down theory regarded the primitive birds and their closest theropod ancestors as arboreal. Before the motion of flapping evolved, these animals could jump from branch to branch/ground, by spreading their winged forelimbs to increase the lift and to slow their descent speed; later, wing-assisted gliding occurred, then incipient motion of wingbeating was acquired, and finally powered flight evolved (Chatterjee and Templin 2012). The arboreal theory has received increasing support from recent fossil evidence and becomes the most popular hypothesis about the origin of flight. Chatterjee and his colleagues have elaborated the arboreal theory using biomechanical methods (see review in Chatterjee and Templin 2012) and proposed a series of transitional stages by which the powered avian flight could have arisen. We suggest interested readers refer their studies for detailed account.

While we do not intend to present a complete description of that hypothesis, some most recent discoveries could shed new lights on this issue. Xu et al. (2015) described a new scansoriopterygid taxon *Yi qi*, from the Middle-Late Jurassic Tiaojishan Formation, Hebei Province, China (Fig. 1.9). *Yi* preserves many striking features, which is impossible to imagine prior to its discovery. There is a slightly curved, distally tapered, rod-like structure, associated with and extending from the wrist to the body. By using energy dispersive spectrometry, that rod-like structure was confirmed as a bone or a skeletal element composed of calcified cartilage (Xu et al. 2015). Equivalent structure is unknown in any other dinosaurs. Xu et al. (2015) suggested

Fig. 1.9 Photograph and life reconstruction of the enigmatic scansoriopterygid *Yi qi* from the Middle-Late Jurassic Tiaojishan Formation, Hebei Province, China. *Yi qi* possesses feathered wings with membranous component (photo by Xing Xu; life reconstruction by Brian Choo)

that the rod-like element (termed styliform element) was likely similar to the enlarged carpally situated elements in some petauristines, such as the Japanese giant flying squirrel *Petaurista leucogenys*. In addition to the feathers, *Yi* preserves membranous soft tissues around hands, the styliform element, and other parts of the body (Fig. 1.9). Among other amniotes that have the styliform element, it is invariably responsible for supporting an aerodynamic membrane that contributes to powered flight or gliding (Schutt and Simmons 1998; Thorington et al. 1998; Bennett 2007). Following that, it is parsimonious to interpret the same function for the styliform element in *Yi* (Xu et al. 2015). Although the flight apparatus of *Yi* cannot be reconstructed with certainty because of the poor preservation, a range of possible flight apparatus configuration for *Yi* was proposed in that work. It demonstrates that *Yi* could use wings with membranous component to fly or glide (a locomotion style unknown among dinosaurs but resembling that of pterosaurs, bat, and some gliding mammals; Fig. 1.9; Xu et al. 2015). Apparently, *Yi* is highly derived in volant capability and provides little information about the evolutionary acquisition of avian flight. However, it further corroborates the notion that numerical flight-related morphological changes have been experimented with by multiple lineages close to the origin of birds (Chatterjee and Templin 2007; Xu et al. 2014).

1.5 Insights About Respiratory System of Mesozoic Birds

Among living vertebrate animals, birds have a derived and unique respiratory system characterized by the high efficient flow-through ventilation and a high compliance air-sac system, providing birds with the most efficient respiratory apparatus (Maina 2005; Duncker 2013) (Chap. 8). The air sacs are of central importance to avian respiratory system (McLelland 1989). On one hand, they serve as air storage and pump a continuous unidirectional airflow through the lungs (Duncker 1972); on the other hand, they help regulate the body temperature and protect the internal organs during flight (Gill 2007). The avian skeletons display a numerous adaptive modifications related to that system (Duncker 2013). Notably, the postcranial skeletons are variably penetrated and pneumatized by air sacs, a feature unique to birds with few exceptions in other modern vertebrates (Duncker 1978; O'Connor 2004).

Recent fossil evidences have documented occurrences of the postcranial skeletal pneumaticity in several lineages of theropods, e.g., *Sinraptor*, *Allosaurus*, *Tyrannosaurus*, *Majungatholus*, and *Rahonavis* (Forster et al. 1998; Ruben et al. 1999; Vickers-Rich et al. 2002; Schachner et al. 2009; Benson et al. 2012), suggesting that the modern bird-like respiratory system evolved before the origin of birds (O'Connor and Claessens 2005; O'Connor 2006; Schachner et al. 2009). Correlations between regional pneumaticity of postcranial skeletons and specific air sacs are well known in living birds (O'Connor 2009), and that correlation is likely true between non-avian theropods and birds regarding the axial pneumaticity (O'Connor and Claessens 2005). Therefore, it is possible to infer the existence of the specific air sac in stem-group birds by looking at the patterns of the pneumaticity. Unfortunately, the bones of birds are hollow and fragile, and in most case the early birds are two-dimensionally preserved resultant from postmortem compression. Consequently, the pneumatic foramen on the bone surface is difficult to identify, needless to say the fine pneumatic diverticula organization within the bone.

In *Archaeopteryx*, pneumaticity has been recognized in cervical and anterior thoracic vertebrae in the Berlin, Eichstätt, and London specimens and the pelvis of the London specimen (Britt et al. 1998; Christiansen and Bonde 2000), indicating the presence of the cervical and abdominal air sacs. *Jeholornis* shows pneumaticity in its thoracic vertebrae and possibly the sternum; the lateral trabeculae of the sternum on both sides are perforated by round fenestrae (Fig. 1.2e), which were interpreted as perforated by the clavicular air sac (Zhou and Zhang 2003a). Comparable pneumatic foramen is absent from all the known sternal materials of Cretaceous birds. The lack of pneumaticity of the sternum cannot be solely attributed to preservational artifact, given that the sternum is well preserved in many specimens. If the pneumaticity of the sternum in *Jeholornis* holds true, it represents an autapomorphy for that taxon, and the clavicular or the cranial thoracic air sac was not so well expanded to invade the sternum in most other Cretaceous birds as they do in living birds; alternatively, the foramen in *Jeholornis* may not be related to air sac but serves other function, e.g., reducing the body weight. Aside from the above fossil evidences, the record of unequivocal pneumaticity of postcranial skeletons in Mesozoic birds is still rare (Sereno et al. 2002). Noticeable, an enantiornithine specimen preserves the earliest record of pneumaticity in limb bones among Mesozoic birds (Chiappe and Walker 2002). It is a right humerus (PVL-4022) from the Late Cretaceous El Brete of Argentina. On the caudal surface of the proximal end, a foramen is developed ventral to the ventral tubercle (Chiappe and Walker 2002), corresponding well to the pneumotricipital fossa in living birds. The pneumotricipital fossa is penetrated by the diverticulum of the clavicular air sac (Baumel and Witmer 1993), suggesting the same structure in place. Despite the sparse and temporally restricted fossil records, the occurrences of the cervical, clavicular, and abdominal air sacs in different lineages of basal birds and theropods indicate that early birds had already evolved some features of modern birds' respiratory system in a phylogenetic context (Britt et al. 1998; Christiansen and Bonde 2000). The study of respiratory system in early birds can be further advanced by using new techniques, e.g., high-resolution computed tomography (CT), to uncover the degree of the pneumaticity of postcranial skeletons and the internal architecture of the pneumatic foramen.

1.6 Conclusion

Birds have evolved on the planet over 150 million years, and their evolutionary success has been of considerable attention. Our understanding of the origin of birds has been greatly refined by unprecedented discoveries of basal birds and their theropod ancestors over the last two decades. It is generally accepted that birds originated approximately 150 Ma from within theropods. Many features previously considered unique to modern birds, in fact, had already been acquired by multiple lineages of Mesozoic birds or even their closest theropod ancestors. The Mesozoic Era is a critical interval that records the first phase of radiation in avian evolution and shows that the bauplan of modern birds are archived through piecemeal acquisition of key features. Despite these achievements that have already been made, questions still remain such as the origin of flight, why Enantiornithes perished and only Ornithuromorpha survived at the end-Cretaceous mass extinction, and the evolution of biologically modern birds, which call for future analyses that combine information from paleontological, functional, paleogeographical, and genomic studies.

Acknowledgments We are grateful to Xing Xu, Gerald Mayr, and Oliver Rauhut for sharing specimen photographs and Brain Choo for the illustration of *Yi qi*. This study was supported by the National Natural Science Foundation of China (91514302, 41502002) and the Youth Innovation Promotion Association (CAS, 2016073).

References

Alvarez J, Meseguer J, Meseguer E, Perez A. On the role of the alula in the steady flight of birds. Ardeola. 2001;48:161–73.

Baumel JJ, Witmer LM. Osteologia. In: Baumel JJ, King AS, Breazile JE, Evans HE, Vanden Berge JC, editors. Handbook of avian anatomy: Nomina anatomica avium. 2nd ed. Cambridge: Nuttall Ornithological Club; 1993. p. 45–132.

Bell A, Chiappe LM. A species-level phylogeny of the Cretaceous Hesperornithiformes (Aves: Ornithuromorpha): implications for body size evolution amongst the earliest diving birds. J Syst Palaeontol. 2016;14:239–51.

Bell AK, Chiappe LM, Erickson GM, Suzuki S, Watabe M, Barsbold R, Tsogtbaatar K. Description and ecologic analysis of *Hollanda luceria*, a Late Cretaceous bird from the Gobi Desert (Mongolia). Cretac Res. 2010;31:16–26.

Bennett SC. Articulation and function of the pteroid bone of pterosaurs. J Vertebr Paleontol. 2007;27:881–91.

Benson RBJ, Butler RJ, Carrano MT, O'Connor PM. - Air-filled postcranial bones in theropod dinosaurs: physiological implications and the reptile-bird transition. Biol Rev. 2012;87:168–93.

Bock W. The arboreal origin of avian flight. Mem Calif Acad Sci. 1986;8:57–72.

Britt BB, Makovicky PJ, Gauthier J, Bonde N. Postcranial pneumatization in *Archaeopteryx*. Nature. 1998;395:374–6.

Brusatte SL, O'Connor JK, Jarvis ED. The origin and diversification of birds. Curr Biol. 2015;25:R888–98.

Bundle MW, Dial KP. Mechanics of wing-assisted incline running (WAIR). J Exp Biol. 2003;206:4553–64.

Burgers P, Chiappe LM. The wing of *Archaeopteryx* as a primary thrust generator. Nature. 1999;399:60–2.

Burnham DA, Feduccia A, Martin LD, Falk AR. Tree climbing-a fundamental avian adaptation. J Syst Palaeontol. 2011;9:103–7.

Chatterjee S. The rise of birds: 225 million years of evolution. 2nd ed. Baltimore: Johns Hopkins University Press; 2015.

Chatterjee S, Templin RJ. The flight of *Archaeopteryx*. Naturwissenschaften. 2003;90:27–32.

Chatterjee S, Templin RJ. Feathered coelurosaurs from China: new light on the arboreal origin of avian flight. In: Currie PJ, Koppelhaus EB, Shugar MA, Wright JL, editors. Feathered dragons: studies on the transition from dinosaurs to birds. Bloomington: Indiana University Press; 2004a. p. 1–64.

Chatterjee S, Templin RJ. Posture, locomotion, and paleoecology of pterosaurs. Geol Soc Am. 2004b;376:1–64.

Chatterjee S, Templin RJ. Biplane wing planform and flight performance of the feathered dinosaur *Microraptor gui*. Proc Natl Acad Sci U S A. 2007;104:1576–80.

Chatterjee S, Templin RJ. Palaeoecology, aerodynamics, and the origin of avian flight. In: Talent JA, editor.

Earth and life, international year of planet Earth. New York: Springer; 2012. p. 585–612.
Chiappe LM. The first 85 million years of avian evolution. Nature. 1995;378:349–55.
Chiappe LM. Late Cretaceous birds of southern South America: anatomy and systematics of Enantiornithes and *Patagopteryx deferrariisi*. Münchner Geowissenschaftliche Abhandlungen. 1996;30:203–44.
Chiappe LM. Osteology of the flightless *Patagopteryx deferrariisi* from the Late Cretaceous of Patagonia (Argentina). In: Chiappe LM, Witmer LM, editors. Mesozoic birds: above the heads of dinosaurs. Berkeley, CA: University of California Press; 2002. p. 281–316.
Chiappe LM, Walker CA. Skeletal morphology and systematics of the Cretaceous Euenantiornithes (Ornithothoraces: Enantiornithes). In: Chiappe LM, Witmer LM, editors. Mesozoic birds: above the heads of dinosaurs. California: University of California Press; 2002. p. 240–67.
Chiappe LM, Ji S, Ji Q, Norell MA. Anatomy and systematics of the Confuciusornithidae (Theropoda: Aves) from the Late Mesozoic of northeastern China. Bull Am Mus Nat Hist. 1999;242:1–89.
Chiappe LM, Norell M, Clark J. A new skull of *Gobipteryx minuta* (Aves: Enantiornithes) from the Cretaceous of the Gobi desert. Am Mus Novit. 2001;3346:1–15.
Chiappe LM, Zhao B, O'Connor JK, Gao C, Wang X, Habib M, Marugan-Lobon J, Meng Q, Cheng X. A new specimen of the Early Cretaceous bird *Hongshanornis longicresta*: insights into the aerodynamics and diet of a basal ornithuromorph. Peer J. 2014;2:e234.
Christiansen P, Bonde N. Axial and appendicular pneumaticity in *Archaeopteryx*. Proc R Soc Lond. 2000;267:2501–5.
Claramunt S, Cracraft J. A new time tree reveals Earth history's imprint on the evolution of modern birds. Sci Adv. 2015;1:e1501005.
Clarke JA. Morphology, phylogenetic taxonomy, and systematics of *Ichthyornis* and *Apatornis* (Avialae: Ornithurae). Am Mus Nat Hist. 2004;286:1–179.
Clarke JA, Zhou Z, Zhang F. Insight into the evolution of avian flight from a new clade of Early Cretaceous ornithurines from China and the morphology of *Yixianornis grabaui*. J Anat. 2006;208:287–308.
Close RA, Vickers-Rich P, Trusler P, Chiappe LM, O'Connor JK, Rich TH, Kool L, Komarower P. Earliest Gondwanan bird from the Cretaceous of southeastern Australia. J Vertebr Paleontol. 2009;29:616–9.
Cracraft J. The origin and early diversification of birds. Paleobiology. 1986;12:383–99.
de Souza Carvalho I, Novas FE, Agnolin FL, Isasi MP, Freitas FI, Andrade JA. A Mesozoic bird from Gondwana preserving feathers. Nat Commun. 2015;6:7141.
Dial KP. Wing-assisted incline running and the evolution of flight. Science. 2003;299:402–4.
Duncker HR. Structure of avian lungs. Respir Physiol. 1972;14:44–63.
Duncker HR. General morphological principles of amniotic lungs. In: Piiper J, editor. Respiratory function in birds, adult and embryonic. Berlin: Springer; 1978. p. 2–22.
Duncker HR. The lung air sac system of birds: a contribution to the functional anatomy of the respiratory apparatus. Berlin-Heidelberg: Springer; 2013.
Dyke GJ, Nudds RL. The fossil record and limb disparity of enantiornithines, the dominant flying birds of the Cretaceous. Lethaia. 2009;42:248–54.
Elzanowski A. Archaeopterygidae (Upper Jurassic of Germany). In: Chiappe LM, Witmer LM, editors. Mesozoic birds: above the heads of dinosaurs. California: University of California Press; 2002. p. 129–59.
Elzanowski A, Wellnhofer P. Cranial morphology of *Archaeopteryx*: evidence from the seventh skeleton. J Vertebr Paleontol. 1996;16:81–94.
Ericson PGP, Anderson CL, Britton T, Elzanowski A, Johansson US, Källersjö M, Ohlson JI, Parsons TJ, Zuccon D, Mayr G. Diversification of Neoaves: integration of molecular sequence data and fossils. Biol Lett. 2006;2:543–7.
Feduccia A. The problem of bird origins and early avian evolution. J Ornithol. 2001;142:139–47.
Feduccia A. 'Big bang' for tertiary birds? Trends Ecol Evol. 2003;18:172–6.
Forster CA, Sampson SD, Chiappe LM, Krause DW. The theropod ancestry of birds: new evidence from the Late Cretaceous of Madagascar. Science. 1998;279:1915–9.
Forster CA, Chiappe LM, Krause DW, Sampson SD. *Vorona berivotrensis*, a primitive bird from the Late Cretaceous of Madagascar. In: Chiappe LM, Witmer LM, editors. Mesozoic birds: above the heads of dinosaurs. Berkeley, CA: University of California Press; 2002. p. 268–80.
Foth C, Tischlinger H, Rauhut OWM. New specimen of *Archaeopteryx* provides insights into the evolution of pennaceous feathers. Nature. 2014;511:79–82.
Fröbisch J, Reisz RR. The Late Permian herbivore *Suminia* and the early evolution of arboreality in terrestrial vertebrate ecosystems. Proc R Soc Lond. 2009;283:1–8.
Galton PM, Martin LD. *Enaliornis*, an Early Cretaceous Hesperornithiform bird from England. In: Chiappe LM, Witmer LM, editors. Mesozoic birds: above the heads of dinosaurs. Berkeley, CA: University of California Press; 2002. p. 317.
Gao C, Chiappe LM, Zhang F, Pomeroy DL, Shen C, Chinsamy A, Walsh MO. A subadult specimen of the Early Cretaceous bird *Sapeornis chaoyangensis* and a taxonomic reassessment of sapeornithids. J Vertebr Paleontol. 2012;32:1103–12.
Gatesy SM, Dial KP. Locomotor modules and the evolution of avian flight. Evolution. 1996;50:331–40.

Gatesy SM, Middleton KM. Bipedalism, flight, and the evolution of theropod locomotor diversity. J Vertebr Paleontol. 1997;17:308–29.

Gatesy SM, Middleton KM. Skeletal adaptations for flight. In: Hall BK, editor. Fins into limbs: evolution, development, and transformation. Chicago: University of California Press; 2007. p. 269–83.

Gauthier J, Gall LF. Phylogenetic relationships among coelurosaurian theropods. In: Gauthier J, Gall LF, editors. New perspectives on the origin and early evolution of birds. New Haven: Yale University Press; 2001. p. 49–67.

Gauthier J, Padian K. Phylogenetic, functional, and aerodynamic analyses of the origin of birds and their flight. In: Hecht JH, Ostrom GV, editors. The beginnings of birds. Eichstatt: Freunde des Jura-Museum; 1985. p. 185–97.

Gill FB. Ornithology. 3rd ed. New York: W.H. Freeman; 2007.

Gregory JT. The jaws of the Cretaceous toothed birds, *Ichthyornis* and *Hesperornis*. Condor. 1952;54:73–88.

Hackett SJ, Kimball RT, Reddy S, Bowie RCK, Braun EL, Braun MJ, Chojnowski JL, Cox WA, Han K, Harshman J, Huddleston CJ, Marks BD, Miglia KJ, Moore WS, Sheldon FH, Steadman DW, Witt CC, Yuri T. A phylogenomic study of birds reveals their evolutionary history. Science. 2008;320:1763–8.

Harshman J, Braun EL, Braun MJ, Huddleston CJ, Bowie RCK, Chojnowski JL, Hackett SJ, Han K, Kimball RT, Marks BD, Miglia KJ, Moore WS, Reddy S, Sheldon FH, Steadman DW, Steppan SJ, Witt CC, Yuri T. Phylogenomic evidence for multiple losses of flight in ratite birds. Proc Natl Acad Sci U S A. 2008;105:13462–7.

He H, Wang X, Jin F, Zhou Z, Wang F, Yang L, Ding X, Boven A, Zhu R. The $^{40}Ar/^{39}Ar$ dating of the early Jehol Biota from Fengning, Hebei Province, northern China. Geochem Geophys Geosyst. 2006;7:Q04001.

Heers AM, Dial KP. From extant to extinct: locomotor ontogeny and the evolution of avian flight. Trends Ecol Evol. 2012;27:296–305.

Hopson JA. Ecomorphology of avian and nonavian theropod phalangeal proportions: implications for the arboreal versus terrestrial origin of bird flight. In: Gauthier J, Gall LF, editors. New perspectives on the origin and early evolution of birds. New Haven: Yale University; 2001. p. 211–35.

Hou L, Zhou Z, Martin LD, Feduccia A. A beaked bird from the Jurassic of China. Nature. 1995;377:616–8.

Hou L, Matin LD, Zhou Z, Feduccia A, Zhang F. A diapsid skull in a new species of the primitive bird *Confuciusornis*. Nature. 1999;399:679–82.

Hou L, Chiappe LM, Zhang F, Chuong CM. New Early Cretaceous fossil from China documents a novel trophic specialization for Mesozoic birds. Naturwissenschaften. 2004;91:22–5.

Hu D, Hou L, Zhang L, Xu X. A pre-*Archaeopteryx* troodontid theropod from China with long feathers on the metatarsus. Nature. 2009;461:640–3.

Huxley TH. On the animals which are most nearly intermediate between birds and reptiles. Ann Mag Nat Hist. 1868;4:66–75.

Hwang SH, Norell MA, Ji Q, Gao K. New specimens of *Microraptor zhaoianus* (Theropoda: Dromaeosauridae) from northeastern China. Am Mus Novit. 2002;381:1–44.

Jarvis ED, Mirarab S, Aberer AJ, Li B, Houde P, Li C, Ho SYW, Faircloth BC, Nabholz B, Howard JT, Suh A, Weber CC, da Fonseca RR, Li J, Zhang F, Li H, Zhou L, Narula N, Liu L, Ganapathy G, Boussau B, Bayzid MS, Zavidovych V, Subramanian S, Gabaldón T, Capella-Gutiérrez S, Huerta-Cepas J, Rekepalli B, Munch K, Schierup M, Lindow B, Warren WC, Ray D, Green RE, Bruford MW, Zhan X, Dixon A, Li S, Li N, Huang Y, Derryberry EP, Bertelsen MF, Sheldon FH, Brumfield RT, Mello CV, Lovell PV, Wirthlin M, Schneider MPC, Prosdocimi F, Samaniego JA, Velazquez AMV, Alfaro-Núñez A, Campos PF, Petersen B, Sicheritz-Ponten T, Pas A, Bailey T, Scofield P, Bunce M, Lambert DM, Zhou Q, Perelman P, Driskell AC, Shapiro B, Xiong Z, Zeng Y, Liu S, Li Z, Liu B, Wu K, Xiao J, Yinqi X, Zheng Q, Zhang Y, Yang H, Wang J, Smeds L, Rheindt FE, Braun M, Fjeldsa J, Orlando L, Barker FK, Jønsson KA, Johnson W, Koepfli K-P, O'Brien S, Haussler D, Ryder OA, Rahbek C, Willerslev E, Graves GR, Glenn TC, McCormack J, Burt D, Ellegren H, Alström P, Edwards SV, Stamatakis A, Mindell DP, Cracraft J, Braun EL, Warnow T, Jun W, Gilbert MTP, Zhang G. Whole-genome analyses resolve early branches in the tree of life of modern birds. Science. 2014;346:1320–31.

Jin F, Zhang F, Li Z, Zhang J, Li C, Zhou Z. On the horizon of *Protopteryx* and the early vertebrate fossil assemblages of the Jehol Biota. Chin Sci Bull. 2008;53:2820–7.

Kurochkin EN. A true carinate bird from lower Cretaceous deposits in Mongolia and other evidence of early Cretaceous birds in Asia. Cretac Res. 1985;6:271–8.

Lee MSY, Worthy TH. Likelihood reinstates *Archaeopteryx* as a primitive bird. Biol Lett. 2011;12:1–6.

Lee MSY, Cau A, Naish D, Dyke GJ. Sustained miniaturization and anatomical innovation in the dinosaurian ancestors of birds. Science. 2014;345:562–6.

Lee S, Kim J, Park H, Jabłoński PG, Choi H. The function of the alula in avian flight. Sci Rep. 2015;5:9914.

Lefèvre U, Hu D, Escuillié F, Dyke G, Godefroit P. A new long-tailed basal bird from the lower Cretaceous of north-eastern China. Biol J Linn Soc. 2014;113:790–804.

Li Z, Zhou Z, Wang M, Clarke JA. A new specimen of large-bodied basal enantiornithine *Bohaiornis* from the Early Cretaceous of China and the inference of feeding ecology in Mesozoic birds. J Paleontol. 2014;88:99–108.

Longrich NR, Tokaryk T, Field DJ. Mass extinction of birds at the Cretaceous–Paleogene (K–Pg) boundary. Proc Natl Acad Sci. 2011;108(37):15253–7.

Longrich NR, Vinther J, Meng Q, Li Q, Russell AP. Primitive wing feather arrangement in *Archaeopteryx lithographica* and *Anchiornis huxleyi*. Curr Biol. 2012;22:2262–7.

Maina J. The lung-air sac system of birds: development, structure, and function. Berlin: Springer; 2005.

Marsh OC. Odontornithes: a monograph on the extinct toothed birds of North America. Rep Geol Expl Fortieth Parallels. 1880;7:1–201.

Martin L. The origin of birds and of avian flight. In: Johnston R, editor. Current ornithology, vol. 1. New York: Plenum; 1983. p. 105–29.

Martin LD. A new hesperornithid and the relationships of the Mesozoic birds. Trans Kans Acad Sci. 1984;87:141–50.

Mayr G. Paleogene fossil birds. Berlin-Heidelberg: Springer; 2009.

Mayr G. Metaves, Mirandornithes, Strisores and other novelties – a critical review of the higher-level phylogeny of neornithine birds. J Zool Syst Evol Res. 2011;49:58–76.

Mayr G, Pohl B, Peters DS. A well-preserved *Archaeopteryx* specimen with theropod features. Science. 2005;310:1483–6.

Mayr G, Pohl B, Hartman S, Peters DS. The tenth skeletal specimen of *Archaeopteryx*. Zool J Linn Soc Lond. 2007;149:97–116.

McLelland J. Anatomy of the lungs and air sacs. In: King AS, McLelland J, editors. Form and function in birds. London: Academic; 1989. p. 221–79.

Nopcsa F. Ideas on the origin of flight. Proc Zool Soc Lond. 1907;77(1):223–36.

Nopsca F. On the origin of flight in birds. Proc Zool Soc Lond. 1923;1923:463–77.

Norell MA, Makovicky PJ. Dromaeosauridae. In: Weishampel DB, Dodson P, editors. The Dinosauria. 2nd ed. Berkeley, CA: University of California Press; 2004. p. 196–209.

Novacek MJ. 100 million years of land vertebrate evolution: the Cretaceous-Early Tertiary transition. Ann Mo Bot Gard. 1999;86:230–58.

O'Connor PM. Pulmonary pneumaticity in the postcranial skeleton of extant Aves: a case study examining Anseriformes. J Morphol. 2004;261:141–61.

O'Connor PM. Postcranial pneumaticity: an evaluation of soft-tissue influences on the postcranial skeleton and the reconstruction of pulmonary anatomy in archosaurs. J Morphol. 2006;267:1199–226.

O'Connor PM. Evolution of archosaurian body plans: skeletal adaptations of an air-sac-based breathing apparatus in birds and other archosaurs. J Exp Zool. 2009;311A:629–46.

O'Connor JK, Chiappe LM. A revision of enantiornithine (Aves: Ornithothoraces) skull morphology. J Syst Palaeontol. 2011;9:135–57.

O'Connor PM, Claessens LPAM. Basic avian pulmonary design and flow-through ventilation in non-avian theropod dinosaurs. Nature. 2005;436:253–6.

O'Connor JK, Zhou Z. Early evolution of the biological bird: perspectives from new fossil discoveries in China. J Ornithol. 2015;156:333–42.

O'Connor JK, Wang X, Chiappe LM, Gao C, Meng Q, Cheng X, Liu J. Phylogenetic support for a specialized clade of Cretaceous enantiornithine birds with information from a new species. J Vertebr Paleontol. 2009;29:188–204.

O'Connor JK, Chiappe LM, Gao C, Zhao B. Anatomy of the Early Cretaceous enantiornithine bird *Rapaxavis pani*. Acta Palaentol Pol. 2011a;56:463–75.

O'Connor JK, Sun C, Xu X, Wang X, Zhou Z. A new species of *Jeholornis* with complete caudal integument. Hist Biol. 2011b;24:29–41.

O'Connor JK, Zhou Z, Xu X. Additional specimen of *Microraptor* provides unique evidence of dinosaurs preying on birds. Proc Natl Acad Sci U S A. 2011c;108:19662–5.

O'Connor JK, Zhang Y, Chiappe LM, Meng Q, Li Q, Liu D. A new enantiornithine from the Yixian Formation with the first recognized avian enamel specialization. J Vertebr Paleontol. 2013;33:1–12.

Ostrom JH. Osteology of *Deinonychus antirrhopus*, an unusual theropod from the Lower Cretaceous of Montana. Bull Peabody Mus Nat Hist. 1969;30:1–165.

Ostrom JH. *Archaeopteryx* and the origin of flight. Q Rev Biol. 1974;49:27–47.

Ostrom JH. Bird flight: how did it begin? Did birds begin to fly "from the trees down" or "from the ground up"? Reexamination of *Archaeopteryx* adds plausibility to an "up from the ground" origin of avian flight. Am Sci. 1979;67:46–56.

Padian K, Chiappe LM. The origin of birds and their flight. Sci Am. 1998;278:28–37.

Pennycuick CJ. Mechanical constraints on the evolution of flight. Mem Calif Acad Sci. 1986;8:83–98.

Phil S. Scapular orientation in theropods and basal birds, and the origin of flapping flight. Acta Palaeontol Pol. 2006;51:305–13.

Poore SO, Sanchez-Haiman A, Goslow GE. Wing upstroke and the evolution of flapping flight. Nature. 1997;387:799–802.

Prum RO, Berv JS, Dornburg A, Field DJ, Townsend JP, Lemmon EM, Lemmon AR. A comprehensive phylogeny of birds (Aves) using targeted next-generation DNA sequencing. Nature. 2015;526:569–73.

Pu H, Chang H, Lü J, Wu Y, Xu L, Zhang J, Jia S. A new juvenile specimen of *Sapeornis* (Pygostylia: Aves) from the Lower Cretaceous of Northeast China and allometric scaling of this basal bird. Paleontol Res. 2013;17(1):27–8.

Pycraft WP. On the morphology and phylogeny of the Palaeognathae (Ratitae Crypturi) and Neognathae (Carinatae). Trans Zool Soc Lond. 1900;15:149–290.

Roff DA. The evolution of flightlessness: is history important? Evol Ecol. 1994;8:639–57.

Ruben JA, Dal Sasso C, Geist NR, Hillenius WJ, Jones TD, Signore M. Pulmonary function and metabolic physiology of theropod dinosaurs. Science. 1999;283:514–6.

Sanz JL, Chiappe LM, Perez-Moreno BP, Buscalioni AD, Moratalla JJ, Ortega F, Poyato-Ariza FJ. An Early Cretaceous bird from Spain and its implications for the evolution of avian flight. Nature. 1996;382:442–5.

Schachner ER, Lyson TR, Dodson P. Evolution of the respiratory system in nonavian theropods: evidence from rib and vertebral morphology. Anat Rec. 2009;292:1501–13.

Schutt JWA, Simmons NB. Morphology and homology of the chiropteran calcar, with comments on the phylogenetic relationships of *Archaeopteropus*. J Mamm Evol. 1998;5:1–32.

Sereno PC, Rao C, Li J. *Sinornis santensis* (Aves: Enantiornithes) from the Early Cretaceous of northeastern China. In: Chiappe LM, Witmer LM, editors. Mesozoic birds: above the heads of dinosaurs. Berkeley, CA: University of California Press; 2002. p. 184–208.

Thorington RW, Darrow K, Anderson CG. Wing tip anatomy and aerodynamics in flying squirrels. J Mammal. 1998;79:245–50.

Turner AH, Makovicky PJ, Norell MA. A review of dromaeosaurid systematics and paravian phylogeny. Bull Am Mus Nat Hist. 2012;371:1–206.

Vickers-Rich P, Chiappe LM, Kurzanov S. The enigmatic birdlike dinosaur *Avimimus portentosus*. In: Chiappe L, Witmer LM, editors. Mesozoic birds: above the heads of dinosaurs. California: University California Press; 2002. p. 65–86.

Walker A. New subclass of birds from the Cretaceous of South America. Nature. 1981;292:51–3.

Wang M. Taxonomical revision, ontogenetic, ecological and phylogenetic analyses of Enantiornithes (Aves: Ornithothoraces) of China. Dissertation. University of Chinese Academy of Sciences. 2014.

Wang M, Lloyd GT. Rates of morphological evolution are heterogeneous in Early Cretaceous birds. Proc R Soc Biol B. 2016;283:1828.

Wang M, Zhou Z. A new adult specimen of the basalmost ornithuromorph bird *Archaeorhynchus spathula* (Aves: Ornithuromorpha) and its implications for early avian ontogeny. J Syst Palaeontol. 2016; doi:10.1080/14772019.2015.1136968.

Wang M, O'Connor JK, Zhou Z. A new robust enantiornithine bird from the Lower Cretaceous of China with scansorial adaptations. J Vertebr Paleontol. 2014;34:657–71.

Wang M, Li D, O'Connor JK, Zhou Z, You H. Second species of enantiornithine bird from the Lower Cretaceous Changma Basin, northwestern China with implications for the taxonomic diversity of the Changma avifauna. Cretac Res. 2015a;55:56–65.

Wang M, Zheng X, O'Connor JK, Lloyd GT, Wang X, Wang Y, Zhang X, Zhou Z. The oldest record of Ornithuromorpha from the Early Cretaceous of China. Nat Commun. 2015b;6:6987.

Wang M, Zhou Z, Sullivan C. A fish-eating enantiornithine bird from the Early Cretaceous of China provides evidence of modern avian digestive features. Curr Biol. 2016a;26:1170–6.

Wang M, Wang X, Wang Y, Zhou Z. A new basal bird from China with implications for morphological diversity in early birds. Sci Rep. 2016b;6:19700.

Wang M, Zhou Z, Zhou S. A new basal ornithuromorph bird (Aves: Ornithothoraces) from the Early Cretaceous of China with implication for morphology of early Ornithuromorpha. Zool J Linn Soc Lond. 2016c;176:207–23.

Wellnhofer P. *Archaeopteryx*. Münich: Der Urvogel von Solnhofen; 2008.

Wellnhofer P. A short history of research on *Archaeopteryx* and its relationship with dinosaurs. Geol Soc Lond. 2010;343:237–50.

Williston S. Are birds derived from dinosaurs. Kansas City Rev Sci. 1879;3:457–60.

Xu X, Mackem S. Tracing the evolution of avian wing digits. Curr Biol. 2013;23:R538–44.

Xu X, Zhou Z, Wang X. The smallest known non-avian theropod dinosaur. Nature. 2000;408:705–8.

Xu X, Zhou Z, Wang X, Kuang X, Zhang F, Du X. Four-winged dinosaurs from China. Nature. 2003;421:335–40.

Xu X, Zheng X, You H. Exceptional dinosaur fossils show ontogenetic development of early feathers. Nature. 2010;464:1338–41.

Xu X, You H, Du K, Han F. An *Archaeopteryx*-like theropod from China and the origin of Avialae. Nature. 2011;475:465–70.

Xu X, Zhou Z, Dudley R, Mackem S, Chuong CM, Erickson GM, Varricchio DJ. An integrative approach to understanding bird origins. Science. 2014;346:1253293.

Xu X, Zheng X, Sullivan C, Wang X, Xing L, Wang Y, Zhang X, O'Connor JK, Zhang F, Pan Y. A bizarre Jurassic maniraptoran theropod with preserved evidence of membranous wings. Nature. 2015;521:70–3.

You H, Lamanna MC, Harris JD, Chiappe LM, O'Connor JK, Ji S, Lü J, Yuan C, Li D, Zhang X, Lacovara KJ, Dodson P, Ji Q. A nearly modern amphibious bird from the Early Cretaceous of northwestern China. Science. 2006;312:1640–3.

Zhang Y, Tian X. Analyses of diversity of avian tarsometatarsus shape and its function. Sichuan J Zool. 2006;25:703–9.

Zhang F, Zhou Z. A primitive enantiornithine bird and the origin of feathers. Science. 2000;290:1955–9.

Zhang F, Zhou Z, Hou L, Gu G. Early diversification of birds: evidence from a new opposite bird. Chin Sci Bull. 2001;46:945–9.

Zhang F, Zhou Z, Xu X, Wang X. A juvenile coelurosaurian theropod from China indicates arboreal habits. Naturwissenschaften. 2002;89:394–8.

Zhang F, Zhou Z, Dyke GJ. Feathers and 'feather-like' integumentary structures in Liaoning birds and dinosaurs. Geol J. 2006;41:395–404.

Zhang F, Zhou Z, Xu X, Wang X, Sullivan C. A bizarre Jurassic maniraptoran from China with elongate ribbon-like feathers. Nature. 2008a;455:1105–8.

Zhang F, Zhou Z, Benton MJ. A primitive confuciusornithid bird from China and its implications for early avian flight. Sci China Ser A. 2008b;51:625–39.

Zheng X, Martin LD, Zhou Z, Burnham DA, Zhang F, Miao D. Fossil evidence of avian crops from the Early Cretaceous of China. Proc Natl Acad Sci U S A. 2011;108:15904–7.

Zheng X, O'Connor JK, Huchzermeyer F, Wang X, Wang Y, Wang M, Zhou Z. Preservation of ovarian follicles reveals early evolution of avian reproductive behaviour. Nature. 2013;495:507–11.

Zheng X, O'Connor JK, Wang X, Wang M, Zhang X, Zhou Z. On the absence of sternal elements in *Anchiornis* (Paraves) and *Sapeornis* (Aves) and the complex early evolution of the avian sternum. Proc Natl Acad Sci U S A. 2014a;111:13900–5.

Zheng X, O'Connor JK, Huchzermeyer F, Wang X, Wang Y, Zhang X, Zhou Z. New specimens of *Yanornis* indicate a piscivorous diet and modern alimentary canal. PLoS One. 2014b;9:e95036.

Zhou Z. The origin and early evolution of birds: discoveries, disputes, and perspectives from fossil evidence. Naturwissenschaften. 2004a;91:455–71.

Zhou Z. Gastroliths in *Yanornis*: an indication of the earliest radical diet-switching and gizzard plasticity in the lineage leading to living birds? Naturwissenschaften. 2004b;91:455–71.

Zhou Z, Farlow JO. Flight capability and habits of *Confuciusornis*. In: Gauthier J, Gall LF, editors. New perspectives on the origin and early evolution of birds. New Haven: Yale University Press; 2001. p. 237–54.

Zhou Z, Wang Y. Vertebrate diversity of the Jehol Biota as compared with other lagerstätten. Sci China Earth Sci. 2010;53:1894–907.

Zhou Z, Zhang F. Two new ornithurine birds from the Early Cretaceous of western Liaoning, China. Chin Sci Bull. 2001;46:1258–64.

Zhou Z, Zhang F. A long-tailed, seed-eating bird from the Early Cretaceous of China. Nature. 2002a;418:405–9.

Zhou Z, Zhang F. Largest bird from the Early Cretaceous and its implications for the earliest avian ecological diversification. Naturwissenschaften. 2002b;89:34–8.

Zhou Z, Zhang F. *Jeholornis* compared to *Archaeopteryx*, with a new understanding of the earliest avian evolution. Naturwissenschaften. 2003a;90:220–5.

Zhou Z, Zhang F. Anatomy of the primitive bird *Sapeornis chaoyangensis* from the Early Cretaceous of Liaoning, China. Can J Earth Sci. 2003b;40:731–47.

Zhou Z, Zhang F. Discovery of an ornithurine bird and its implication for Early Cretaceous avian radiation. Proc Natl Acad Sci U S A. 2005;102:18998–9002.

Zhou Z, Zhang F. Mesozoic birds of China—a synoptic review. Vertebr Palasiat. 2006a;44:74–98.

Zhou Z, Zhang F. A beaked basal ornithurine bird (Aves, Ornithurae) from the Lower Cretaceous of China. Zool Scr. 2006b;35:363–73.

Zhou Z, Clarke J, Zhang F. Insight into diversity, body size and morphological evolution from the largest Early Cretaceous enantiornithine bird. J Anat. 2008;212:565–77.

Zhou Z, Zhang F, Li Z. A new basal ornithurine bird (*Jianchangornis microdonta* gen. et sp. nov.) from the Lower Cretaceous of China. Vertebr Palasiat. 2009;47:299–310.

Zhou Z, Li Z, Zhang F. A new Lower Cretaceous bird from China and tooth reduction in early avian evolution. Proc R Soc B. 2010;277:219–27.

Zhou S, Zhou Z, O'Connor JK. Anatomy of the basal ornithuromorph bird *Archaeorhynchus spathula* from the Early Cretaceous of Liaoning. China J Vertebr Paleontol. 2013;33:141–52.

Zhou S, Zhou Z, O'Connor JK. A new piscivorous ornithuromorph from the Jehol Biota. Hist Biol. 2014a;26:608–18.

Zhou S, O'Connor JK, Wang M. A new species from an ornithuromorph (Aves: Ornithothoraces) dominated locality of the Jehol Biota. Chin Sci Bull. 2014b;59:5366–78.

Zusi RL. Patterns of diversity in the avian skull. In: Hanken J, Hall BK, editors. The skull. Chicago, IL: University of Chicago Press; 1993. p. 391–437.

The Avian Lingual and Laryngeal Apparatus Within the Context of the Head and Jaw Apparatus, with Comparisons to the Mammalian Condition: Functional Morphology and Biomechanics of Evaporative Cooling, Feeding, Drinking, and Vocalization

2

Dominique G. Homberger

Abstract

The lingual and laryngeal apparatus are the mobile and active organs within the oral cavity, which serves as a gateway to the respiratory and alimentary systems in terrestrial vertebrates. Both organs play multiple roles in alimentation and vocalization besides respiration, but their structures and functions differ fundamentally in birds and mammals, just as the skull and jaws differ fundamentally in these two vertebrate classes. Furthermore, the movements of the lingual and laryngeal apparatus are interdependent with each other and with the movements of the jaw apparatus in complex and little-understood ways. Therefore, rather than updating the existing numerous reviews of the diversity in lingual morphology of birds, this chapter will concentrate on the functional-morphological interdependences and interactions of the lingual and laryngeal apparatus with each other and with the skull and jaw apparatus. It will:

1. Briefly review the salient features of the mammalian head as a baseline against which to understand the uniqueness of the avian head
2. Describe general morphological features of the lingual and laryngeal apparatus within the context of the skull and jaw apparatus
3. Contrast some fundamental functional-morphological differences that exist among the jaw, lingual and laryngeal apparatus of birds
4. Provide models of the movements of the various parts of the lingual and laryngeal apparatus based on biomechanical analyses
5. Integrate these models with behaviors in thermoregulation, feeding, drinking, and vocalization
6. Briefly demonstrate how detailed morphological and functional analyses can be tested and expanded by using 3D visualization and animation
7. Place the provided data in an evolutionary framework

Keywords

Larynx • Tongue • Jaw • Vocalization • Feeding • Biomechanics

D.G. Homberger (✉)
Department of Biological Sciences, Louisiana State University, Baton Rouge, LA 70803-1715, USA
e-mail: zodhomb@lsu.edu

J.N. Maina (ed.), *The Biology of the Avian Respiratory System*, DOI 10.1007/978-3-319-44153-5_2

Contents

Abbreviations (Including Some Selected Synonymies)

Al.ling	*Ala linguae*
Al.lingc	*Ala linguae caudalis*
Al.lingr	*Ala linguae rostralis*
Ap.ling	*Apex linguae*
AR	*Arytenoideum*
AR.cart	Cartilaginous portion of the *Arytenoideum*
Ar.prgl	*Area preglottalis*, lingual base (Zweers and Berkhoudt 1987)
Art.	*Articulatio*, joint
Art.ar	*Art. intra-arytenoidea*
Art.cb	*Art. cerato-basihyalis*
Art.cf	*Art. craniofacialis*
Art.crp	*Art. crico-procricoidea*
Art.ecb	*Art. epi-ceratobranchialis*
Art.eph	*Art. epi-pharyngobranchialis*
Art.icr	*Art. intracricoidea*
Art.icrd	*Art. intracricoidea dorsalis*
Art.icrdv	*Art. intracrioidea ventrodorsalis*
Art.jr	*Art. jugo-rostralis*
Art.par	*Art. processo-arytenoidea*
Art.pgb	*Art. paraglosso-basihyalis*
Art.ppr	*Art. pterygo-palatorostralis*
Art.pr	*Art. palato-rostralis*
Art.pra	*Art. procrico-arytenoidea*
Art.qj	*Art. quadrato-jugalis*
Art.qm	*Art. quadrato-mandibularis*
Art.qpt	*Art. quadrato-pterygoidea*
Art.qqj	*Art. quadrato-quadratojugalis*
Art.qso	*Art. quadrato-squamoso-otica*
BH	*Basihyale*
BHD	*Basihyoideum* (mammals)
C.ling	*Corpus linguae*
CAN	Canine tooth (mammals)
Cart.ar	*Cartilago arytaenoidea* (mammals)
Cart.cr	*Cartilago cricoidea* (mammals)
Cart.thy	*Cartilago thyroidea* (mammals)
CASC	Subcutaneous cervical air sac of the cervico-cephalic air sac system
Cav.lrx	*Cavum laryngis*, laryngeal chamber
Cav.nas	*Cavum nasale*
Cav.or	*Cavum oris*, oral or mouth cavity
Cav.sbl	Cavum sublinguale
Cav.syn	*Cavitas synovialis fasciae vaginalis*
Cav.tymp	*Cavum tympanicum*
CB	*Ceratobranchiale*
Cer	*Cera*, cere
Cerb	*Cerebrum*
Cerbl	*Cerebellum*
CH	Choana
CHD	*Ceratohyoideum* (mammals)

CL	*Clavicula*, Furcula
Cond.mq	*Condylus mandibularis quadrati*
Cond.occ	*Condylus occipitalis*
Condd.occ	*Condyli occipitales* (mammals)
Cot.qm	*Cotyla quadratica mandibulae*
CR	*Cricoideum*
CRd	*Cricoideum dorsale*
CRv	*Cricoideum ventrale*
SK	Skull, braincase, *Cranium*
Cutl	*Cuticula cornificata linguae*, the lingual nail
EB	*Epibranchial*
EH	*Epihyoideum* (mammals)
em	Epimysium (of *M. branchiomandibularis*)
EPGL	Epiglottis (mammals)
Fac.art.ar	*Facies articularis intra-arytenoidea*
Fac.art.bc	*Facies articularis basihyalis ceratobranchialis*
Fac.art.buh	*Facies articularis basihyo-urohyalis*
Fac.art.cb	*Facies articularis ceratobranchialis basihyalis*
Fac.art.mq	*Facies articularis mandibularis quadrati*
Fac.art.par	*Facies articularis processus arytenoidei*
Fac.art.phe	*Facies articularis pharyngobranchialis epibranchialis*
Fac.art.qm	*Facies articularis quadratica mandibulae*
F.cerv	*Fascia cervicalis*
F.sbl	*Fascia sublingualis*
F.vag	*Fascia vaginalis hyoidei*, hyoid sheath
F.vag.pv	*Fascia vaginalis hyoidei parietalis et visceralis*
F.vagp	*Fascia vaginalis hyoidei parietalis*
F.vagv	*Fascia vaginalis hyoidei visceralis*
FM	*Foramen magnum*
FR	Filing ridges on internal surface of projecting upper bill tip in parrots and some cockatoos
GL	*Glottis*
Gl.ling	*Glandula lingualis*; *Gll. linguales rostrales* (McLelland 1993)
Gl.md	*Glandula mandibularis*; *Gll. mandibulares caudales* (McLelland 1993)

Gl.mx	*Glandula maxillaris*
Gl.pregl	*Glandula preglottalis*; *Glandulae linguales caudales* (McLelland 1993)
Gl.phrx	*Glandula pharyngealis*; *Gll. sphenopterygoideae* (Zweers 1982a, b)
Gl.sbl	*Glandula sublingualis; Gll. mandibulares rostrales* (McLelland 1993)
Gl.pl	*Glandula palatina*
HP	Hard, or secondary, palate (mammals)
IM.cnstr.crar	Insertion of *M. constrictor glottidis cricoarytenoideus*
IM.cnstr.iarr	Insertion of *M. constrictor glottidis interarytenoideus rostralis*
IM.cnstr.iars	Insertion of *M. constrictor glottidis interarytenoideus superficialis*
IM.dlgl	Insertion of *M. dilator glottidis*
INC	Incisor tooth (mammals)
INGL	*Ingluvies*, crop
INT	Integument (skin)
J	*Os jugale*
Lab.gl	*Labium glottidis*
Lab.or	*Labium oralis* (mammals)
Lb.pl	*Lobus palatinus*
Lig.iph	*Ligamentum interparahyale*
Lig.crar	*Ligamentum crico-arytenoideum*
Lig.nc	*Ligamentum nodulo-ceratobranchiale*
Lig.po	*Ligamentum postorbitale*
Lphrx	*Laryngopharynx* (mammals)
LP	lips, *Labia oris* (mammals)
LX	Larynx
M.ame	*M. adductor mandibulae externus*
M.amec	*M. adductor mandibulae externus caudalis*
M.amerl	*M. adductor mandibulae externus rostralis lateralis*
M.amert	*M. adductor mandibulae externus rostralis temporalis*
M.amev	*M. adductor mandibular externus ventralis*
M.bm	*M. branchiomandibularis*
M.bmr	*M. branchiomandibularis rostralis*
M.bmc	*M. branchiomandibularis caudalis*
M.cg	*M. ceratoglossus*

M.cgl	*M. ceratoglossus lateralis*	M.th	*M. tracheohyoideus*
M.ch	*M. ceratohyoideus*	M.thc	*M. tracheohyoideus caudalis*
M.cnstr	*M. constrictor glottidis*	M.thr	*M. tracheohyoideus rostralis*
M.cnstr. crar	*M. constrictor glottidis cricoarytenoideus*	M.tl	*M. tracheolateralis*
		M.tll	*M. tracheolateralis lateralis*
M.cnstr. iarc	*M. constrictor glottidis interarytenoideus caudalis*	M.tlm	*M. tracheolateralis medialis*
		M.tlu	*M. tracheolateralis urohyalis*
M.cnstr. iarr	*M. constrictor glottidis interarytenoideus rostralis*	MD	*Os mandibulare*, mandible
		Med.obl	*Medulla oblongata*
M.cnstr. iars	*M. constrictor glottidis interarytenoideus superficialis*	Med.sp	*Medulla spinalis*
		Mem.pl	*Membrana connectiva palatina* (palatine joint membrane between the corneous and soft parts of the palate)
M.cnstr. icr	*M. constrictor glottidis intercricoideus*		
M.crh	*M. cricohyoideus*	MH	Mesohyoideum
M.crhd	*M. cricohyoideus dorsalis*	Mm.lrx	Laryngeal muscles
M.crhdi	*M. cricohyoideus dorsalis intermedius*	Mm.ling	Lingual muscles
		Mns.lrx	*Mons laryngealis*
M.crhdp	*M. cricohyoideus dorsalis profundus*	MOL	Molar tooth (mammals)
		N	*Naris*
M.crhds	*M. cricohyoideus dorsalis superficialis*	NCASC	Passageway between naris and cervical air sac CASC
M.crhl	*M. cricohyoideus lateralis*	ND	*Nodulus* (sesamoid bone on the urohyal)
M.crhm	*M. cricohyoideus medialis*		
M.crhv	*M. cricohyoideus ventralis*	Neck	Neck (vertebral column and neck musculature)
M.crpp	*M. cricopapillaris profundus*		
M.crps	*M. cricopapillaris superficialis*	Nphrx	*Nasopharynx* (mammals)
M.dlgl	*M. dilator glottidis*	OC	Optic chiasma
M.dt	*M. dermotemporalis*	OE	Esophagus
M.em	*M. ethmomandibularis*	OM.cnstr. crar	Origin of *M. constrictor glottidis cricoarytenoideus*
M.dm	*M. depressor mandibulae*		
M.ggl	*M. genioglossus*	OM.cnstr. iarc	Origin of *M. constrictor glottidis interarytenoideus caudalis*
M.ho	*M. hyoglossus obliquus*		
M.hol	*M. hyoglossus obliquus lateralis*	OM.cnstr. iarr	Origin of *M. constrictor glottidis interarytenoideus rostralis*
M.hom	*M. hyoglossus obliquus medialis*		
M.hgr	*M. hyoglossus rostralis*	OM.cnstr. icr	Origin of *M. constrictor glottidis intercricoideus*
M.mgl	*M. mesoglossus*		
M.mhr	*M. mylohyoideus rostralis*	OM.dlgl	Origin of *M. dilator glottidis*
M.mhc	*M. mylohyoideus caudalis*	Ophrx	*Oropharynx* (mammals)
M.pq	*M. protractor quadrati*	Papp.ch	*Papillae choanales*
M.pss	*M. pseudotemporalis superficialis*	Papp.gl	*Papillae glottidiales*
M.pt	M. pterygoideus	Papp.lrx	*Papillae laryngeales*; ventral pharyngeal scrapers (Zweers and Berkhoudt 1987)
M.ptl	M. pterygoideus lateralis		
M.pvl	*M. pterygoideus ventralis lateralis*		
M.retr.pl	*M. retractor pterygopalatini*	Papp.ling	*Papillae linguales*
M.sgl	*M. supraglossus*	Papp.phrx	*Papillae pharyngeales*
M.sh	*M. serpihyoideus*	Papp.pl	*Papillae palatinae*
M.st	*M. stylohyoideus*	PASP	Parapatagial air sac of the pulmonary air sac system
M.sth	*M. sternohyoideus* (mammals)		

PB	*Pharyngobranchiale*
PC	*Procricoideum*
PG	*Paraglossum*
PIT	Pituitary gland
PL	*Os palatinum*
Pl.corn	*Palatum corneum*
Plut.ch	*Pluteum choanae*, lateral choanal shelf
PM	Premolar tooth (mammals)
Proc.ar	*Processus arytenoideus*
Proc.lat	*Processus lateralis quadrati*
Proc.mq	*Processus mandibularis quadrati*
Proc.mst	*Processus mastoideus* (mammals)
Proc.orb	*Processus orbitalis quadrati*
Proc.ot	*Processus oticus quadrati*
Proc.ph	*Processus parahyalis*
Proc.po	*Processus postorbitalis*
Proc.rtr	*Processus retroarticularis mandibulae*
Proc. uh	*Processus urohyalis*
Proc.zyg	*Processus zygomaticus*
PT	*Os pterygoideum*
Q	*Os quadratum*
R.mx	*Rostrum maxillare*, maxilla, upper beak
R.sph	*Rostrum sphenoidale*
Ram.md	*Ramus mandibularis*
Rh.md	*Rhamphotheca mandibularis*
Rh.mx	*Rhamphotheca maxillaris*
Rim.inf	*Rima infundibuli*, entrance to the Eustachian tubes
Sin.f	*Sinus frontalis* (mammals)
Sin.sph	*Sinus sphenoidalis* (mammals)
SPL	Soft palate (mammals)
STH	*Stylohyoideum* (mammals)
SX	Syrinx
T	Tongue (mammals)
T.md	*Tomium mandibulare*, cutting edge of the mandibular rhamphotheca
TB	Tympanic bulla (mammals)
TMH	*Tympanohyoideum* (mammals)
Tor.pl	*Torus palatinus*
TR	Trachea
TRh	Tracheal half rings
TYH	*Thyrohyoideum* (mammals)
TS	Transverse step of the corneous palate (*Gradus transversus palati cornei*)
VERT	Vertebral column
VF	Vocal fold, *Plica vocalis* (mammals)

2.1 Introduction

It has been firmly established how the processes and underlying structures of pulmonary ventilation and respiration of birds differ fundamentally from those of mammals. What is much less well studied is how the accessory organs of the respiratory system (e.g., the tongue and larynx within the oral cavity) compare in these two vertebrate classes.

The lingual and laryngeal apparatus of birds differ fundamentally from those of mammals in their structure and function. These differences are a consequence of the fundamental differences in the skull and jaw apparatus because of the separate evolutionary paths of the therapsid and sauropsid lineages probably dating back to the Devonian. Nevertheless, the uniqueness of the avian condition can be appreciated best when it is contrasted with the mammalian condition. This chapter will, therefore, first summarize the better-known mammalian configurations of the tongue and larynx within the context of the head and jaw apparatus as a baseline against which the subsequent general overview of the surface features of the avian configuration is compared. An overview of the kinesis of the avian skull and jaw apparatus will provide the framework for some examples of the functional morphology of the lingual and laryngeal apparatus to illustrate the breadth of possible adaptive expressions and to form the basis for a discussion of some behaviors that employ the lingual and laryngeal apparatus.

In general, an understanding of the interplay of the movements of the jaw, tongue, and larynx of birds has been sought through physiological and experimental techniques (e.g., Zweers 1974, 1982a, b, c; Zweers et al. 1995; Suthers and Zollinger 2008), and such studies will be reviewed. In addition, this chapter seeks a more direct understanding through anatomical, biomechanical, and imaging approaches to achieve integrated insights into the functions, biological roles, adaptations, and evolutionary histories of morphological constructions and configurations. A comparison of salient mammalian and avian features will close the circle started with the first section on mammals, and the chapter will conclude with a discussion of the evolutionary implications to guide future research.

A review of the vast literature on the morphology of the avian skull, jaw, and tongue is beyond the scope of this chapter, which, therefore, focuses on a selection of more recent, relevant, or lesser-known references. The figures are based on unpublished originals or have been redrawn and adapted for consistency of style and information. In the text, anglicized anatomical terms are used for legibility wherever appropriate, but the figures are labeled with abbreviations or acronyms of Latin anatomical terms, which are listed separately as well as with every figure caption to facilitate the analysis of the figures.

2.2 General Aspects of the Mammalian Head

Mammals are characterized by an akinetic skull[1] in which the temporomandibular jaw joint is the only diarthrodial, or synovial, joint,[2] which serves as the pivot around which the lower jaw, or mandible, moves relative to the upper jaw, which is part of the face (Fig. 2.1a). The jaws of most mammals bear teeth for prehension or mastication, although teeth are reduced or absent in mammals that specialize on small prey items that form malleable boluses (e.g., baleen whales or Mysticeti), on soft earthworms [some rodents (Balete et al. 2007; Esselstyn et al. 2012)], or on ants and termites and, thereby, also pick up small stones [e.g., anteaters (Vermilingua), numbats (*Myrmecobius fasciatus*), pangolins (Pholidota)].

The oral cavity is bounded rostrally and laterally by the upper and lower rows of teeth and by soft lips, which can seal off the mouth cavity to retain its moisture or, in infants, help form a seal around the nipple of the mother during suckling. The oral cavity is lined by a soft-cornified mucosal epithelium that needs to be kept moist. The ceiling of the oral cavity, or palate, consists rostrally of the hard palate, which is supported by the bony secondary palate formed by the maxillary and palatine bones. Caudally, the muscular soft palate extends the palate far back; its caudal margin demarcates the caudally joining pharynx with the choana, or internal naris, which is located above the larynx in the resting position when the mouth is closed. Together, the hard and soft palate separate the oral cavity from the nasal cavity, which is very large in mammals (Fig. 2.1b). The single opening to the Eustachian tubes is located in the ceiling of the nasopharynx above the soft palate. The floor of the oral cavity is formed, in effect, by the fleshy tongue and, except in adult humans, the larynx.

The hyoid skeleton of the lingual apparatus is reduced and consists of the small hyoid bone, which is suspended from the tympanic area of the skull by paired chains of hyoid ossicles, as in carnivores, ungulates, and Xenarthra (armadillos, sloths, and anteaters) (Fig. 2.1a), or paired ligaments, as in rabbits and primates (Naples 1986, 1999; Nickel et al. 1986; Walker and Homberger 1993; Hiiemae 2000; Homberger and Walker 2004). These paired suspensory limbs underlie the fleshy palatoglossal arches, which frame the entrance to the pharynx, the *Isthmus faucium*, and constrain the size of the food items that can pass into the pharynx and be swallowed (Schummer et al. 1979; Dyce et al. 1987; Homberger 1994a, 1999; Homberger and Walker 2004).

The body of the tongue is fleshy-muscular and devoid of bone (Fig. 2.1b); it functions as a

[1] The term akinetic for the mammal skull originated at a time when it was believed that the skull, excluding the lower jaw, was a rigid entity. Since then it has been increasingly understood that the sutures between the individual cranial and facial bones provide significant flexibility, which is necessary not only for the growth and maintenance of the cranial bones but also for balancing the forces that are exerted on the skull during mastication (see, e.g., Herring and Teng 2000).

[2] Besides the joints of the incus with the stapes and malleus in the middle ear.

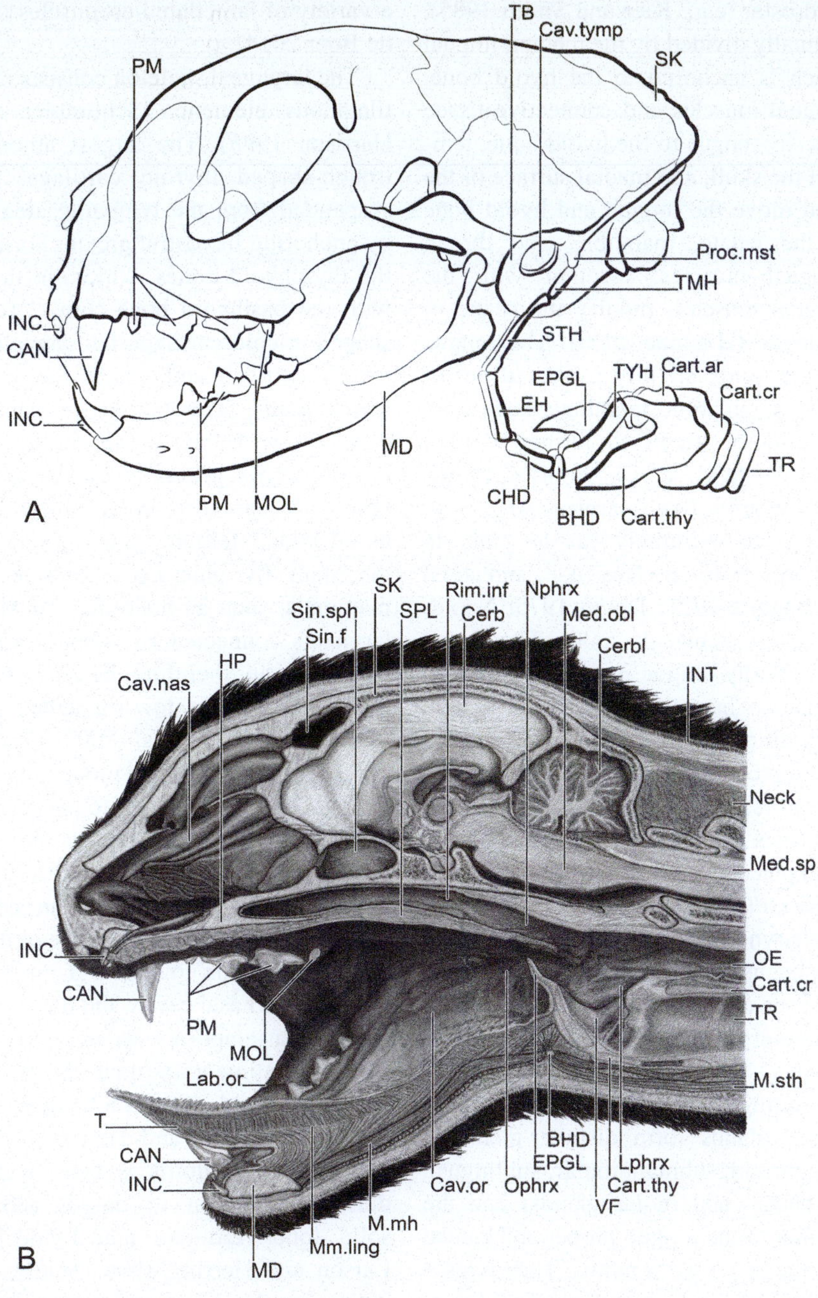

Fig. 2.1 Salient features of the akinetic skull and soft tissue configuration of the head of a cat (*Felis catus*). (**a**) Lateral view of the skeletal elements of the akinetic skull, jaw, hyoid, and larynx. (**b**) Longitudinal section through the head to highlight the secondary palate, fleshy tongue, the apposition of the larynx to the choana, teeth, and the crossing of the air and food passages. BHD = basihyoid bone; CAN = canine tooth; Cart.ar = arytenoid cartilage; Cart.cr = cricoid cartilage; Cart.thy = thyroid cartilage; Cav.nas = nasal cavity; Cav.or = oral cavity; Cav.tymp = tympanic cavity; Cerb = *Cerebrum*; Cerbl = *Cerebellum*; CHD = ceratohyoid bone; EH = epihyoid bone; EPGL = epiglottis; HP = hard palate; INC = incisor tooth; INT = integument (skin); Lab.or = lips; Lphrx = laryngopharynx; M.mh = *M. mylohyoideus*; M.sth = *M. sternohyoideus*; MD = mandible; Med.obl = *Medulla oblongata*; Med.sp = spinal cord; Mm.ling = lingual muscles; MOL = molar tooth; Neck = neck; Nphrx = nasopharynx; OE = esophagus; Ophrx = oropharynx; PM = premolar tooth; Proc.mst = mastoid process; Rim.inf = entrance to Eustachian tubes; Sin.f = frontal sinus; Sin.sph = sphenoid sinus; SK = skull (braincase); SPL = soft palate; STH = stylohyoid bone; T = tongue; TB = tympanic bulla; TYH = thyrohyoid bone; TMH = tympanohyoid bone; TR = trachea; VF = vocal fold [Adapted from Homberger and Walker (2004)]

muscular hydrostat (e.g., Kier and Smith 1985). It is longitudinally divided by the median lingual septum, which is anchored to the hyoid bone. Extrinsic lingual muscles and connective tissue anchor the fleshy tongue to the hyoid bone, temporal bone of the skull, and medial surface of the mandible and move the tongue and hyoid bone relative to the palate, mandible, and throat. Intrinsic lingual muscles, together with the extrinsic lingual muscles, modify the shape of the fleshy tongue (Hiiemae 2000). The tongue can generally be only slightly projected out of the mouth, although some mammals have long tongues that can be projected farther to grab grass or leaves (e.g., cows, giraffes, black rhinoceroses) or drink (e.g., carnivores), or can be projected even extremely far to pick up termites and ants from crevices (e.g., anteaters) (Doran and Baggett 1971; Doran 1975; Naples 1999; Reiss 2000; Schwenk 2000). The dorsal surface of the fleshy tongue is strewn with epithelial papillae to increase the grip with food items for deglutition. It also bears a large number of taste receptors and nerve endings as mechanoreceptors. During the act of swallowing, the tongue pushes an aliquot of liquid, an appropriately small piece of food, or a masticated food bolus past the *Isthmus faucium* into the pharynx where the pharyngeal musculature squeezes it into the esophagus (Hiiemae 2000; Matsuo and Palmer 2008; see also below).

Various salivary glands produce saliva that is serous, mucous, or a mixture of serous and mucous depending on the glands and species. Many salivary glands form distinct glandular bodies (e.g., parotid, submandibular, sublingual, zygomatic, buccal and molar glands) and are located external to the tongue proper and caudal or ventral to the mandibular ramus. They release their saliva through ducts on the sublingual floor of the oral cavity underneath the tip of the free tongue or into the oral vestibule between the tooth rows and cheeks. Small salivary glands are spread over the mucosal lining of the oral cavity and tongue, as well as around the gustatory vallate papillae (von Ebner's glands). Some serous saliva contains enzymes to predigest food. Mechanoreception is based on nerve endings and a variety of lamellated corpuscles (Haggard and de Boer 2014).

The laryngeal skeleton consists of several cartilaginous elements (Schummer et al. 1979; Harrison 1995). The largest among them, the trough-shaped thyroid cartilage, is indirectly suspended from the tympanic area of the skull by anchoring its rostral greater horns (cornua) to the thyrohyoid bones, which, in turn, articulate with the basihyoid bone (Fig. 2.1a). The ring-shaped cricoid cartilage fits caudally within the thyroid cartilage and connects to it *via* its caudal lesser horns. The paired arytenoid cartilages articulate with the cricoid cartilage. The mammalian larynx generally moves with the hyoid bone, but can also be moved relative to the hyoid bone, sternum, pharynx, and the base of the skull by pharyngeal muscles. In certain mammals, such as male deer or large cats, the larynx is connected by ligaments to the hyoid bone and can, therefore, be retracted toward the thorax to generate low-frequency vocalizations (Fitch 2006; Frey et al. 2007; Frey and Gebler 2010). Intrinsic muscles move the arytenoid and cricoid cartilages relative to the thyroid cartilage to open and close the glottis and modulate the tension of the vocal folds (Titze 1994). Except in adult humans,[3] the larynx can be raised and suspended from the caudal edge of the soft palate by the epiglottis to create a direct connection between the nasal cavity and the trachea, thereby allowing a temporary separation of the airways from the alimentary tract (Dyce et al. 1987; Schummer et al. 1979; Laitman and Reidenberg 1993a, b; Larson and Herring 1996; Hiiemae 2000). The epiglottis is bent over the larynx during deglutition of liquids and boluses of solid food (Laitman and Reidenberg 1993a; Larson and Herring 1996; Matsuo and Palmer 2008), but this function is apparently not essential to prevent liquids and food particles from entering the trachea (Laitman and Reidenberg 1993a; Leder et al. 2010).

[3] For a discussion of the capacity for articulated speech in adult humans, see Sect. 2.9.

2.3 General Aspects of the Avian Head

Birds are characterized by a kinetic skull and jaw apparatus that differ in fundamental aspects from the akinetic skull and jaw of mammals and, therefore, require fundamentally different configurations of the tongue and larynx and their suspensory apparatus (Fig. 2.2). The avian skull comprises a multielement upper jaw apparatus and a mandible, both of which are moved by an intermediary bone, the quadrate, which articulates with the skull (see Sect. 2.4). The toothless jaws of extant birds are covered by hard-cornified rhamphothecae that are richly endowed with a variety of mechanoreceptors (see below).

Paired (external) nares are located on the dorsal side of the upper beak near its articulation with the cranium, except in kiwis (*Apteryx* spp.), which sport the nares at the tip of their upper beak, and in some diving birds, such as cormorants and darters (Phalacrocoracidae) and gannets and boobies (Sulidae), in which they are replaced by openings at the caudal end of the gape. The nares lead into nasal cavities that fill part of the upper beak and vary in size and in the number and complexity of the turbinates (conchae) among birds (Bang and Wenzel 1985; Porter and Witmer 2016; Danner et al. 2017).

The oral cavity[4] is bounded rostrally and laterally by the edges of the maxillary and mandibular rhamphotheca. These hard-cornified edges of the beak are largely congruent with each other in most birds, so that the oral cavity can be sealed off if the palate is covered by soft-cornified epithelium. In filter-feeding birds, these rhamphothecal edges bear rows of cornified lamellae (e.g., Goodman and Fisher 1962; Zweers 1974; Zweers et al. 1977, 1995; Kooloos et al. 1989; Jackowiak et al. 2011). In some birds, such as toucans, the rhamphothecal edges are fashioned into sawblade-like corneous teeth that are pointing rostrally (Bühler 1995). In some birds (e.g., parrots, cockatoos, open-bill storks), a gap remains between the upper and lower beaks even when the beak is closed, but parrots and cockatoos (Psittaciformes) can close off their oral cavity by fitting their tongues within the arch created by fleshy palatal lobes lateral to the palatal torus (see below). In birds without patent external nares (see above), such gaps serve as alternative openings for breathing air.

The ceiling of the oral cavity, or palate, varies by species and diet. The rostral part of the palate is supported by the upper beak and is lined by soft-cornified epithelium in most birds, such as pigeons (Zweers 1982a), flamingos (Zweers et al. 1995), geese and ducks (Nickel et al. 1977; Zweers et al. 1977), chickens and other Galliformes (Nickel et al. 1977; Sağsöz et al. 2013), raptors (Erdoğan et al. 2012a), and ratites (Gussekloo and Bout 2005). In certain birds, such as parrots and cockatoos (Homberger 1980, 2001; Redd et al. 2012; Fig. 2.2) and seed-eating passerines (Bowman 1961; Ziswiler 1965, 1969; Ziswiler et al. 1972; Heidweiller and Zweers 1990; Genbrugge et al. 2012; Soons et al. 2012; van Hemert et al. 2012), the rostral part of the palate that is supported by the upper beak is covered by a hard-cornified epithelium that is continuous with the maxillary rhamphotheca and forms a ridged corneous palate against which seeds and other hard materials can be cut or crushed with the lower beak. In birds with a corneous palate, a soft-cornified, pleated palatine joint membrane creates the transition to the caudal part of the palate that is lined by a soft-cornified epithelium and forms a soft palate that is not supported by a bony shelf (Figs. 2.2 and 2.3a). It is stretched when the upper beak is raised (Homberger 1980). Rostrally, the soft palate forms a palatal torus, or cushion (Figs. 2.2 and 2.3a). In many birds, including parrots and cockatoos, a maxillary salivary gland opens just caudal to this torus. Caudally, the soft palate splits into usually caudally diverging paired palatal shelves that frame the choana, or internal naris, and are rimmed by caudally pointing choanal papillae. They end in a row of palatal

[4] The anatomical terms follow the Nomina Anatomica Avium (Baumel et al. 1993); hence, oral cavity (*cavum oris*) is used for what recent physiological-experimental papers have termed "oropharyngeal" or "oropharyngeal-esophageal" cavity (for a review, see Riede et al. 2013).

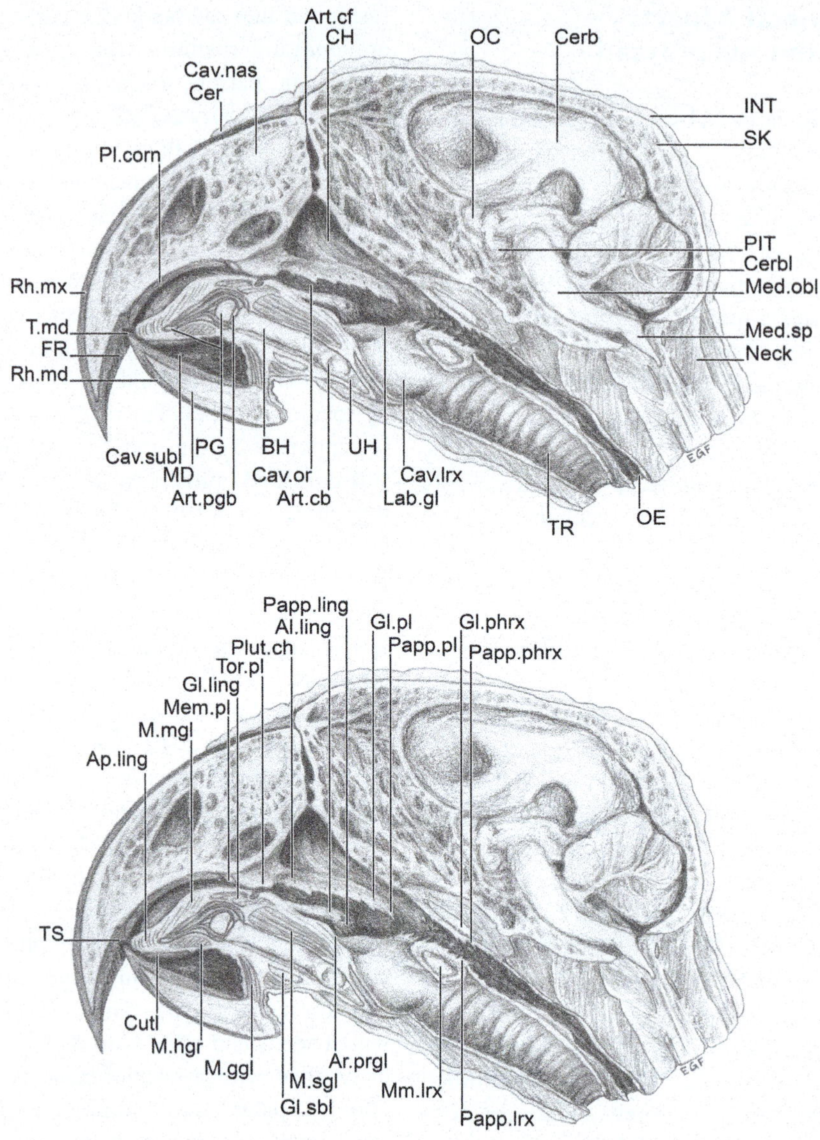

Fig. 2.2 Salient features of the head of birds in a longitudinal section of an African Grey Parrot (*Psittacus erithacus*) in resting position to highlight the long choana, the lingual tip apposed to the corneous palate near the transverse step, the lingual body apposed to the rostral part of the choana, the lingual papillae aligned with the palatal papillae, the laryngeal mound apposed to the caudal part of the choana, and the laryngeal papillae aligned with the pharyngeal papillae, the absence of teeth, and the crossing of the air and food passages. The figure is reproduced twice to accommodate all the labels. Al.ling = lingual wing; Ap.ling = lingual tip; Ar.prgl = preglottal area; Art.cb = cerato-basihyal joint; Art.cf = craniofacial joint; Art.pgb = paraglosso-basihyal joint; Cav.lrx = laryngeal chamber; Cav.or = oral cavity; Cav.sbl = sublingual cavity; Cav.nas = nasal cavity; Cer = cere; Cerb = *Cerebrum*; Cerbl = *Cerebellum*; CH = choana; Cutl = lingual nail; FR = filing ridges; Gl.ling = lingual salivary gland; Gl.phrx = pharyngeal salivary gland; Gl.pl = palatal mucous gland; INT = integument (skin); Lab.gl = glottal lips; M.ggl = *M. genioglossus*; M.hgr = *M. hyoglossus rostralis*; M. mgl = *M. mesoglossus*; M.sgl = *M. supraglossus*; MD = mandible; Med.obl. = *Medulla oblongata*; Med.sp = spinal cord; Mem.pl = palatine joint membrane; Mm.lrx = laryngeal muscles; Neck = neck; OC = optic chiasma; OE = esophagus; Papp.ling = lingual papillae; Papp.lrx = laryngeal papillae; Papp.phrx = pharyngeal papillae; Papp.pl = palatal papillae; PG = paraglossum; PIT = pituitary gland; Pl.corn = corneous palate; Plut.ch = lateral choanal shelf; Rh.md = mandibular rhamphotheca; Rh.mx = maxillary rhamphotheca; SK = skull (braincase); T.md = cutting edge of the mandibular rhamphotheca; Tor.pl = palatal torus; TR = trachea; TS = transverse step; UH = urohyal (Drawn by Ellen Farrar from several sections)

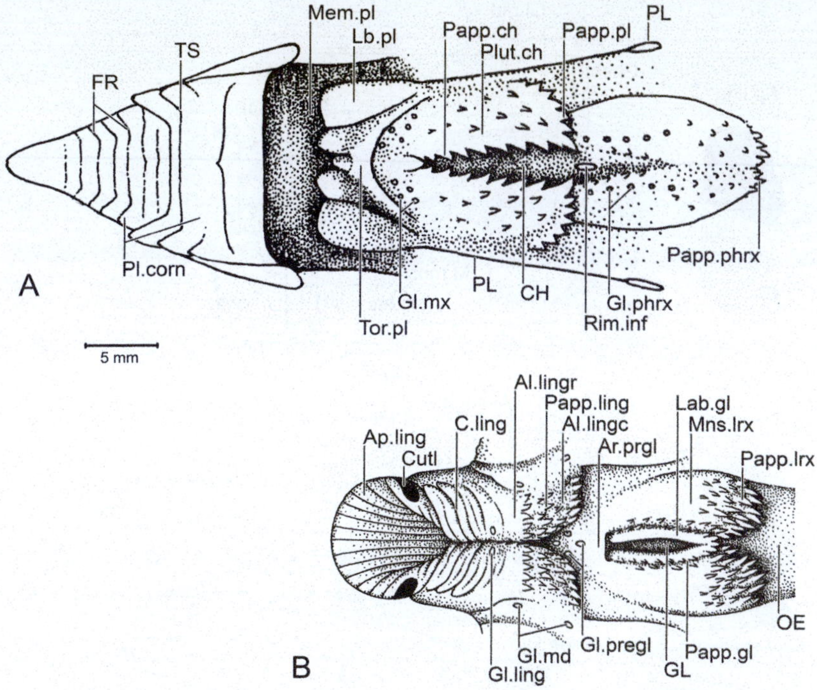

Fig. 2.3 The surface features of the ceiling and floor of the oral cavity of a budgerigar (*Melopsittacus undulatus*). (**a**) Ventral view of the corneous and soft palate, choana, and oral ceiling. (**b**) Dorsal view of the lingual body, preglottal area, and laryngeal mound. Al.lingc = caudal part of lingual wing; Al.lingr = rostral part of lingual wing; Ap.ling = lingual tip; Ar.prgl = preglottal area; C.ling = lingual body; CH = choana; Cutl = lingual nail; FR = filing ridges; GL = glottis; Gl.ling = lingual salivary gland; Gl.md = mandibular salivary gland; Gl.mx = maxillary salivary gland; Gl.phrx = pharyngeal salivary gland; Gl.pregl.= preglottal salivary gland; Lab.gl = glottal lips; Lb.pl = palatal lobe; Mem.pl = palatine joint membrane; Mns.lrx = laryngeal mound; OE = esophagus; Papp.ch = choanal papillae; Papp.gl = glottal papillae; Papp.ling = lingual papillae; Papp.lrx = laryngeal papillae; Papp.phrx = pharyngeal papillae; Papp.pl = palatal papillae; PL = palatine bone; Pl.corn = corneous palate; Plut.ch = lateral choanal shelf; Rim.inf = entrance to the Eustachian tubes; TS = transverse step; Tor.pl = palatal torus [Adapted from Homberger (1980)]

papillae and contain rows of mucous palatal glands that open individually on their dorsal (nasal) side (Figs. 2.2 and 2.3a). In some birds, such as ratites, the choana is shortened (Gussekloo and Bout 2005; Gussekloo 2006) apparently through an elongation of the soft palate. Hence, the paired nares communicate with the oral cavity through the generally elongated choana (Fig. 2.2), which cannot be closed (*contra* White 1969). It remains to be confirmed whether the choana can be temporarily closed by mucus during swallowing as suggested by Zweers (1992).

The entrance to the Eustachian tubes is located in the center of the oral ceiling at the level of the paired transverse rows of palatal papillae (Figs. 2.2 and 2.3a). The caudal ceiling of the oral cavity is lined by a soft-cornified epithelium, which is underlain by the pharyngeal salivary glands and is delimited caudally by rows of pharyngeal papillae (Figs. 2.2 and 2.3a).

The floor of the oral cavity is formed, in effect, by the lingual body, the preglottal area, and the laryngeal mound [Figs. 2.3b and 2.4b; see also McLelland (1979) and Zweers and Berkhoudt (1987) for a general description]. The dorsal surface of the lingual body is covered by a thick soft-cornified epithelium (Homberger 1980, 1986; Homberger and Brush 1986; for additional references, see, e.g., Bragulla and Homberger 2009; Erdoğan and Iwasaki 2014; Rico-Guevara 2014; Abumandour and

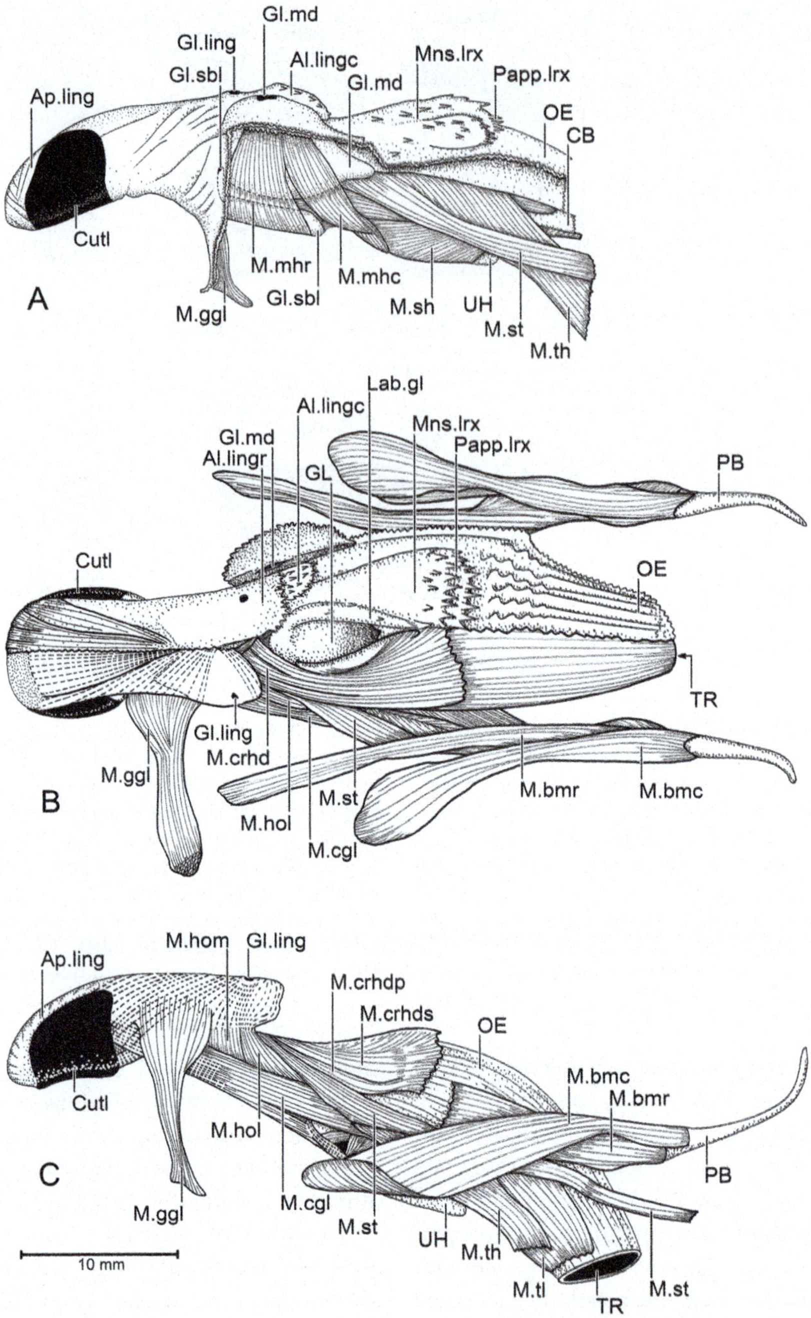

Fig. 2.4 The surface structures of the tongue and laryngeal mound in an African Grey Parrot (*Psittacus erithacus*) and their relationships to underlying muscles and salivary glands. (**a**) Lateral view with the surface epithelium cut along its attachment to the lateral walls of the oral cavity. (**b**) Dorsal view with the surface epithelium removed on the *left side*. (**c**) Lateral view with the surface epithelium removed except at the lingual tip. Al.lingc = caudal lingual wing; Al.lingr = rostral lingual wing; Ap.ling = lingual tip; CB = ceratobranchial; Cutl = lingual nail; GL = glottis; Gl.ling = lingual salivary gland; Gl.md = mandibular salivary gland; Gl.sbl = sublingual salivary gland; M.bmc = *M. branchiomandibularis caudalis*; M.bmr = *M. branchiomandibularis rostralis*; M.cgl = *M. ceratoglossus lateralis*; M.crhd = *M. cricohyoideus dorsalis*; M.crhdp = *M. cricohyoideus dorsalis profundus*; M.crhds = M. *cricohyoideus dorsalis superficialis*; M.ggl = *M. genioglossus*; M.hol = *M. hyoglossus obliquus lateralis*;

El-Bakary 2016; Skieresz-Szewczyk and Jackowiak 2016).

Whereas the tongues of most birds are shaped to fit within the confines of the mandible, special structures have evolved in some birds for food gathering or processing. The sides and dorsal surfaces of the tongues of filter-feeding birds, such as ducks, geese, and flamingos, bear cornified papillae (Zweers et al. 1977, 1995; Jackowiak et al. 2011; Skieresz-Szewczyk and Jackowiak 2016). The lateral and ventral sides of the lingual tip are covered by a cuff-like hard-cornified, but resilient cuticle, the lingual nail (Susi 1969; Homberger 1980, 1986; Homberger and Brush 1986; Jackowiak and Godynicki 2005; Jackowiak et al. 2011; Skieresz-Szewczyk and Jackowiak 2016), which bends apart when the lingual tip is inflated and flattened or pressed against the palate or food items, and returns the lingual tip to its resting condition by resilience (Homberger 1986, 1988a; see Sect. 2.5.2). In some birds (e.g., passerine birds, hummingbirds, toucans, woodpeckers), the lingual nail is elongated beyond the fleshy lingual tip and forms a narrow, stiff brush (e.g., Bock 1972, 1985a, b; Rico-Guevara and Rubega 2011; Rico-Guevara 2014; Rico-Guevara et al. 2015). The lingual tip of seed-eating passerine birds is reinforced and thickened by the underlying neomorphic bone, fat bodies, and hydraulic tissues and can be changed from flat to spoon-shaped (Bock and Morony 1978; Ziswiler 1979). The lingual tip of some nectar- and pollen-feeding parrots and other birds bears epithelial papillae of various lengths (Güntert and Ziswiler 1972; Homberger 1980; Paton and Collins 1989). In some birds, such as ratites, hornbills (Bucerotidae), storks (Ciconiidae), pelicans, and cormorants, which swallow large prey items whole, the lingual body is reduced (McLelland 1979; Tomlinson 2000; Gussekloo 2006; Jackowiak et al. 2006; Jackowiak and Ludwig 2008; Baussart and Bels 2011).

Caudally, the lingual body splits into a pair of lingual wings (Figs. 2.3b and 2.4b). Their rostral portions are supported by the bodies of the lingual salivary glands (Homberger 1980, 1986), and their caudal portions are strewn with lingual papillae. The lingual wings are flanked by the mandibular salivary glands and overhang the caudally extending preglottal area (lingual base), which is covered by a soft epithelium and supported by the preglottal salivary glands. Caudally, the preglottal area is followed by the laryngeal mound with a longitudinal glottis in its center (Figs. 2.3b and 2.4a, b). It bears glottal papillae along the glottal lips and laryngeal papillae along its caudal end bordering the esophagus. The glottis leads into a laryngeal chamber that continues into the characteristically long trachea (Fig. 2.2), which in some birds may be elongated by loops and even longitudinally subdivided (McLelland 1989).

Salivary glands are mostly mucous, occasionally seromucous, and vary in their size and distribution by species and their diet. They are numerous in granivorous and insectivorous species, such as parrots and cockatoos, passerine birds, woodpeckers, chickens, pigeons, raptors, and ducks and geese (Foelix 1970; Ziswiler et al. 1972; Bock et al. 1973; Nickel et al. 1977; McLelland 1979, 1993; Homberger 1980, 1986, 1988a; Zweers 1982a, Homberger and Meyers 1989; Heidweiller and Zweers 1990; Jackowiak and Godynicki 2005; Jackowiak et al. 2011; Erdoğan et al. 2012b; Sağsöz et al. 2013; Erdoğan and Iwasaki 2014; Abumandour and El-Bakary 2016; Skieresz-Szewczyk and Jackowiak 2016), but reduced in birds that eat fish and other aquatic prey (e.g., Jackowiak et al. 2006). In addition to producing saliva to moisten the oral epithelium, salivary glands also serve as

Fig. 2.4 (continued) M.hom = *M. hyoglossus obliquus medialis*; M.mhc = *M. mylohyoideus caudalis*; M.mhr = *M. mylohyoideus rostralis*; M.sh = *M. serpihyoideus*; M. st = *M. stylohyoideus*; M.th = *M. tracheohyoideus*; M.tl = *M. tracheolateralis*; Mns.lrx = laryngeal mound; OE = esophagus; Papp.lrx = laryngeal papillae; PB = pharyngobranchial; TR = trachea; UH = urohyal [Adapted from Homberger (1986)]

hydrostatic skeletons (Homberger 1986; Homberger and Meyers 1989).

The location and density of taste receptors vary by species and diet; in general, they are less numerous in birds than in mammals and are concentrated around the openings of salivary glands (Berkhoudt 1977, 1985; Zweers 1982a; Zweers and Berkhoudt 1987; Evans and Martin 1993; Zweers et al. 1994, 1995; Mason and Clark 2000; Erdoğan et al. 2012a, b; Erdoğan and Iwasaki 2014; Crole and Soley 2015; Abumandour and El-Bakary 2016; Skieresz-Szewczyk and Jackowiak 2016). Taste acuity varies greatly among species (Berkhoudt 1985; Powell et al. 2017). In contrast, the beak, palate, and tongue are generally richly endowed with a variety of touch receptors with distinct modalities (Ziswiler and Trnka 1972; Berkhoudt 1976, 1980; Zweers et al. 1977, 1995; Bock and Morony 1978; Krulis 1978; Homberger 1980, 2001, 2003; Zweers 1982a; Gerritsen and Sevenster 1985; Gottschaldt 1985; Gerritsen and Meiboom 1986; Homberger and Brush 1986; Heidweiller and Zweers 1990; Zweers et al. 1994; Cunningham et al. 2010; Jackowiak et al. 2011; Crole and Soley 2014; Skieresz-Szewczyk and Jackowiak 2016).

The hyoid skeleton is well developed in birds in contrast to that of mammals. It forms the skeletal support of the entire lingual body and tip and supports the laryngeal apparatus (see Sect. 2.6). A pair of hyoid horns articulate with the lateral sides of basihyal-urohyal and are invested in connective tissue sheaths, the hyoid sheaths, which are attached to the fasciae of the jaw muscles or throat (Figs. 2.5 and 2.6). This hyoid suspension construction enables some birds to protract their tongues far beyond the tip of the beak, as in lorikeets (Homberger 1980), woodpeckers (Leiber 1907; Bock 1999a), and hummingbirds (Rico-Guevara and Rubega 2011; Rico-Guevara 2014; Rico-Guevara et al. 2015). The hyoid skeleton is suspended between the mandibular rami by fasciae and extrinsic muscles. The laryngeal skeleton supports the laryngeal mound and rests on the hyoid skeleton (see Sect. 2.6).

In the resting position of the linguo-laryngeal apparatus relative to the closed beak and palate, the rows of palatal and laryngeal papillae are aligned so that the laryngeal mound and its glottis are positioned underneath the widest part of the choana, and the lingual body is placed against and closes off the rostral narrower part of the choana (Fig. 2.2a; see also Zweers 1982a). The laryngeal mound of birds that lost their lingual body (e.g., ratites) is positioned underneath the shortened choana (see Gussekloo and Bout 2005; Gussekloo 2006). The dorsoventral position of the linguo-laryngeal apparatus relative to the mandible varies among birds and is correlated with the position of the hyoid sheath (see Sect. 2.5.2). In parrots and cockatoos, the lingual body is elevated above the sublingual floor, and the hyoid sheaths are tucked high up between the jaw muscles (Fig. 2.5a). In many birds, such as the chicken, the lingual body rests on the sublingual floor on the mandibular symphysis, and the hyoid sheaths are attached on the ventral side of the *M. depressor mandibulae* (Homberger and Meyers 1989). In many other birds that swallow large prey or food items whole (e.g., cormorants, pelicans, herons), the linguo-laryngeal apparatus is situated well below the mandible, and the hyoid sheaths are incorporated into the gular wall (Fig. 2.6).

2.4 The Jaw Apparatus of Birds: Components and Movements of the Upper Jaw Apparatus

The bones of the braincase, or cranium, of birds are fused in the adult, but the elements of the upper jaw apparatus are joined by diarthrodial, syndesmotic, or synostotic articulations and moved by a very complex jaw musculature (for reviews, see Bowman 1961; Dzerzhinsky 1972; Richards and Bock 1973; Bock 1985c; Bühler 1981; Baumel and Witmer 1993; Zusi 1993; see also Bock 1964, 1972; Zweers 1974; Bhattacharyya 1988; Homberger 2003, 2006; Nuijens and Zweers 1997; van Gennip 1986; Dawson et al. 2011; Bhullar et al. 2016; Claes et al. 2016; Olsen and Westneat 2016). The paired quadrate bones are directly or indirectly linked to all skeletal elements of the head and play a central role in all movements of the upper

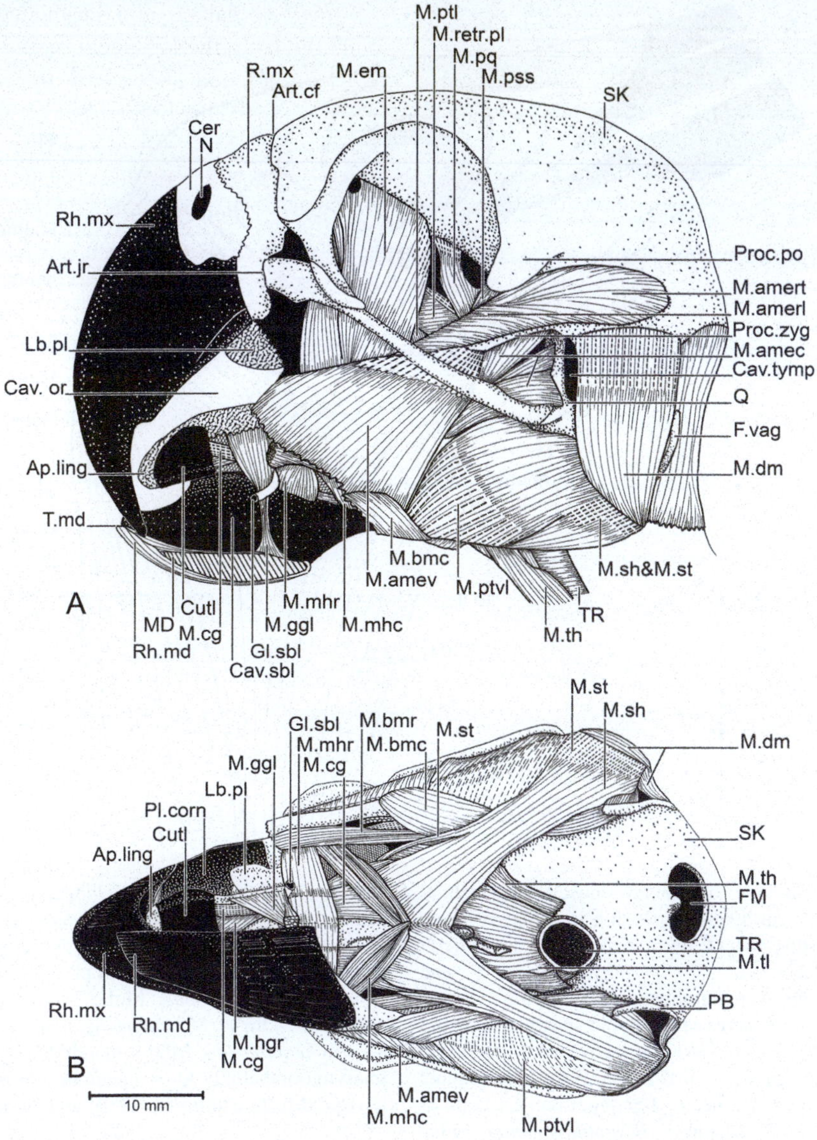

Fig. 2.5 Jaw musculature and tight hyoid suspension in an African Grey Parrot (*Psittacus erithacus*); the *left half* of the lower beak was removed to reveal the tongue. (**a**) Lateral view of the head to show how the lingual apparatus is framed laterally by the mandibular rami and jaw musculature. (**b**) Ventral view of the head to show the extrinsic lingual muscles; the hyoid sheath was removed to reveal the pharyngobranchial. Ap.ling = lingual tip; Art.cf = craniofacial joint; Art.jr = jugorostral joint; Cav.or = oral cavity; Cav.sbl = sublingual cavity; Cav.tymp = tympanic (middle ear) cavity; Cer = cere; Cutl = lingual nail; F.vag = hyoid sheath; FM = *foramen magnum*; Gl.sbl = sublingual salivary gland; Lb.pl = palatal lobe; M.amec = *M. adductor mandibulae externus caudalis*; M.amerl = *M. adductor mandibulae externus lateralis*; M.amert = *M. adductor mandibulae rostralis temporalis*; M.amev = *M. adductor mandibulae externus ventralis*; M.bmc = *M. branchiomandibularis caudalis*; M.bmr = *M. branchiomandibularis rostralis*; M.cg = *M. ceratoglossus*; M.dm = *M. depressor mandibulae*; M. em = *M. ethmomandibularis*; M.ggl = *M. genioglossus*; M. hgr = *M. hyoglossus rostralis*; M.mhc = *M. mylohyoideus caudalis*; M.mhr = *M. mylohyoideus rostralis*; M.pq = *M. protactor quadrati*; M.pss = *M. pseudotemporalis superficialis*; M.ptl = *M. pterygoideus lateralis*; M.ptvl = *M. pterygoideus ventralis lateralis*; M.retr.pl = *M. retractor pterygopalatini*; M.sh = *M. serpihyoideus*; M.st = *M. stylohyoideus*; M.th = *M. tracheohyoideus*; M. tl = *M. tracheolateralis*; MD = mandible; N = naris; PB = pharyngobranchial; Pl.corn = corneous palate; Proc.po = postorbital process; Proc.zyg = zygomatic process; Q = quadrate bone; R.mx = maxillary rostrum (upper beak); Rh.md = mandibular rhamphotheca; Rh.mx = maxillary rhamphotheca; SK = skull; T.md = cutting edge of the mandible; TR = trachea [Adapted from Homberger (1986)]

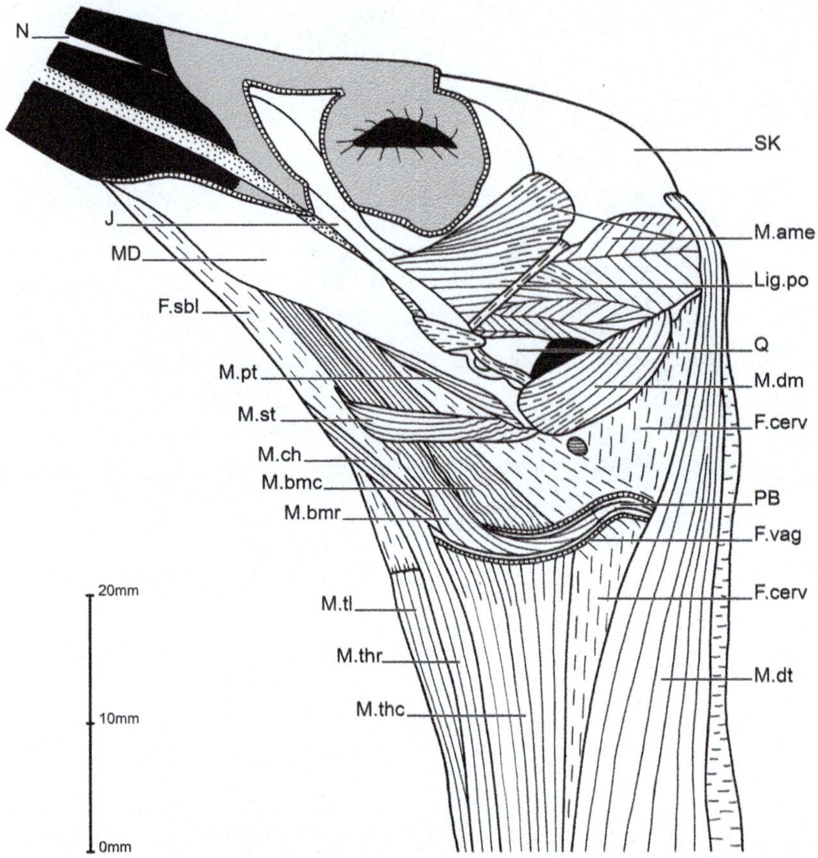

Fig. 2.6 Lateral view of the jaw muscles, postorbital ligament, and loose hyoid suspension in a Snowy Egret (*Egretta garzetta*); the *M. constrictor colli cervicalis et intermandibularis*, *M. serpihyoideus*, and *M. mylohyoideus* were removed to reveal the position and muscular attachments of the hyoid sheath; the rostral end of the *M. tracheolateralis* is still covered by the *fascia sublingualis*. F.cerv = cervical fascia; F.sbl = sublingual fascia; F.vag = hyoid sheath; J = jugal bone; Lig.po = postorbital ligament; M.ame = *M. adductor mandibulae externus*; M.bmc = *M. branchiomandibularis caudalis*; M.bmr = *M. branchiomandibularis rostralis*; M.ch = *M. ceratohyoideus*; M.dm = *M. depressor mandibulae*; M.dt = *M. dermotemporalis*; M.pt = *M. pterygoideus*; M. st = *M. stylohyoideus*; M.thc = *M. tracheohyoideus caudalis*; M.thr = *M. tracheohyoideus rostralis*; M.tl = *M. tracheolateralis*; MD = mandible; N = naris; PB = pharyngobranchial; Q = quadrate bone; SK = skull [Adapted from Cummins (1986) and Homberger (1999)]

and lower jaws by forming four main joints (Figs. 2.7, 2.8, 2.9, 2.10 and 2.11). The dorsal otic process forms the diarthrodial quadrato-squamoso-otic joint with usually two condyles to fit in the prootic-opisthotic and squamosal articular facets on the skull (Baumel and Witmer 1993). The ventral mandibular process forms the diarthrodial quadrato-mandibular joint with three condyles in most birds (Fig. 2.11a) to fit in the corresponding articular cotyla on the caudal end of the mandibular ramus. In parrots and cockatoos, the mandibular process bears a single, elongated, and diagonally oriented condyle which fits in the corresponding articular cotyla of the mandible to form the medial part of the quadrato-mandibular joint (Fig. 2.9). The rostro-medio-dorsally pointing condyle of the quadrate bone forms the diarthrodial quadrato-pterygoid joint with the articular cotyla of the pterygoid bone (Figs. 2.7a and 2.11a). The lateral process of the quadrate bone bears a rostrally oriented cotyla, which forms a diarthrodial or

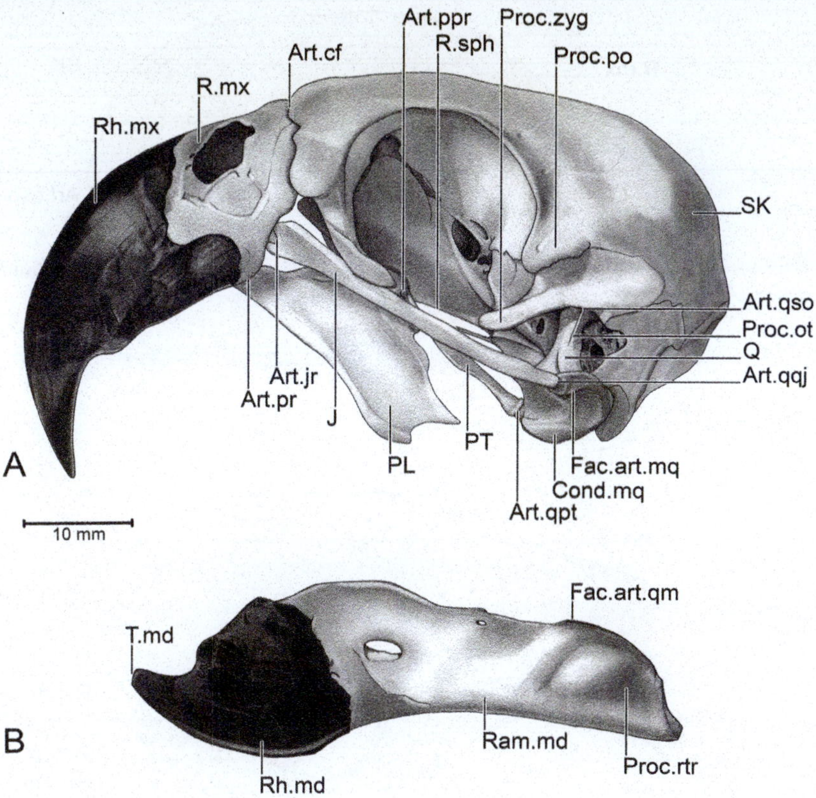

Fig. 2.7 Lateral views of the kinetic skull and jaw apparatus of an African Grey Parrot (*Psittacus erithacus*). (**a**) The skull and upper jaw apparatus. (**b**) The mandible. Art. cf = craniofacial joint; Art.jr = jugorostral joint; Art.ppr = pterygo-palatorostral joint; Art.pr = palatorostral joint; Art.qqj = quadrato-quadratojugal joint; Art.qpt = quadrato-pterygoid joint; Art.qso = quadrato-squamoso-otic joint; Cond.mq = mandibular condyle of the quadrate bone; Fac.art.mq = mandibular articular facet of the quadrate bone; Fac.art.qm = quadrate articular facet of mandible; J = jugal bone; Proc.ot = otic process of quadrate bone; Proc.po = postorbital process; Proc.rtr = retroarticular process of the mandible; Proc.zyg = zygomatic process; PL = palatine bone; PT = pterygoid bone; Q = quadrate bone; R.mx = maxillary rostrum (upper beak); R.sph = *Rostrum sphenoidale*; Ram.md = mandibular ramus; Rh.md = mandibular rhamphotheca; Rh. mx = maxillary rhamphotheca; SK = skull; T.md = cutting edge of the mandibular rhamphotheca

syndesmotic quadrato-quadratojugal joint with the corresponding condyle of the quadratojugal bone (Baumel and Witmer 1993; Figs. 2.7a and 2.11a, b). In parrots and cockatoos, the ventral surface of the lateral process bears also an articular facet to fit the articular facet on the elevated and broadened edge of the caudal end of the mandibular ramus bordering the articular cotyla of the mandible for the quadrate bone and, thereby, forms the lateral part of the quadrato-mandibular joint (Figs. 2.7 and 2.9).

The paired pterygoid bones point rostro-medio-dorsally to articulate with the paired palatine bones; together, they form the compound pterygo-palatorostral joint with the *Rostrum sphenoidale*, the longitudinal crest on the ventral surface of the cranium (Figs. 2.7a, 2.9a and 2.11a). The paired palatine bones connect rostrally through diarthroses in parrots and cockatoos (Fig. 2.9a) or synostoses in many birds (Fig. 2.11a) to the caudo-ventral edge of the upper beak[5] (maxilla). The quadratojugal bones do the same more laterally (Figs. 2.7a, 2.9a and 2.11a, b). The upper beak also

[5] The adult upper beak is formed from various embryonic elements (for details, see Bühler 1981).

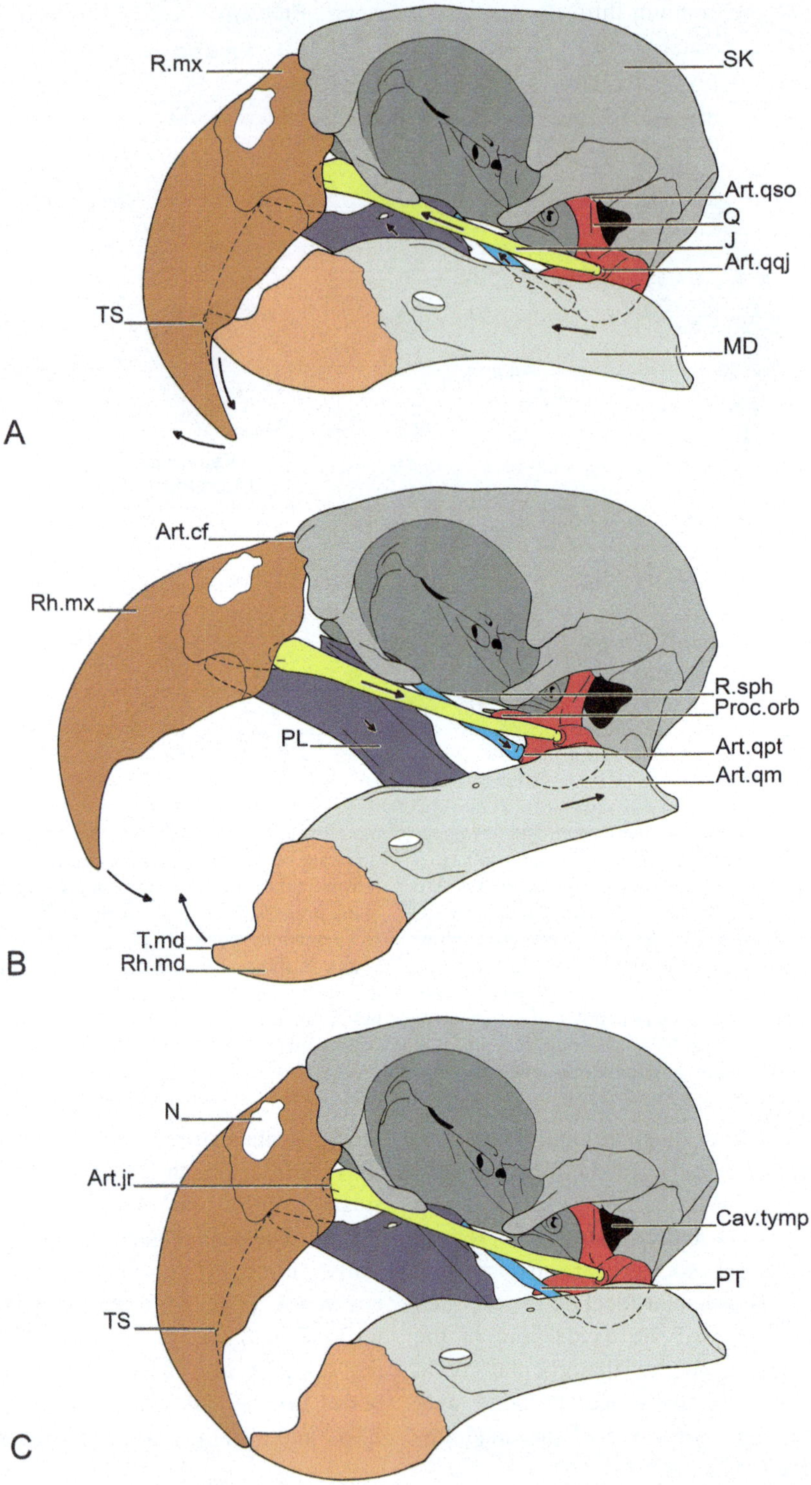

Fig. 2.8 Graphic models of lateral views of the uncoupled cranial kinesis in an African Grey Parrot (*Psittacus erithacus*) as a representative of a Psittaciformes species with a psittacine beak. (**a**) Resting, closed-beak position with lowered upper beak and raised lower beak; *arrows* indicate the movements to open the beak. (**b**) Open-beak position with raised upper beak and lowered lower beak; *arrows* indicate the

articulates with the cranium through the craniofacial joint[6] (Figs. 2.7a, 2.8 and 2.11b).

A forward rotation of the quadrate around the quadrato-squamoso-otic joint pushes the pterygo-palato-quadratojugal complex rostrally, thereby forcing the upper beak to rotate upward around the craniofacial joint (Figs. 2.8a, b and 2.11b, c; see also Bock 1964; Bühler 1981; Gussekloo et al. 2001). Because the quadrate bones rotate along a slight medio-rostral to latero-caudal axis, the distance between the quadrato-mandibular joints diminishes slightly, which allows the pterygoid bones to reduce the angle they enclose at the pterygo-palatorostral joint, which increases their longitudinal force components during the raising of the upper beak. This slight narrowing of the distance between the quadrato-mandibular joints is possible in most birds that raise the maxilla only slightly. In parrots and cockatoos, whose maxillary excursion and quadrate rotation can be significant, the simple, rounded condyles of the ventral processes of the quadrate bones rotate towards the rostral end of the quadrate articular facets of the mandible where the distance between the joints is narrower (Fig. 2.8 and 2.9). To depress the upper beak, the quadrate bones rotate backward, the distance between them increases slightly, the pterygoid and palatine bones are retracted, while the angle between the rostral ends of the pterygoid bones increase slightly, and the maxilla is depressed.

The jaw muscles of birds are much more complex than those of mammals as a consequence of their kinetic skull and upper jaw apparatus with its multiple joints. Nevertheless, their basic functions in opening and closing the beak can be analyzed in principle, while also keeping in mind that they are quite complex so that their interplay can result in finely tuned and complex movements especially in birds that intraorally process their food before swallowing it (e.g., seed-shelling parrots, cockatoos, and passerine birds). Depending on the species, the movements of the mandible and upper jaw are independent from or linked to each other.

2.4.1 The Uncoupled Jaw Movements of Parrots and Cockatoos

Birds with an uncoupled jaw apparatus (e.g., parrots and cockatoos) can move their upper and lower beaks independently from each other (Fig. 2.8). The hallmarks of an uncoupled jaw apparatus are the absence of a postorbital ligament between the postorbital process of the skull and the mandible (see Sect. 2.4.2; see also Korzun et al. 2008) and the presence of simple articular facets of the quadrato-mandibular joints (Homberger 2003). In parrots and cockatoos with a psittacine beak in a resting and closed position, the upper beak is depressed, and the mandible is raised so that the transverse cutting edge of the mandibular rhamphotheca is apposed to the transverse step on the corneous palate (Figs. 2.2 and 2.8a; Homberger 2003, 2006). The tip of the upper beak projects over the lower beak, and the tip of the tongue touches the corneous palate where it meets the lower beak (Fig. 2.2).

Fig. 2.8 (continued) movements to close the beak. (c) Pincered-beak position with lowered upper beak and lowered lower beak. Art.cf = craniofacial joint; Art.jr = jugorostral joint; Art.qm = quadrato-mandibular joint; Art.qpt = quadrato-pterygoid joint; Art.qqj = quadrato-quadratojugal joint; Art.qso = quadrato-squamoso-otic joint; Cav.tymp = tympanic (middle ear) cavity; J = jugal bone; MD = mandible; N= naris; PL = palatine bone; Proc.orb = orbital process of the quadrate bone; PT = pterygoid bone; Q = quadrate bone; R.mx = maxillary rostrum (upper beak); R.sph = *Rostrum sphenoidale*; Rh.md = mandibular rhamphotheca; Rh.mx = maxillary rhamphotheca; SK = skull; T.md = cutting edge of mandibular rhamphotheca; TS = transverse step [Adapted from Homberger (2003)]

[6] The exact place and structure of this joint varies among birds (for details, see Bühler 1981).

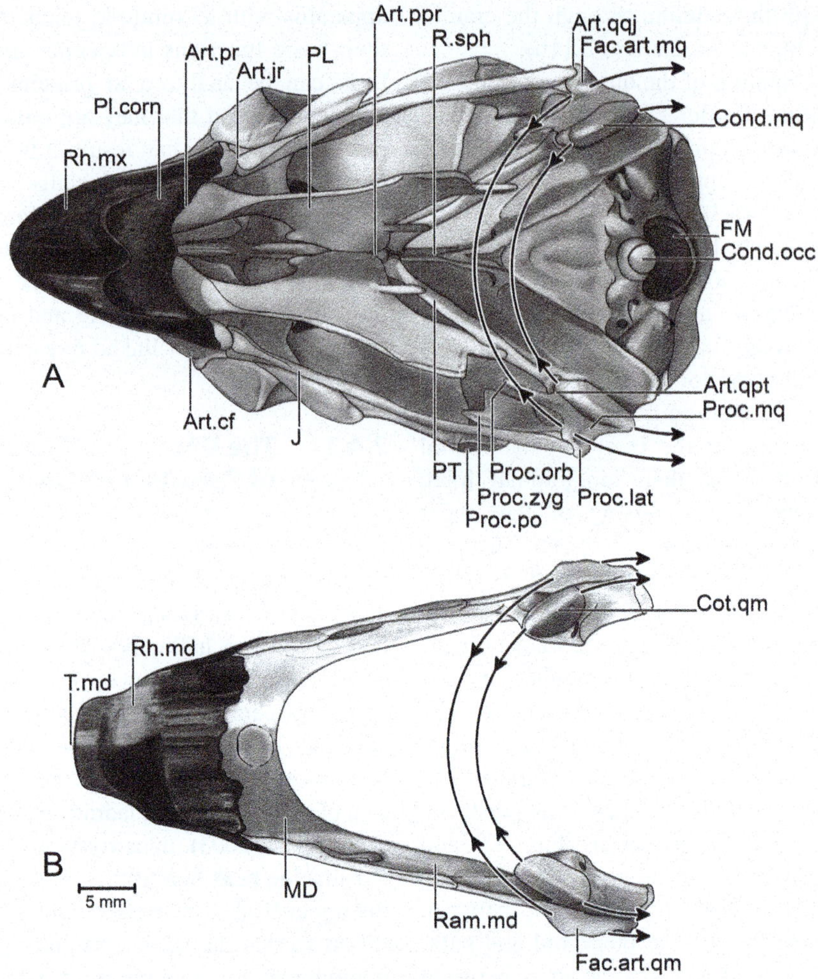

Fig. 2.9 The quadrato-mandibular joint of an African Grey Parrot (*Psittacus erithacus*) enabling lateral rotational movements of the mandible as indicated by *arrows* and *semicircular lines*. (**a**) Ventral view of the skull and upper jaw apparatus. (**b**) Dorsal view of the mandible. Art.cf = craniofacial joint; Art.jr = jugorostral joint; Art.ppr = pterygo-palatorostral joint; Art.pr = palatorostral joint; Art.qpt = quadrato-pterygoid joint; Art.qqj = quadrato-quadratojugal joint; Cond.mq = mandibular condyle of the quadrate bone; Cond.occ = occipital condyle; Cot.qm = quadrate socket of the mandible; Fac.art.mq = mandibular articular facet of the quadrate bone; Fac.art.qm = quadrate articular facet of the mandible; FM = *Foramen magnum*; J = jugal bone; MD = mandible; PL = palatine bone; Pl.corn = corneous palate; Proc.lat = lateral process of the quadrate bone; Proc.mq = mandibular process of the quadrate bone; Proc.orb = orbital process of the quadrate bone; Proc.po = postorbital process; Proc.zyg = zygomatic process; PT = pterygoid bone; R.sph = *Rostrum sphenoidale*; Ram.md = mandibular ramus; Rh.md = mandibular rhamphotheca; Rh.mx = maxillary rhamphotheca; T.md = cutting edge of mandibular rhamphotheca [Adapted from Homberger (2003)]

To open the beak widely (Fig. 2.8b), the small one-joint *M. protractor quadrati*, which originates from the skull and inserts on the orbital process of the quadrate bone (Figs. 2.5a, 2.8b and 2.9a), protracts the quadrate and, thereby, raises the upper beak (see Sect. 2.4). The mandible is depressed by the larger two-joint *M. depressor mandibulae*, which originates from the skull and inserts on the retroarticular process of the mandible (Figs. 2.5a and 2.7b) and, thereby, not only depresses the mandible but also supports the protraction of the quadrate.

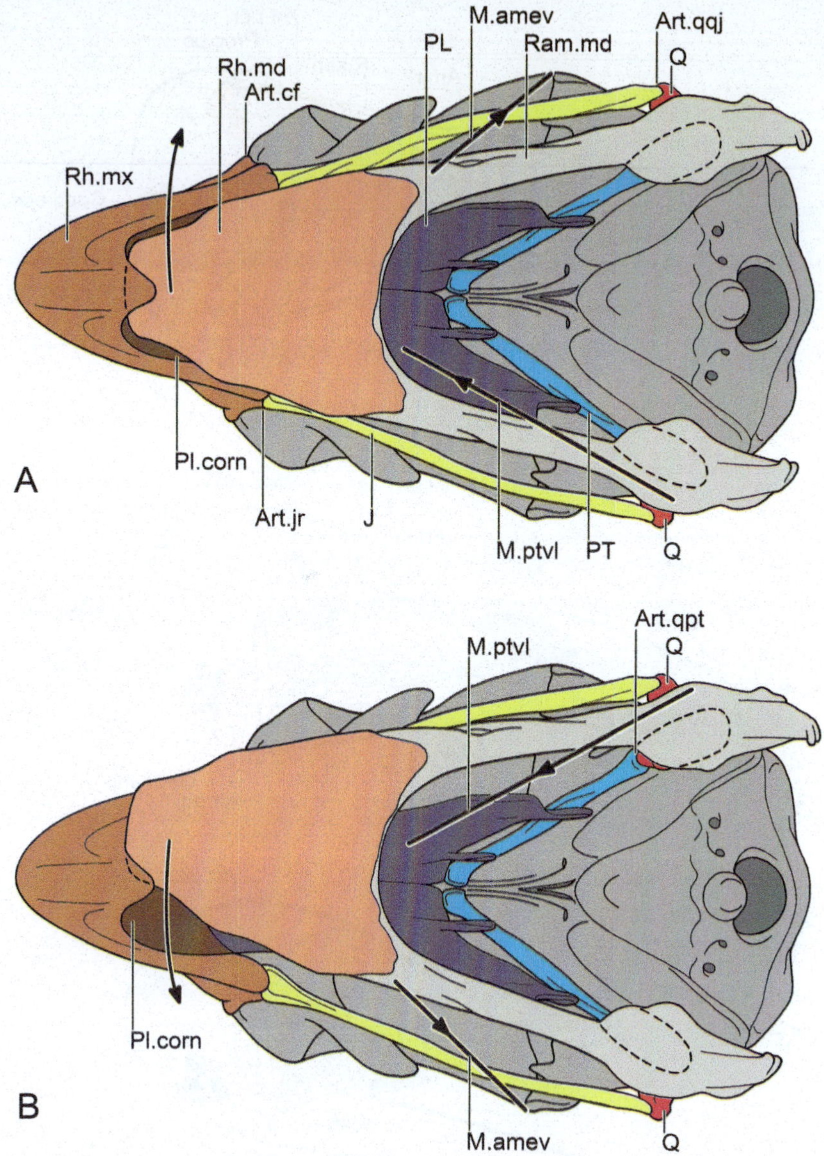

Fig. 2.10 Graphic model of ventral views of lateral movements of the mandible relative to the skull and upper jaw apparatus of an African Grey Parrot (*Psittacus erithacus*) based on rotational movements at the quadrato-mandibular joints. (**a**) Mandible aligned with the upper beak; *arrows* indicate the movements and muscular pulls necessary to deflect the mandible to the *left*. (**b**) Mandible deflected to the *left* through rotation on the mandibular condyle of the quadrate bone; *arrows* indicate the movements and muscular pulls necessary to align the mandible with the skull. Art.cf = cracio-facial joint; Art.jr = jugorostral joint; Art.qpt = quadrato-pterygoid joint; Art.qqj = quadrato-quadratojugal joint; J = jugal bone; M. amev = *M. adductor mandibulae externus ventralis*; M.ptvl = *M. pterygoideus ventralis lateralis*; PL = palatine bone; Pl.corn = corneous palate; PT = pterygoid bone; Q = quadrate bone; R.sph = *Rostrum sphenoidale*; Ram.md = mandibular ramus; Rh.md = mandibular rhamphotheca; Rh.mx = maxillary rhamphotheca [(**b**) Adapted from Homberger (2003)]

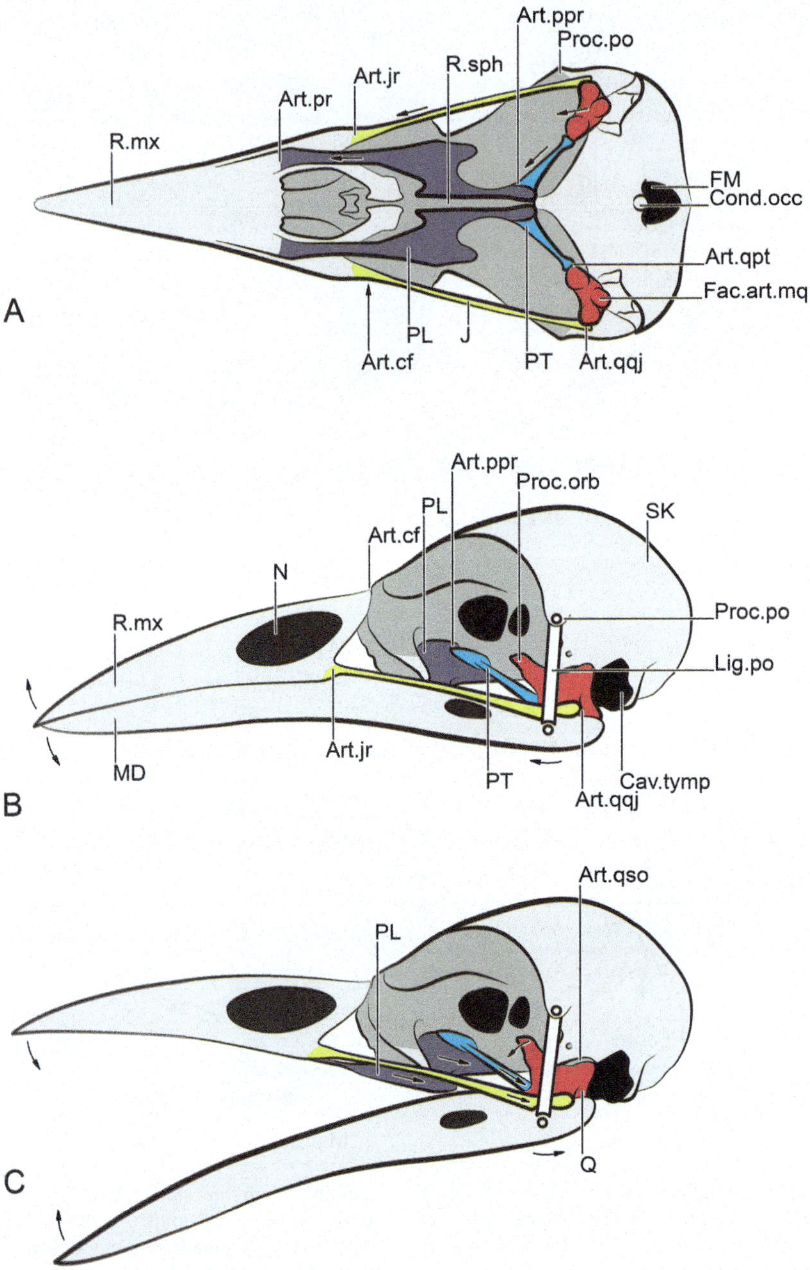

Fig. 2.11 Semidiagrammatic graphic model of the coupled cranial kinesis with a collagenous postorbital ligament in an American Crow (*Corvus brachyrhynchos*). (**a**) Ventral view of the skull and upper jaw apparatus; the *arrows* on the *left side* indicate the movements for raising the upper beak. (**b**) Lateral view of the closed-beak resting position; *arrows* indicate the movements for opening the beak. (**c**) Lateral view of the open-beak position with a raised upper beak linked to a lowered mandible; the *arrows* indicate the movements for closing the beak. Art. cf = craniofacial joint; Art.jr = jugorostral joint; Art.ppr = pterygo-palatorostral joint; Art.pr = palatorostral joint; Art.qpt = quadrato-pterygoid joint; Art.qqj = quadrato-quadratojugal joint; Art.qso = quadrato-squamoso-otic joint; Cav.tymp = tympanic (middle ear) cavity; Cond.occ = occipital condyle; Fac. art.mq = mandibular articular facet of the quadrate bone; FM = *Foramen magnum*; J = jugal bone; Lig.po = postorbital ligament; N= naris; PL = palatine bone; Proc.orb = orbital process of the quadrate bone; Proc.po = postorbital process; PT = pterygoid bone; Q = quadrate; R.sph = *Rostrum sphenoidale*; MD = mandible; R. mx = maxillary rostrum (upper beak); SK = skull [Redrawn and adapted from Bock (1964, 1999b) by Ellen Farrar]

Many more muscles are involved in closing the beak than in opening it. Most closing jaw muscles are two- or multi-joint muscles with several subdivisions characterized by different attachments on the skull and mandible and various degrees of pinnation (for details, see Homberger 2003). Two main muscle groups depress the upper beak and raise the lower beak at the same time. The *M. adductor mandibulae externus* originates from the temporal region of the skull and inserts externally on the rostral end of the mandibular ramus (Figs. 2.5 and 2.10). It raises and retracts the mandible while at the same time supporting the backward rotation of the quadrate bone and, thereby, the retraction of the upper beak (Fig. 2.10b). The large *M. pterygoideus ventralis lateralis* originates on the palato-pterygoid complex of the upper beak apparatus and inserts on the external side of the caudal end of the mandibular ramus after having slung itself around the ventral edge of the mandibular ramus (Figs. 2.5 and 2.10). It retracts the palato-pterygoid complex while exerting a dorso-rostral pull on the mandible (Fig. 2.10b).

Two muscles specifically depress the upper beak or raise the mandible and are responsible for increasing the bite force. The *M. retractor pterygopalatini* is a one-joint muscle originating from the basioccipital region of the skull and inserting on the palato-pterygoid complex (Fig. 2.5a). It depresses the upper beak by retracting the palato-pterygoid complex and, at the same time, rotating the quadrate bone backward. The *M. ethmomandibularis* is unique to parrots and cockatoos; it originates from the interorbital septum and inserts internally on the rostral end of the mandibular ramus (Fig. 2.5a). It raises the mandible with considerable mechanical advantage because of its attachments that are as far from the quadrato-mandibular joint as possible.

In order for the tips of the upper and lower beak to meet in a pincerlike fashion to grab items, the maxilla and mandible need to move independently by depressing the maxilla while keeping the mandible depressed (Fig. 2.8c). This motion can be achieved through a contraction of the one-joint *M. retractor pterygopalatini* and a simultaneous contractions of the *M. depressor mandibulae* and of the small one-joint *M. pseudotemporalis profundus*, which connects the mandible and orbital process of the quadrate bone and counterbalances the tendency of the quadrate bone from rotating forward when the *M. depressor mandibulae* contracts.

Parrots can also protract their mandible beyond the upper bill tip (Fig. 2.5).

2.4.2 The Coupled Jaw Movements and Postorbital Ligament of Most Birds

The movements and muscles of the upper jaw apparatus and mandible in most birds are very similar to those described above for parrots and cockatoos (see also Fig. 2.6). However, most birds, in contrast to parrots and cockatoos, are characterized by a compound articulation of the quadrato-mandibular joint with two to three articular facets (Fig. 2.11a) and a collagenous postorbital ligament between the postorbital process and the external side of the mandibular ramus at the level of the quadrato-mandibular joint (Figs. 2.6 and 2.11b, c). This ligament links the depression of the lower mandible to a forward rotation of the quadrate and the concomitant raising of the upper beak *via* the forward motion of the palato-pterygoid complex (Fig. 2.11b, c; see Bock 1964, 1999b). This ligament was described in passerine birds (Bock 1964, 1974, 1985c; Bock et al. 1973; Richards and Bock 1973; Bhattacharyya 1982; Kalyakin and Dzerzhinsky 1997; Nuijens and Bout 1998; Nuijens et al. 2000; Genbrugge et al. 2011; Zubkova and Korzun 2014), barbets (Trunov et al. 1996), woodpeckers (Bock 1999a), ducks (Goodman and Fisher 1962; Zweers 1974, 1992; Nickel et al. 1977), pigeons (Zweers 1982a; van Gennip 1986; Bhattacharyya 1989, 1997, 1998, 2013), chickens (Nickel et al. 1977; van den Heuvel 1991), cranes (Fisher and Goodman 1955), herons (Cummins 1986; Fig. 2.6), and many other avian groups (Nuijens and Bout 1998). However, the presence and morphology of the postorbital ligament can vary even among related species, and the degree to which this ligament links the beak movements has been shown to vary to a certain degree, but the causes for these variations are not yet understood (Bock 1999b; Nuijens and Bout 1998), and experimental studies did not clarify the function of this ligament (van den Heuvel 1991; Zweers 1992; Nuijens et al. 2000; but see Sect. 2.4.3).

2.4.3 Lateral Mobility of the Avian Mandible

The mandible is a rigid unit in birds that use their beaks to process their food intraorally (e.g., parrots and cockatoos, granivorous passerine birds), hammer food on a substrate (e.g., nuthatches, corvid jays, woodpeckers), or spear prey (e.g., herons). The mandibular rami may, however, contain synostotic bending zones that allow them to bend and thereby widen the space between them to catch flying insects or swallow large prey items (Bühler 1981) or scoop large amounts of water in pelicans (Meyers and Myers 2005).

Parrots and cockatoos have the capacity to move their mandible from side to side because of the lack of a postorbital ligament and their specialized quadrato-mandibular jaw joints which lie on a virtual arc (Fig. 2.9). The mandibular tip can move from side to side while the beak is being closed, such as during their bill-honing behavior or for fine positional adjustments when opening seeds or nuts (Fig. 2.8; Homberger and Ziswiler 1972; Homberger 2001, 2003, 2006). The crosswise arrangement of two jaw-closing muscles enables them to play synergistic roles: The *M. adductor mandibulae externus* pulls the mandibular ramus latero-dorso-caudally, while the *M. pterygoideus ventralis lateralis* pulls the mandible medio-dorso-apically (Figs. 2.5a and 2.10; see also Sect. 2.4.1).

Pelicans also lack a postorbital ligament and have the capacity to bow their mandibular rami outward when scooping up water with fishes (Meyers and Myers 2005), which necessitates a rotation at the quadrato-mandibular joints (Böker 1938).

Many, if not most, birds seem to have no, or only a limited, capacity to move their mandible sideways relative to the upper beak, presumably due to their postorbital ligament and their complex quadrato-mandibular joint (see Nuijens and Bout 1998 for a discussion). A strict alignment of the upper and lower beaks may protect the cornified edges of the hard-cornified and brittle maxillary and mandibular rhamphotheca from splintering. The presence or absence of a postorbital ligament varies in seed-shelling passerine birds and does not seem to correlate with the capacity for lateral movements of the mandible during seed-shelling movements (Nuijens and Bout 1998; Bock 1999b), but Nuijens and Zweers (1997) described that lateral mandibular movements in fringillid finches is correlated with flattened articular facets of the quadrato-mandibular joint to allow lateral rotations of the mandible (see also Sect. 2.7.2.2). Ratites also seem to lack a postorbital ligament, but lateral movements of the mandible were not described for this avian group (see Gussekloo and Bout 2005).

2.5 The Lingual Apparatus of Birds

In birds, the lingual and laryngeal apparatus together form the *de facto* floor of the mouth in birds (Figs. 2.2, 2.3b and 2.4a, b) and play a central role in feeding and drinking (see Sect. 2.7.3). Despite their superficial consolidation and the fact that the laryngeal apparatus rests on and moves relative to the lingual apparatus, they are separate entities, each with its own skeleton and its own extrinsic and intrinsic musculature. To understand their common roles, it is necessary to analyze first their separate functional morphology and biomechanics. The lingual apparatus is directly attached to and suspended only from the mandible (not the skull) through extrinsic muscles and fasciae, presumably to prevent strain when the mandible and upper jaw apparatus move relative to the skull when the beak is opened and closed (see Sect. 2.4).

The hyoid skeleton and musculature have been the object of a number of studies (e.g., Bock 1972, 1978a, 1985b, c, 1999a; Richards and Bock 1973; Zweers 1974, 1982a; Zweers et al. 1977, 1995; Bhattacharyya 1985, 1998; Cummins 1986; Homberger 1986; Homberger and Meyers 1989; Heidweiller and Zweers 1990). Synonymies and homologies are included in the studies by Fisher and Goodman (1955), Zweers (1982a), Homberger (1986), Homberger and Meyers (1989), Baumel and Witmer (1993), and Vanden Berge and Zweers (1993).

2.5.1 The Hyoid Skeleton

The complex hyoid skeleton is suspended from the mandible by fasciae and extrinsic lingual muscles (Homberger 1986; Homberger and Meyers 1989). The bony basihyal and the urohyal with its cartilaginous caudal extension form the longitudinal central axis of the hyoid skeleton (Figs. 2.2, 2.12, 2.13 and 2.14b, c). Apically, it articulates with the paraglossum, which supports the lingual body. The saddle-shaped paraglosso-basihyal joint allows the lingual body to move up and down as well slightly from side to side. The paraglossum is paired in parrots and cockatoos (Figs. 2.12a and 2.13), but fused in most birds (Fig. 2.14b, c). Some birds, such as ratites, hornbills, cormorants, and pelicans, have lost the paraglossum and lingual body, but have retained the remainder of the hyoid skeleton to support and suspend the laryngeal apparatus (see Sect. 2.3 and 2.6).

A pair of hyoid horns articulate with the lateral sides of basihyal-urohyal transition (Figs. 2.2, 2.12a and 2.14b, c). They comprise the bony ceratobranchials, which connect diarthrodially (in parrots) or syndesmotically (in most birds) to

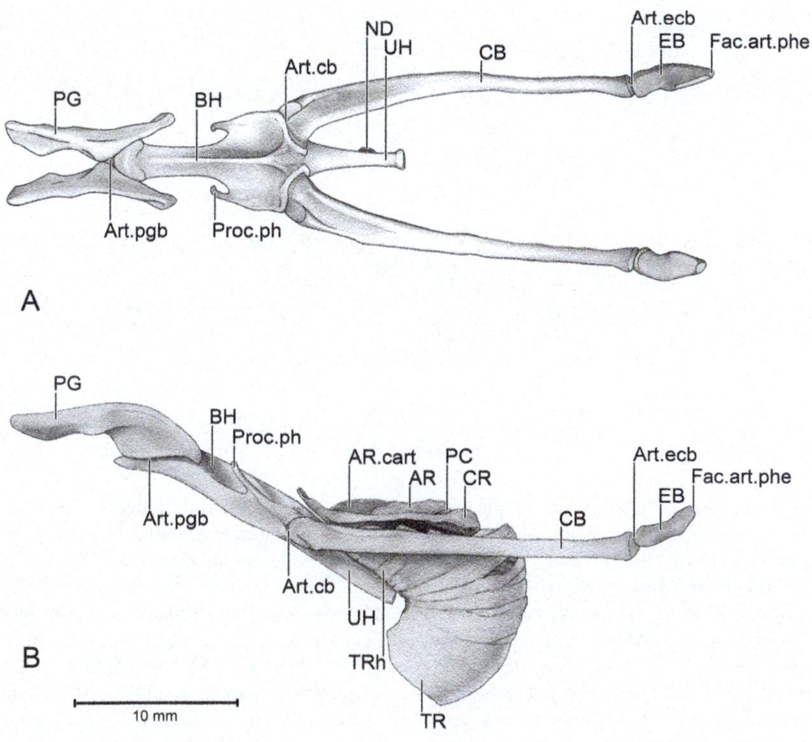

Fig. 2.12 The hyoid and laryngeal skeleton of an African Grey Parrot (*Psittacus erithacus*); the cartilaginous pharyngobranchials and urohyal process are omitted. (**a**) Dorsal view. (**b**) Lateral view with the laryngeal skeleton and trachea in situ; the rostral tip of the paraglossum is in an elevated position. AR = arytenoid; AR.cart = cartilaginous portion of the arytenoid; Art.cb = cerato-basihyal joint; Art.ecb = epi-ceratobranchial joint; Art.pgb = paraglosso-basihyal joint; BH = basihyal; CB = ceratobranchial; CR = cricoid; EB = epibranchial; Fac.art.phe = pharyngobranchial articular facet of the epibranchial; ND = *Nodulus*; PC = procricoid; PG = paraglossum; Proc.ph = parahyal process; TR = trachea; TRh = tracheal half rings; UH = urohyal [(**a**) Adapted from Homberger (1986)]

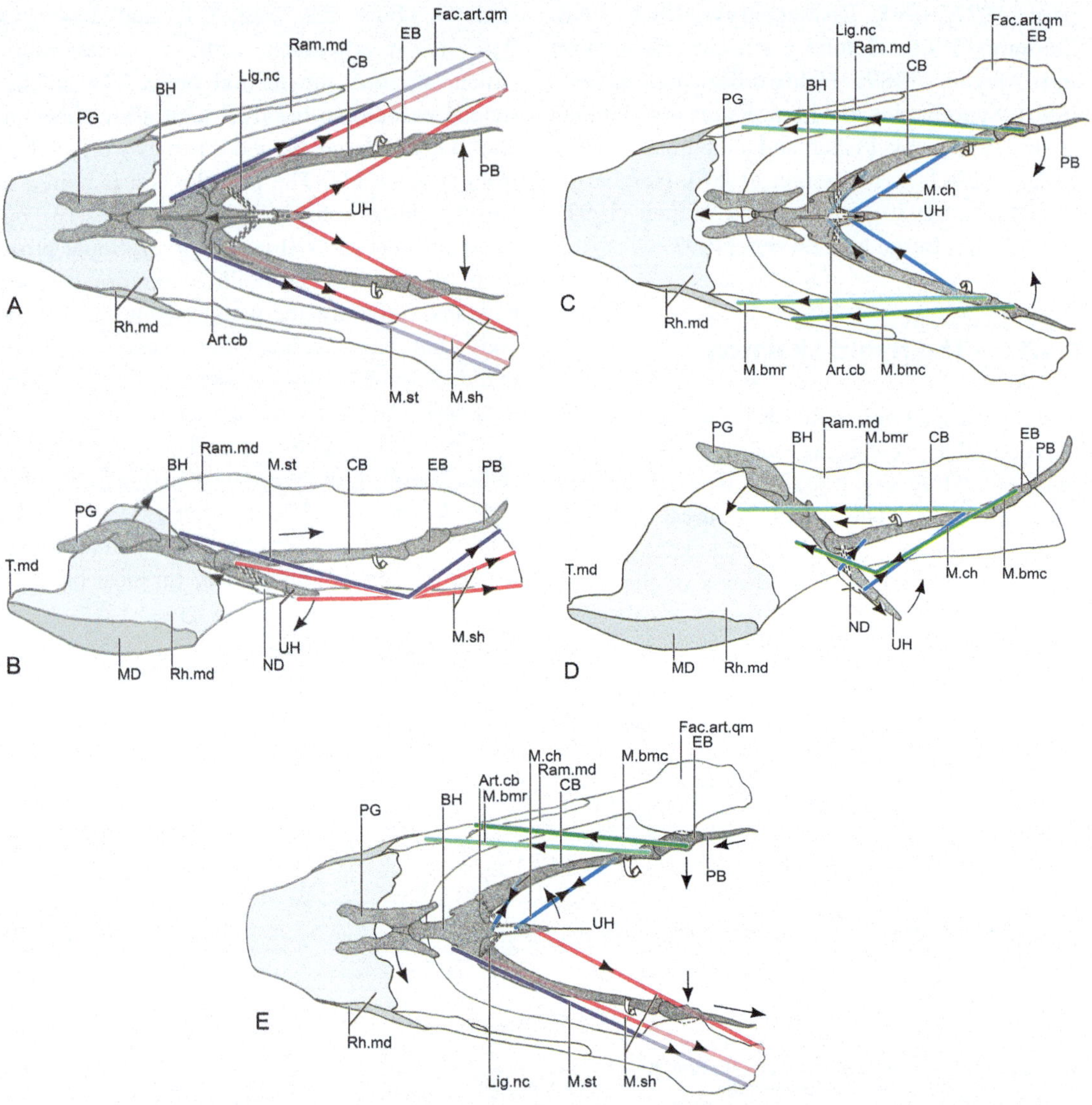

Fig. 2.13 Graphic models of movements of the hyoid skeleton relative to the mandible of an African Grey Parrot (*Psittacus erithacus*); the *arrows* indicate the direction of movements and forces. (**a**)–(**b**) Protraction of the retracted hyoid skeleton. (**a**) Dorsal view. (**b**) Lateral view. (**c**)–(**d**) Retraction of the protracted hyoid skeleton. (**c**) Dorsal view. (**d**) Lateral view. (**e**) Lateral movement of the hyoid skeleton in intermediate position between retracted and protracted positions toward the *left* through a simultaneous unilateral protraction and retraction in dorsal view. Art.cb = ceratobranchial joint; BH = basihyal; CB = ceratobranchial; EB = epibranchial; Fac.art.qm = quadrate articular facet of the mandible; Lig.nc = nodulo-ceratobranchial ligament; M.bmc = *M. branchiomandibularis caudalis*; M.bmr = *M. branchiomandibularis rostralis*; M.ch = *M. ceratohyoideus*; M.sh = *M. serpihyoideus*; M.st = *M. stylohyoideus*; MD = mandible; ND = *Nodulus*; PB = pharyngobranchial; PG = paraglossum; Ram.md = mandibular ramus; Rh.md = mandibular rhamphotheca; UH = urohyal; T.md = cutting edge of mandibular rhamphotheca [(**a**)–(**d**) Adapted from Homberger (1986)]

the bony epibranchials, which in turn are synchondrotically connected to the distal cartilaginous pharyngobranchials (Figs. 2.12, 2.13, and 2.14b, c), so that the distal ends of the hyoid horns are bendable and can conform to the curvature of the hyoid sheath (see Sect. 2.5.1). The complex convex articular facet at the proximal end of the ceratobranchial fits into the equally complex socket

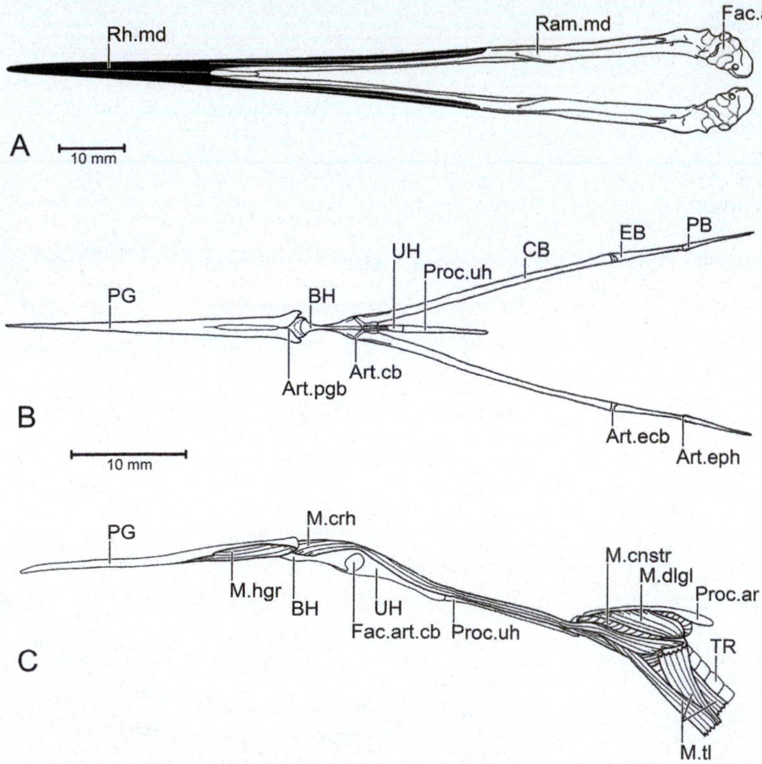

Fig. 2.14 The mandible, hyoid skeleton, and larynx of a Snowy Egret (*Egretta garzetta*); (**a**) and (**b**)–(**c**) have their own scales. (**a**) Dorsal view of the mandible. (**b**) Dorsal view of the hyoid skeleton. (**c**) Lateral view of the lingual and laryngeal apparatus; the hyoid horns are omitted, and the epithelial cover was removed to reveal the laryngeal musculature. Art.cb = cerato-basihyal joint; Art.ecb = epi-ceratobranchial joint; Art.eph = epi-pharyngobranchial joint; Art.pgb = paraglosso-basihyal joint; BH = basihyal; CB = ceratobranchial; EB = epibranchial; Fac.art.cb = ceratobranchial articular facet of the basihyal-urohyal transition; Fac.art.qm = quadrate articular facet of the mandible; M.cnstr = *M. constrictor glottidis*; M.cr = *M. cricohyoideus*; M. dlgl = *M. dilator glottidis*; M.hgr = *M. hyoglossus rostralis*; M.tl = *M. tracheolateralis*; PB = pharyngobranchial; PG = paraglossum; Proc.ar = arytenoid process; Proc.uh = urohyal process; Ram.md = mandibular ramus; Rh.md = mandibular rhamphotheca; TR = trachea; UH = urohyal [(**b**) Adapted from Cummins (1986); (**c**) Adapted from Homberger (1999)]

at the basihyal-urohyal transition to form the diarthrodial cerato-basihyal joint (Homberger 1986; Homberger and Meyers 1989) and directs a complex movement between the two bones. This linkage is supported by the nodulo-ceratobranchial ligament, which glides the sesamoid *Nodulus* along the ventral surface of the urohyal during hyoid movements (see Sect. 2.5.2.1).

The hyoid skeleton varies the proportions of its elements depending on the species. The paraglossum can be elongated to adjust the length of the lingual body to fit within a long beak, as in herons or ducks, or can be lost altogether in birds that possess a loose hyoid suspension and swallow large prey whole (see Sects. 2.5.2.2 and 2.7.3.2). In birds that project their tongues extraorally (e.g., woodpeckers and hummingbirds), the hyoid horns can elongate to such an extent that they are wound around the skull and orbits.

2.5.2 The Hyoid Suspension Apparatus and Hyoid Sheaths

The lingual apparatus is a tensegritous construction (Scarr 2014) in which the hyoid skeleton represents the elements resisting compression, the fasciae are the elements resisting tension, and the extrinsic muscles move the system before it returns to the equilibrium of its resting

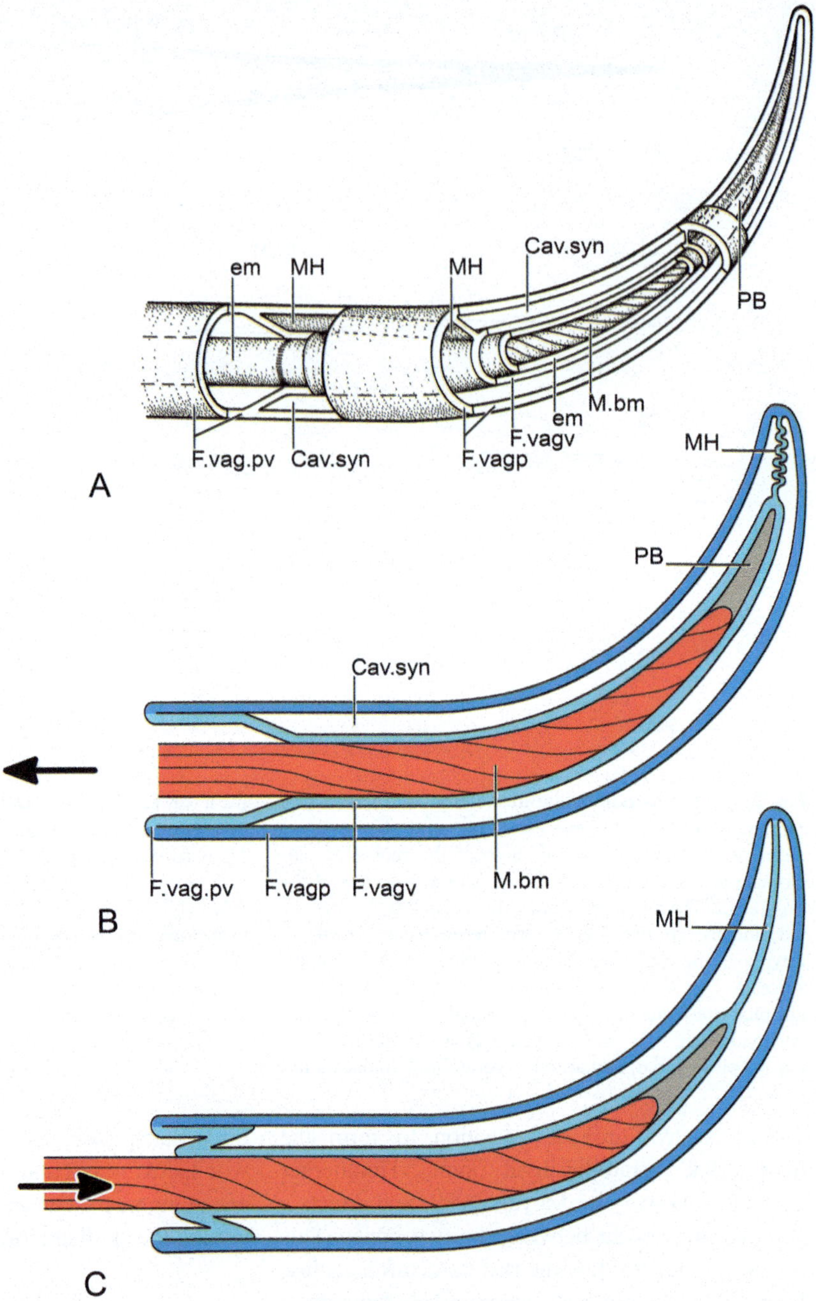

Fig. 2.15 Graphic models of the hyoid sheath in birds with a tight or relaxed hyoid suspension. (**a**) Membranous construction of a hyoid sheath. (**b**) Retracted position of a hyoid horn within its hyoid sheath; the *arrow* indicates the protraction of the hyoid horn into the position shown in (**c**). (**c**) Protracted position of a hyoid horn within its hyoid sheath; the *arrow* indicates the retraction of the hyoid horn into the position shown in (**b**). Cav.syn = synovial cavity of the hyoid sheath; em = epimysium of the branchiomandibular muscle; F.vag.pv = combined parietal and visceral layers of the hyoid sheath; F.vagp = parietal layer of hyoid sheath; F. vagv = visceral layer of the hyoid sheath; M.bm = *M. branchiomandibularis*; MH = mesentery of the hyoid sheath; PB = pharyngobranchial [Adapted from Homberger and Meyers (1989)]

position. Caudally, the hyoid horns are invested and supported by membranous hyoid sheaths (Homberger and Meyers 1989; Homberger 1999), which were previously mentioned, but not described in detail, by Leiber (1907) and Ghetie and Atanasiu (1962). The hyoid sheaths anchor the hyoid horns to the fascial system of the jaw muscles or throat fascia (Figs. 2.5a, 2.6, 2.15, and 2.16) and are analogous to synovial tendon sheaths in the limbs of mammals (see, e.g., Shields 1923; Dyce et al. 1987; Banks 1993; Salmons 1995).

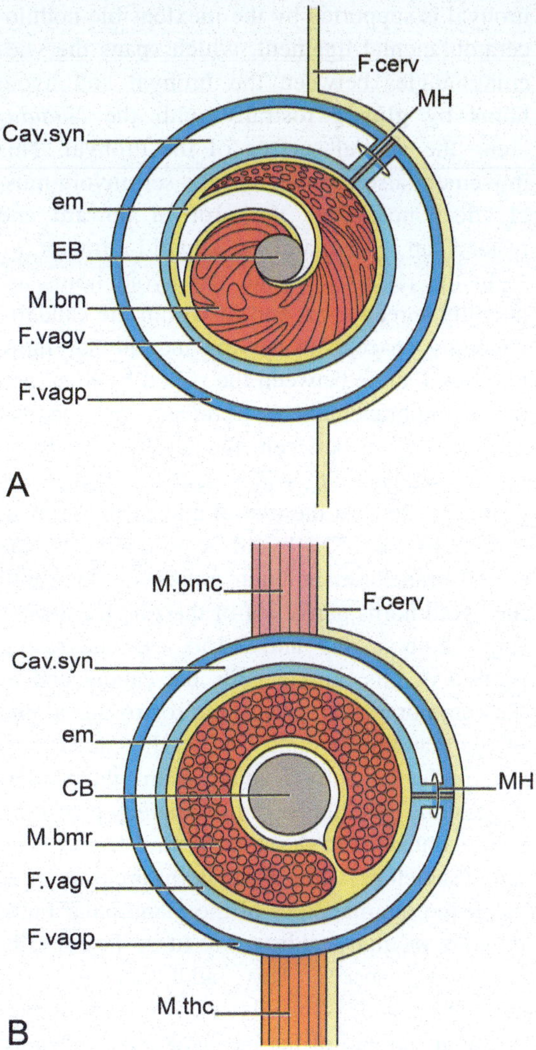

Fig. 2.16 Comparative diagrammatic cross sections through the hyoid suspension apparatus of birds. (**a**) Relaxed hyoid suspension of a chicken (*Gallus gallus*). (**b**) Loose hyoid suspension of a Snowy Egret (*Egretta garzetta*). Cav.syn = synovial cavity of the hyoid sheath; CB = ceratobranchial; EB = epibranchial; em = epimysium of the branchiomandibular muscle; F.cerv = cervical fascia; F.vagp = parietal layer of the hyoid sheath; F. vagv = visceral layer of the hyoid sheath; M.bm = *M. branchiomandibularis*; M.bmc = *M. branchiomandibularis caudalis*; M.bmr = *M. branchiomandibularis rostralis*; M.thc = *M. tracheohyoideus caudalis*; MH = mesentery of the hyoid sheath [(**a**) Redrawn and adapted from Homberger and Meyers (1989); (**b**) redrawn and adapted from Cummins (1986); both by Ellen Farrar)]

The hyoid sheaths envelop the caudal parts of the hyoid horns and consist, in principle, of two membranous layers that are separated by a synovial cavity filled with a thin layer of lubricating synovial fluid (Fig. 2.15a). The internal visceral layer envelops and attaches to the epimysium of the branchiomandibular muscles, which insert on and surround the ceratobranchial and epibranchial bones. This visceral layer folds together along the length of the hyoid sheath and forms the theoretically bilayered mesentery-like mesohyoideum, which separates again into the external parietal layer, thereby enclosing a synovial cavity with lubricating synovial fluid (Figs. 2.15 and 2.16). The mesohyoideum serves as a mesentery through which nerves and blood vessels reach the long hyoid horn and which anchors the caudal end of a hyoid horn.

2.5.2.1 Movements of the Hyoid Skeleton in a Tight or Relaxed Hyoid Suspension

In order to be able to execute precise movements to grasp or process food, the lingual apparatus is endowed with various mechanisms that link particular movements to one another while also allowing a great diversity of movements, but the hyoid skeleton also needs a pivot against which to move. In birds that use their tongues to grab or intraorally process food items (see Sect. 2.7.2), the hyoid sheaths are fastened to the fasciae between the mandibular ramus and throat muscles in parrots and cockatoos (Fig. 2.5) or ventral to the mandibular ramus and jaw muscles in the chicken and passerine birds (Figs. 2.16a and 2.30; Homberger and Meyers 1989). In hummingbirds and woodpeckers, which can project their tongue far out of their beaks, the extremely elongated hyoid horns are housed in hyoid sheath that curve up around the skull, in some species even extending into the nares or circling around the orbits. This kind of hyoid suspension construction holds the hyoid apparatus close to the base of the skull and is, therefore, called a tight hyoid suspension, although the hyoid suspension in chickens, ducks, and passerine birds is less tight and can be called a relaxed hyoid suspension (Figs. 2.16a and 2.30; see also Fig. 30 in Nickel et al. 1977; Table 2.2). The synovial fluid in these hyoid sheaths serves not only to lubricate the two layers of the hyoid sheath while they are displaced against each other but also to equalize the changing volume within of the hyoid sheath during movements of the hyoid horns (Fig. 2.15b, c).

The mechanism underlying the movements of the hyoid skeleton has been analyzed only in parrots, but it is very similar in passerine birds (personal unpublished observations) and in many other birds that possess the same hyoid elements. Here, only the main movements are described. As with the movements of the jaw apparatus, a great variety of movements can be effected by various combinations of synchronized muscle contractions (for details, see Homberger 1986).

The complex movements of the hyoid skeleton at the cerato-basihyal joints are best understood by starting from a protracted position, which may correspond to the resting position of the tongue when the beak is closed in parrots (see also Fig. 2.2): The hyoid horns are medially adducted to fit within the narrowing apical space between the flanking mandibular rami, the dorsal angle between the extended basihyal and the ceratobranchials is wide, the ventromedial angles between the urohyal and the ceratobranchials are narrow, and the sesamoid *Nodulus* is retracted toward the distal end of the urohyal (Figs. 2.12 and 2.13a, b). The hyoid skeleton is retracted by the paired serpihyoid muscles (Figs. 2.4a and 2.5), which pull the ceratobranchials caudolaterally and push the hyoid horns back into their hyoid sheaths (Fig. 2.15b, c). The complex cerato-basihyal joint automatically links the spreading apart of the hyoid horns with their outward axial rotation and, thus, an upward rotation of the rostral end of the basihyal and a downward rotation of the caudal end of the urohyal (Fig. 2.13c, d). The upward rotation of the basihyal is supported by the contraction of the retracting paired stylohyoid muscles, which insert on the parahyal processes on the dorsal side of the basihyal (Figs. 2.4, 2.5 and 2.13b). The rotation of the urohyal is supported by the inextensible nodulo-ceratobranchial ligament, which spans the widening angles between the urohyal and hyoid horns by gliding rostrally with the *Nodulus* along the ventral surface of the urohyal. This movement also ensures that the serpihyoid muscle does not rotate the urohyal upward and instead can serve mainly as a hyoid retractor.

In the retracted hyoid, the hyoid horns are laterally spread apart and fit within the caudally widening space between the mandibular rami, the dorsal angle between the raised basihyal and the ceratobranchials is reduced, the ventromedial angles between the urohyal and the ceratobranchials are wider, and the *Nodulus* is protracted toward the rostral end of the urohyal (Fig. 2.13c, d). The hyoid is protracted by the paired branchiomandibular muscles, which pull the hyoid horns partly out of their hyoid sheaths (Figs. 2.4b, c, 2.5 and 2.15b, c). The rostral portion of this muscle pulls the caudal end of the ceratobranchial toward the rostro-dorsal end of the mandibular ramus, while the caudal portion pulls the epibranchial toward the ventral edge of the mandibular ramus, thereby providing some control as to the dorsoventral level of the hyoid skeleton. As the hyoid horns are protracted along the internal sides of the mandibular rami, they are medially adducted to fit in the rostrally narrowing space. This adduction of the ceratobranchials is linked to a downward rotation of the apical end of the basihyal and an upward rotation of the caudal end of the urohyal (Fig. 2.13a, b). This rotation by the basihyal and urohyal is supported by the intrinsic ceratohyoideus muscles, which reduce the angles between the urohyal and ceratobranchials and pull the *Nodulus* back along the ventral surface of the urohyal (Fig. 2.13c, d).

The hyoid skeleton can also be raised and lowered to certain degrees. In parrots, the rostral part can be raised by a contraction of the mylohyoid muscles, which form a sling underneath the basihyal body and suspend it from the mandibular rami (Figs. 2.4a and 2.5), and lowered by the genioglossus muscle, which originates from the mandibular symphysis and inserts on the paraglossum (Figs. 2.2, 2.4, and

2.5). The hyoid skeleton can be pulled down while it is being retracted by the tracheohyoid muscle (Figs. 2.4a, c and 2.5), which inserts on the parahyal process and originates from the sternum. Various sideways movements of the hyoid skeleton are possible by unilaterally protracting and retracting the hyoid skeleton (Fig. 2.13e).

2.5.2.2 Movements of the Hyoid Skeleton in a Loose Hyoid Suspension

In at least some birds that swallow large prey items whole (e.g., pelicans, cormorants, herons, hornbills; see Sect. 2.7.3.2), the hyoid sheaths are built into the flexible walls (fasciae) of the throat below the mandibular rami (Fig. 2.6), partly because the hyoid skeleton would not fit within the usually long and narrow mandible (Fig. 2.14a). This construction is called the loose hyoid suspension. In these birds, part of the hyoid protractor and retractor muscles have modified their insertions and insert not only on the hyoid skeleton but also to the hyoid sheath (Figs. 2.6 and 2.16b). Therefore, the hyoid skeleton is moved together with the hyoid sheath and throat (cervical) fascia, and the hyoid horns are not moved in and out of the hyoid sheaths (Figs. 2.6 and 2.16b).

In birds, whose hyoid sheaths are built into their throat fascia (Fig. 2.6), the movements of the hyoid skeleton are unlikely to be as differentiated as in birds whose lingual apparatus play a central role in the acquisition and processing of food before deglutition, such as parrots, seed-eating passerines but also ducks and geese, chickens, and pigeons (see Sects. 2.7.2 and 2.7.3.1). In herons, for example, the hyoid horns within the hyoid sheaths are located well below the mandible, because the mandible is very narrow and would not be able to accommodate the hyoid apparatus between its rami as it would in parrots or finches (Figs. 2.6 and 2.14a, b). In addition, some of the hyoid protractor and retractor muscles have changed their insertions: The protracting caudal branchiomandibular muscle inserts on the dorsal side of the hyoid sheath instead of on the epibranchial so that a contraction of the caudal and cranial branchiomandibular muscles pulls up the hyoid horn and sheath simultaneously, and the entire throat fascia is protracted together with the hyoid skeleton. The retracting tracheohyoid muscle splits off a caudal portion that attaches to the ventral side of the hyoid sheath instead of on the hyoid skeleton so that a contraction of the caudal and rostral tracheohyoid muscles pulls down the hyoid horn and sheath simultaneously, and the entire throat fascia is retracted together with the hyoid skeleton.

2.5.3 Movements of the Lingual Body and Lingual Tip

The lingual body is supported by the paraglossum and, at least in parrots, also by the bodies of the lingual salivary glands (Figs. 2.3b and 2.4c). It is moved at the paraglosso-basihyal joint by two antagonistic muscles (Figs. 2.4b, c, 2.5 and 2.12). The one-joint portion of the ceratoglossus muscle rotates the tip of the paraglossum and lingual body downward, while the two-joint portions of this muscle link this downward rotation to movements at the cerato-basihyal joints. The one-joint medial oblique hyoglossus muscle rotates the tip of the paraglossum and lingual body upward, while the two-joint lateral oblique hyoglossus muscle links this movement to the movements at the cerato-basihyal joints (for details, see Homberger 1986). In addition, the lingual tip is modified into a muscular hydrostatic system that is inflated and flattened by the action of special muscles (the supraglossus muscle and the rostral and transverse hyoglossus muscles) and is passively deflated by the resilience of the lingual nail (Homberger 1980, 1986; Homberger and Brush 1986).

The lingual tip of many granivorous passerine birds can be modified from a spoonlike to a flat surface through inflatable cavernous tissues. The thick soft-cornified lingual epithelium contains large numbers of dermal papillae filled with a variety of mechanoreceptors (Ziswiler 1965, 1979; Bock and Morony 1978; Krulis 1978; Heidweiller and Zweers 1990[7]; Zweers et al. 1994). In ploceid and estrildid finches, this

[7] The "seed cup" described by Heidweiller and Zweers (1990) is a different structure between the tip of the fleshy tongue and a rostral projection of the lingual nail.

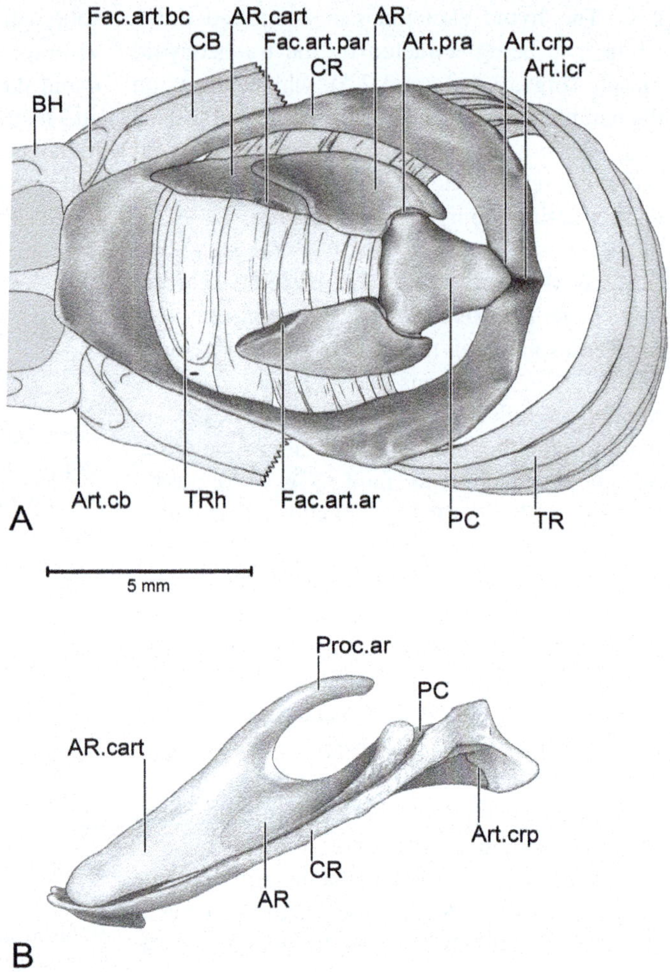

Fig. 2.17 The skeletal elements of the kinetic larynx of an African Grey Parrot (*Psittacus erithacus*). (**a**) Dorsal in situ view; the cartilaginous part of the arytenoid is omitted on the *left* and shriveled on the *right*. (**b**) Lateral view of the isolated laryngeal skeleton. AR = arytenoid; AR.cart = cartilaginous portion of the arytenoid; Art.cb = cerato-basihyal joint; Art.crp = crico-procricoid joint; Art. procrico-arytenoid joint; BH = basihyal; CB = ceratobranchial; CR = cricoid; Fac.art.ar = interface between the cartilaginous and bony portions of the arytenoid; Fac.art.bc = basihyal articular facet of the ceratobranchial; Fac.art.par = interface between the cartilaginous portion of the arytenoid and the arytenoid process; PC = procricoid; Proc.ar = arytenoid process; TR = trachea; TRh = tracheal half rings

seed-cuplike lingual tip is ventrally supported by neomorphic bones (preglossale of Bock and Morony 1978; entoglossum and hypentoglossum of Ziswiler 1979), which articulate with the rostral end of the paraglossum.

2.6 The Laryngeal Apparatus of Birds

The laryngeal mound is part of the *de facto* floor of the oral cavity in birds together with the lingual body and the preglottal area (see Sect. 2.3). In general, the laryngeal skeleton rests on the dorsal side of the basihyal-urohyal to which it is anchored by connective tissue that allows translational back-and-forth movements (Figs. 2.12b, 2.14c, 2.20 and 2.21; see also Zweers and Berkhoudt 1987). In birds that allow extensive laryngeal retractions beyond the caudal end of the urohyal, the larynx is secured by connective tissue to the tip of the cartilaginous process of the urohyal (Figs. 2.14c and 2.30).

The avian larynx essentially consists of at least four basic elements that are part bone and part cartilage and that enclose the laryngeal chamber (Figs. 2.17, 2.18, 2.19, 2.20, 2.21, 2.22, 2.23 and 2.24). The cricoid, whose lateral limbs are syndesmotically joined caudally and form the crico-procricoid diarthrodial joint with the unpaired bony procricoid, is the largest element. A pair of arytenoids forms the sides of the

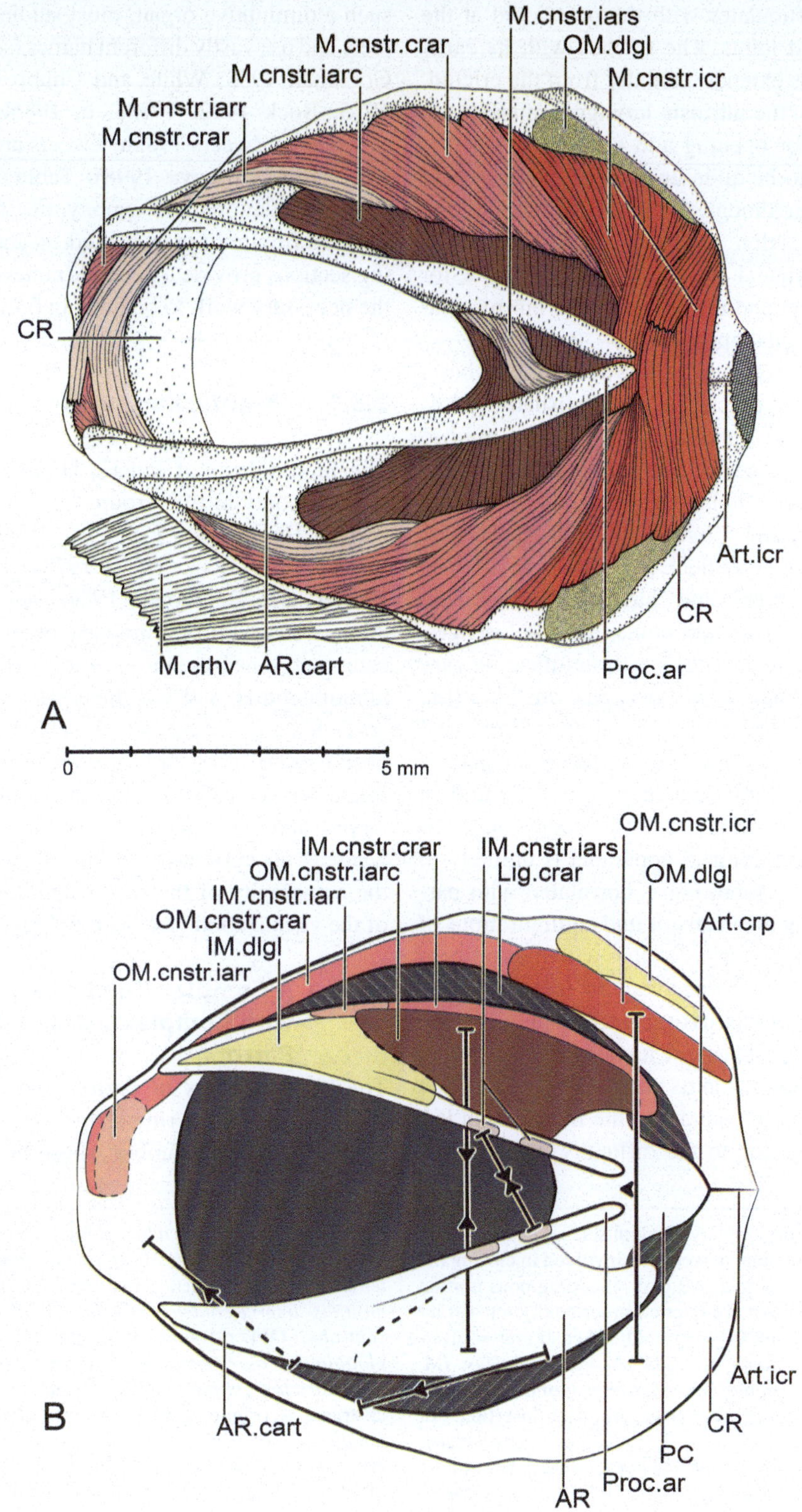

Fig. 2.18 Dorsal view of the intrinsic musculature of the kinetic larynx of an African Grey Parrot (*Psittacus erithacus*); the glottis is maximally opened. (**a**) Constrictor muscles; only one limb of the crosswise-arranged superficial interarytenoid constrictor muscle is shown. (**b**) *Right half*: Attachment sites of the intrinsic

glottis and articulates with the procricoid at the ary-procricoid joints. The trachea with its cartilaginous rings extends caudally from the cricoid.

In general, the intrinsic laryngeal muscles are responsible for opening and closing the glottis; they appear to be quite uniform across a variety of birds even though the configuration of the laryngeal skeleton may vary in details. The extrinsic laryngeal muscles are responsible for the protraction and retraction of the larynx relative to the lingual apparatus.

At the present state of knowledge, two basic types of the larynx can be distinguished in birds [cf. Zweers and Berkhoudt (1987) who tried to distinguish a passerine and a nonpasserine type of larynx]. Akinetic larynges cannot appreciably vary the shape and volume of the laryngeal chamber they enclose (see Sect. 2.6.2); they seem to be common in birds and have been found in chickens, pigeons, and herons. Kinetic larynges, in contrast, have the capacity to significantly vary the size and shape of the laryngeal chamber (see Sect. 2.6.1). These more elaborate laryngeal configurations evolved convergently in parrots and passerine birds and may also be found in hawks and hummingbirds. The degree of complexity of the laryngeal apparatus is unlikely to influence respiration, but is correlated with particular feeding mechanisms and configurations of the hyoid suspension, as well as with the ability to generate articulated vocalizations (see Sect. 2.7).

The avian larynx has received relatively scant interest in the past because it was surmised only to open and close the glottis and to have no function in vocalization in contrast to the mammalian larynx. Probably due to the difficulty in dissecting such a diminutive organ, most studies are incomplete and use vastly different names (see Fisher and Goodman 1955; White and Chubb 1968; White 1975; Bock 1972, 1978a, b, 1985b, c; Zweers et al. 1981; Zweers 1982a; Zweers and Berkhoudt 1987; Bhattacharyya 1998). Table 2.1 provides some correlations and synonymies for laryngeal muscles from more recent studies, and the following sections provide a general framework to guide the necessary work to be done in future.

2.6.1 Kinetic Larynges

The morphology of kinetic larynges have been analyzed only in two groups of birds, namely, in parrots (Homberger 1979, 1997, 1999) and in passerine birds (Bock 1972, 1978a, b, 1985b, c; Zweers and Berkhoudt 1987; Heidweiller and Zweers 1990), but preliminary observations indicate that other avian groups, such as raptors, hummingbirds, and woodpeckers, may also possess kinetic larynges (unpublished personal observations). So far, a correlation has been found between the degree of complexity of the larynx and the complexity of the vocalization, whereas no correlation could be found between the complexity of the syrinx and the complexity of the vocalization (see Sect. 2.7.4).

2.6.1.1 A Complex Kinetic Larynx with Complex Extrinsic Muscles: Parrots

The kinetic larynx of parrots and cockatoos is characterized by a ring-shaped cricoid that sits atop a laryngeal chamber whose floor and sides

Fig. 2.18 (continued) laryngeal muscles. *Left half*: Muscle forces and skeletal movements involved in closing the glottis. AR = arytenoid; AR.cart = cartilaginous portion of the arytenoid; Art.crp = crico-procricoid joint; Art.icr = intracricoid joint; CR = cricoid; IM.cnstr.crar = insertion of *M. constrictor glottidis crico-arytenoideus*; IM.cnstr.iarr = insertion of *M. constrictor glottidis interarytenoideus rostralis*; IM.cnstr.iars = insertion of *M. constrictor glottidis interarytenoideus superficialis*; IM.dlgl = insertion of *M. dilator glottidis*; Lig.crar = crico-arytenoid ligament; M.cnstr.crar = *M. constrictor glottidis crico-arytenoideus*; M.cnstr.iarc = *M. constrictor glottidis interarytenoideus caudalis*; M. cnstr.iarr = *M. constrictor glottidis interarytenoideus rostralis*; M.cnstr.iars = *M. constrictor glottidis interarytenoideus superficialis*; M.cnstr.icr = *M. constrictor glottidis intercricoideus*; M.crhv = *M. cricohyoideus ventralis*; OM.cnstr.crar = origin of *M. constrictor glottidis crico-arytenoideus*; OM.cnstr.iarc = origin of *M.constrictor glottidis interarytenoideus caudalis*; OM. cnstr.iarr = origin of *M. constrictor glottidis interarytenoideus rostralis*; OM.cnstr.icr = origin of *M. constrictor glottidis intercricoideus*; OM.dlgl = origin of *M. dilator glottidis*; PC = procricoid; Proc.ar = arytenoid process

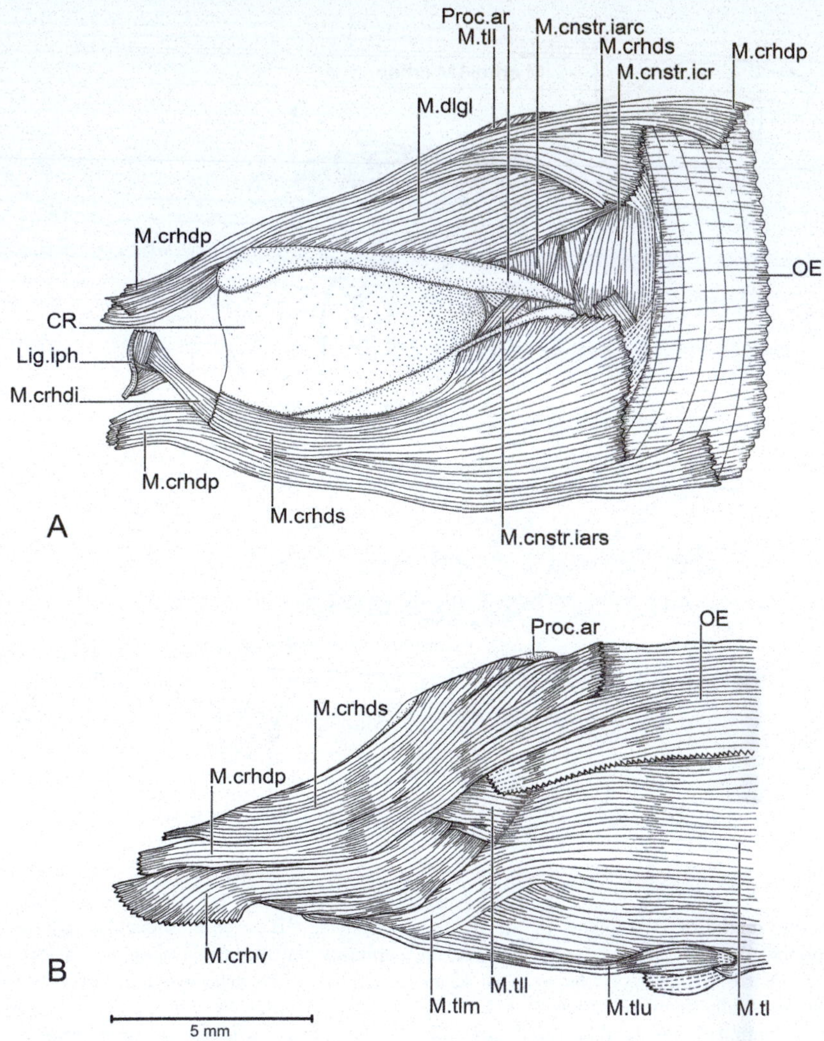

Fig. 2.19 The extrinsic musculature of the kinetic larynx of an African Grey Parrot (*Psittacus erithacus*). (**a**) Dorsal view of the extrinsic protractor muscles of the larynx and parts of the intrinsic laryngeal dilator and constrictor muscles. (**b**) Lateral view of the extrinsic protractor and retractor muscles of the larynx. CR = cricoid; Lig.iph = interparahyal ligament; M.cnstr.iarc = *M. constrictor glottidis interarytenoideus caudalis*; M.cnstr.iars = *M. constrictor glottidis interarytenoideus superficialis*; M.cnstr.icr = *M. constrictor glottidis intercricoideus*; M.crhdi = *M. cricohyoideus dorsalis intermedius*; M.crhdp = *M. cricohyoideus dorsalis profundus*; M.crhds = M. *cricohyoideus dorsalis superficialis*; M.crhv = *M. cricohyoideus ventralis*; M. dlgl = *M. dilator glottidis*; M.tl = *M. tracheolateralis*; M. tll = *M. tracheolateralis lateralis*; M.tlm = *M. tracheolateralis medialis*; M.tlu = *M. tracheolateralis urohyalis*; OE = esophagus; Proc.ar = arytenoid process

are formed by tracheal half rings and by a complex extrinsic musculature (Figs. 2.2, 2.12b, 2.17, 2.18, 2.19, 2.20 and 2.21).

The cricoid is a flat bony ring with a short flat apical plate and lateral limbs that are broadened caudally. Where the latter join syndesmotically with each other caudally, they are sharply bent ventrally and, thereby, create a spur that projects into the laryngeal cavity (Fig. 2.17). The bony procricoid is nestled between the caudal ends of the cricoid limbs and forms with them the diarthrodial crico-procricoid joint. The bony caudal

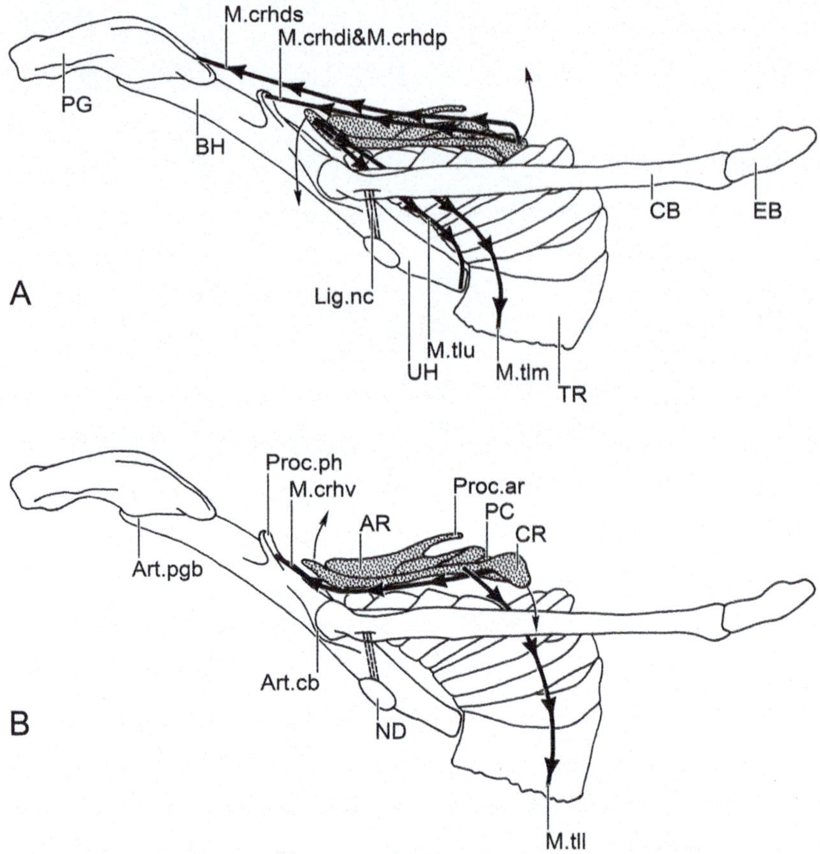

Fig. 2.20 Graphic model of movements of the larynx relative to the trachea through extrinsic laryngeal muscles in an African Grey Parrot (*Psittacus erithacus*). (**a**) Caudally elevated and rostrally depressed cricoid and larynx; *arrows* indicate the action of muscles that would move the cricoid and larynx into the position shown in (**b**). (**b**) Caudally depressed and rostrally elevated cricoid and larynx; the *arrows* indicate the action of muscles that would move the cricoid and larynx into the position shown in (**a**). AR = arytenoid; Art.cb = cerato-basihyal joint; Art.pgb = paraglosso-basihyal joint; BH = basihyal; CB = ceratobranchial; CR = cricoid; EB = epibranchial; Lig.nc = nodulo-ceratobranchial ligament; M.crhds = *M. cricohyoideus dorsalis superficialis*; M. crhdi = *M. cricohyoideus dorsalis intermedius*; M.crhdp = *M. cricohyoideus dorsalis profundus*; M.crhv = *M. cricohyoideus ventralis*; M.tll = *M. tracheolateralis lateralis*; M.tlm = *M. tracheolateralis medialis*; M.tlu = *M. tracheolateralis urohyalis*; ND = *Nodulus*; PC = procricoid; PG = paraglossum; Proc.ar = arytenoid process; Proc.ph = parahyal process; TR = trachea; UH = urohyal [Adapted from Homberger (1999)]

halves of the arytenoids articulate diarthrodially with the rostral end of the procricoid (Fig. 2.17a) and are extended rostrally by the cartilaginous halves that bear the caudally projecting arytenoid processes (Figs. 2.17b, 2.18, 2.19, 2.20 and 2.21). The arytenoids are connected along their entire lateral side to the internal margin of the cricoid by the cricoarytenoid ligaments with which they form the ceiling of the laryngeal chamber (Fig. 2.18b).

Among the intrinsic laryngeal muscles, the parallel-fibered paired dilator muscles of the glottis originate from the caudolateral ends of the cricoid on either side of the crico-procricoid joint and insert on the cartilaginous apical half of the arytenoids (Figs. 2.18 and 2.19a). To open or widen the glottis, they pull apart the arytenoids by rotating them caudolaterally as well as axially at the ary-procricoid joints so that the arytenoids

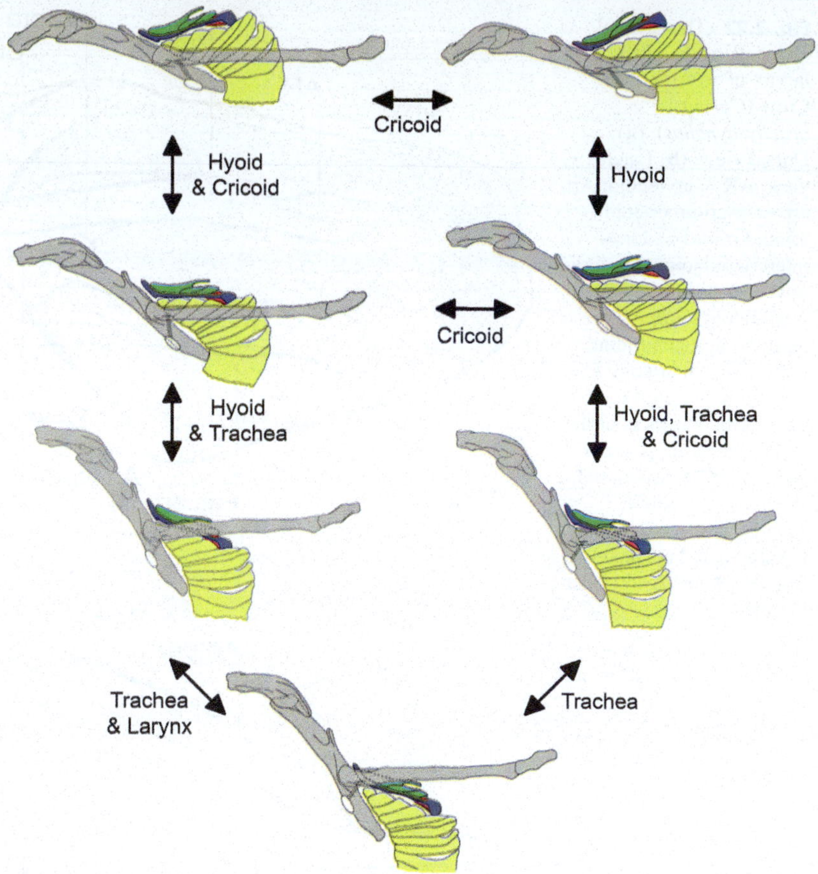

Fig. 2.21 Graphic models showing the effects of diverse combinations of hyoid, tracheal and laryngeal movements on the shape of the laryngeal chamber of an African Grey Parrot (*Psittacus erithacus*). The paraglossum is shown in a fixed position, but its movements would only increase the number of possible permutations of positions that would affect the configuration of the floor of the oral cavity and the oral cavity as a whole. The labeled *arrows* indicate which skeletal elements change their positions between each model. Color codes: blue = cricoid; green = arytenoid; grey = hyoid skeleton; red = procricoid; yellow = trachea

assume a more vertical orientation. At the same time, they elevate the procricoid.

The constrictor muscles of the glottis are more complex than the dilator muscle and consist of several slips with distinctive localized functions (Fig. 2.18). The interarytenoid constrictor muscle comprises three distinct parts and is responsible for adducting the glottal lips. The largest among them, the caudal interarytenoid constrictor muscle, inserts on the dorsal sides of the caudal bony parts of the arytenoids and rotates them medially while also rotating them axially at the procricoarytenoid joints so that they assume a more horizontal orientation. The rostral interarytenoid constrictor muscle originates on the dorsal side of the apical plate of the cricoid from the connective tissue covering the origins of the cricoarytenoid muscles (see below) and inserts on the lateral edges of the arytenoids at the transition between their bony and cartilaginous portions; it adducts the cartilaginous halves of the arytenoids. The superficial interarytenoid constrictor muscles adduct the arytenoid processes. The paired cricoarytenoid constrictor muscles originate from the dorsal sides of the rostral two-thirds of the cricoid limbs and insert along the dorsal sides of the lateral edges of the bony portion of the arytenoids; they pull the arytenoids rostromedially. The caudalmost and most superficial intercricoid muscle originates from the caudal ends of the cricoid limbs and crosses over the procricoid, which it pushes down in antagonism to the elevation by its dilator muscle (Fig. 2.17b).

Hence, the various portions of the constrictor muscle act as antagonists to the relatively simple actions of the dilator muscles in widening the glottis. In addition, the distinct portions of the constrictor muscle can subtly and differentially

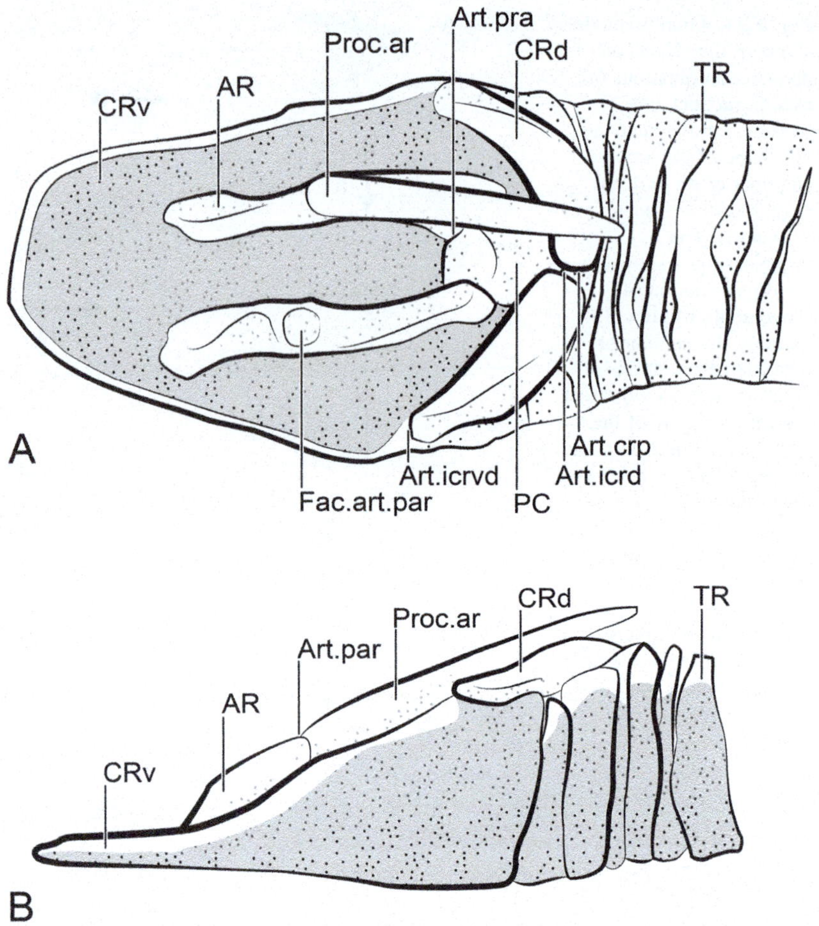

Fig. 2.22 The skeletal elements of the kinetic larynx of an American Crow (*Corvus brachyrhynchos*). (**a**) Dorsal view. (**b**) Lateral view. AR = arytenoid; Art. crp = crico-procricoid joint; Art.icrd = dorsal intercricoid joint; Art. icrvd = ventro-dorsal intercricoid joint; Art.pra = procrico-arytenoid joint; CRd = dorsal cricoid; CRv = ventral cricoid; Fac.art. par = articular facet of the arytenoid for the arytenoid process; PC = procricoid; Proc.ar = arytenoid process; TR = trachea [Redrawn and adapted from Bock (1978) by Ellen Farrar]

affect the shape of the laryngeal chamber, as well as the shape of the glottis and its lips, which is supported for the most part by the cartilaginous portion of the arytenoids.

The extrinsic muscles, which move the larynx relative to the hyoid skeleton and the trachea, are more complex in parrots than in other birds (Figs. 2.19 and 2.20). The protracting cricohyoid muscle also raises the caudal end of the cricoid and comprises a dorsal and a ventral portion (Fig. 2.20a). The dorsal cricohyoid muscle inserts on the connective tissue attaching the laryngeal papillae of the laryngeal mound to the caudal end of the cricoid and, in turn, comprises two to three parts (Figs. 2.2, 2.4b, c, 2.19 and 2.20). Its deep portion originates from the parahyal process of the basihyal, and its occasional middle portion inserts on the interparahyal ligament, which spans the paired parahyal processes, thereby protracting the larynx relative to the basihyal. The superficial portion of the dorsal cricohyoid muscle, however, originates from the paraglossum, thereby linking the protraction of the larynx to the movements of the lingual body at the paraglosso-basihyal joint. The ventral cricohyoid muscle inserts on the ventral side of the caudal end of the cricoid and originates from the base of the parahyal process (Figs. 2.18a and 2.20b), thereby raising the rostral end of the cricoid while protracting it (Fig. 2.20b).

The retracting tracheolateral muscle comprises three portions (Figs. 2.19b and 2.20). Its lateral portion inserts on the dorsal side of the caudal end of the cricoid, whereas its medial

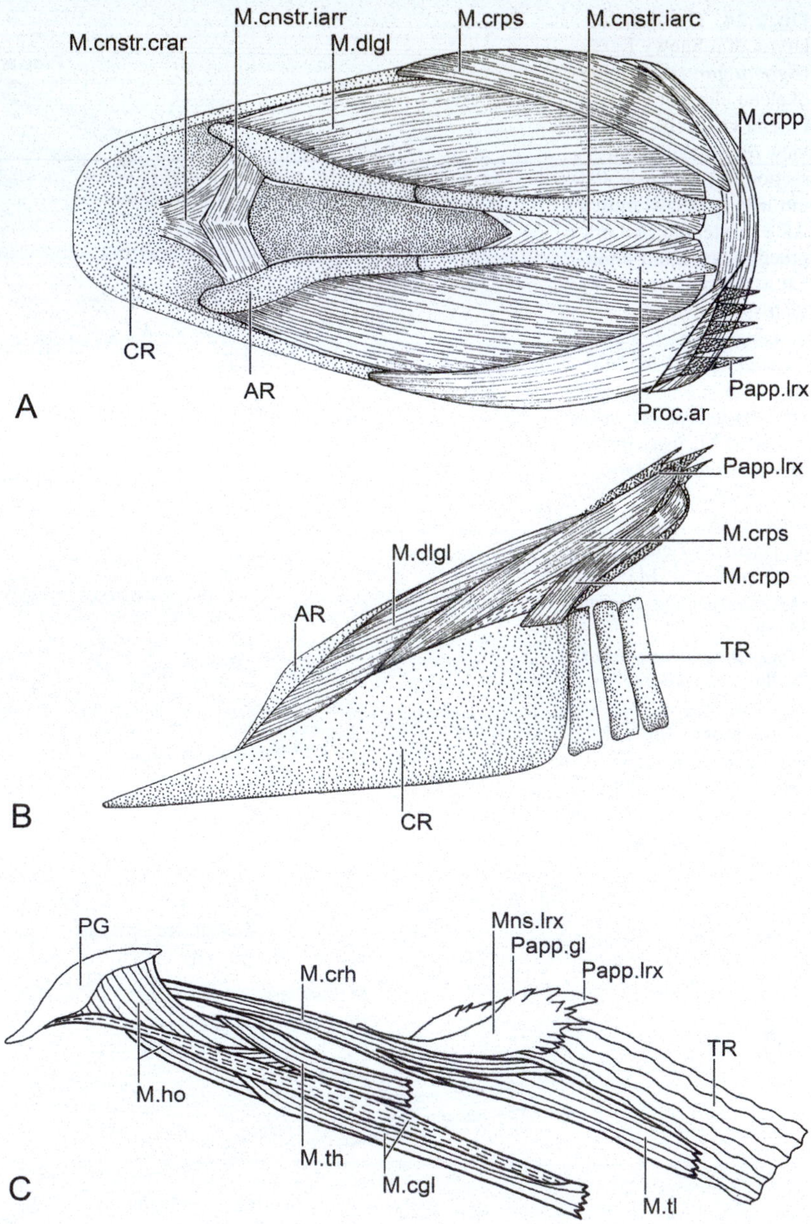

Fig. 2.23 Laryngeal muscles of passerine birds. (**a**, **b**) Intrinsic laryngeal muscles of an American Crow (*Corvus brachyrhynchos*); (**a**, **b**) and (**c**) are not to scale. (**a**) Dorsal view. (**b**) Lateral view. (**c**) Extrinsic muscles of the larynx of a grosbeak (*Coccothraustes* sp.). AR = arytenoid; CR = cricoid; M.cgl = *M. ceratoglossus lateralis*; M.cnstr.crar = *M. constrictor glottidis crico-arytenoideus*; M. cnstr.iarc = *M. constrictor glottidis interarytenoideus caudalis*; M.cnstr.iarr = *M. constrictor glottidis interarytenoideus rostralis*; M.crh = *M. cricohyoideus*; M.crpp = *M. cricopapillaris profundus*; M.crps = *M. cricopapillaris superficialis*; M.dlgl = *M. dilator glottidis*; M.ho = *M. hyoglossus obliquus*; M.th = *M. tracheohyoideus*; M.tl = *M. tracheolateralis*; Mns.lrx = laryngeal mound; Papp.gl = glottal papillae; Papp.lrx = laryngeal papillae; PG = paraglossum; Proc.ar = arytenoid process; TR = trachea [(**a**) and (**b**) adapted from Bock (1978); (**c**) redrawn from W.J. Bock unpublished]

portion inserts on the ventral side of the apical end of the cricoid; both portions originate from fascial tissue at the base of the neck. The short urohyal portion of the tracheolateral muscles originates from the caudal end of the urohyal and inserts on the apical end of the cricoid with the medial portion.

Both the shapes and sizes of the laryngeal chamber and oral cavity are modified by the actions of the extrinsic laryngeal muscles in parrots. When the larynx is pulled rostrally toward or caudally away from the paraglossum by the cricohyoid and tracheolateral muscles, respectively, the configuration of the floor of

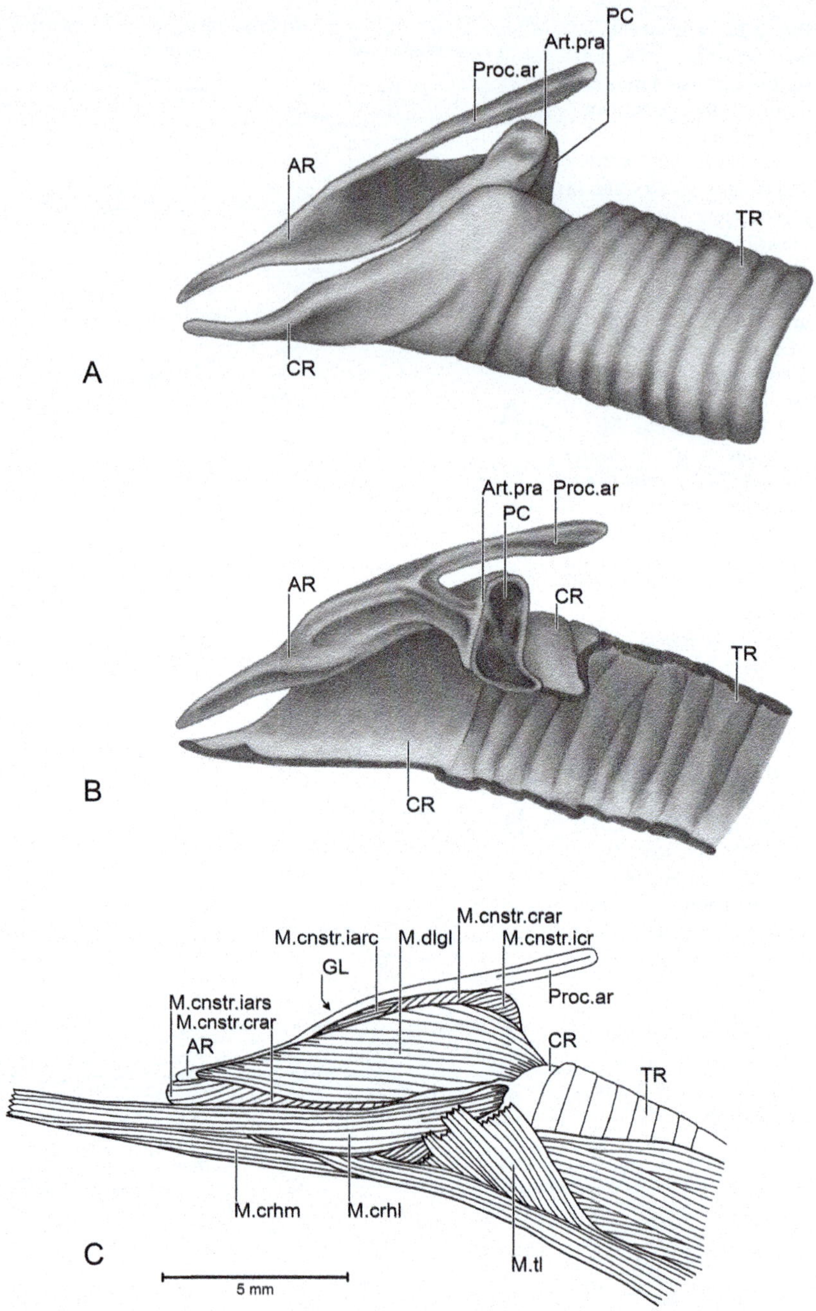

Fig. 2.24 The akinetic larynx of a Snowy Egret (*Egretta garzetta*). (**a–b**) Skeletal elements. (**a**) Lateral view. (**b**) Medial view of a longitudinal section. (**c**) Extrinsic and intrinsic laryngeal muscles. AR = arytenoid; Art.pra = procrico-arytenoid joint; CR = cricoid; GL = glottis; M.cnstr.crar = *M. constrictor glottidis crico-arytenoideus*; M. cnstr.iarc = *M. constrictor glottidis interarytenoideus caudalis*; M.cnstr.iars = *M. constrictor glottidis interarytenoideus superficialis*; M.cnstr.icr = *M. constrictor glottidis intercricoideus*; M.crhl = *M. cricohyoideus lateralis*; M.crhm = *M. cricohyoideus medialis*; M.dlgl = *M. dilator glottidis*; *M.tl* = *M. tracheolateralis*; PC = procricoid; Proc.ar = arytenoid process; TR = trachea [(**c**) Adapted from Homberger (1999)]

the oral cavity changes its configuration and the length of the preglottal area between the lingual body and the laryngeal mound. The configuration of the floor of the oral cavity can also be changed through a pivoting of the cricoid above the laryngeal chamber (Figs. 2.20 and 2.21; see Sect. 2.7.4.1). Simultaneous contractions of the dorsal cricohyoid muscle and the medial and urohyal tracheolateral muscles raise the caudal end of the cricoid and, thus, the laryngeal

Table 2.1 Comparisons and synonymies of the laryngeal muscles of birds

Psittaciformes (Homberger this chapter)	Passeriformes: *Corvus* and various species (e.g., Bock 1972*; 1978a^{+}, b^; 1985b$^{\#}$, c$^{\$}$)	Passeriformes: *Corvus* (Zweers and Berkhoudt 1987); estrildid finches (Heidweiller and Zweers 1990)	Columbiformes: *Columba* (Zweers 1982a*; Zweers et al. 1981^); *Treron* (Bhattacharyya 1998$^{\#}$)	Galliformes (White 1975; White and Chubb 1968)
M. dilator glottidis	*M. thyreoarytenoideus**, *M. dilator glottidis*^	*M. dilator glottidis* (*sive M. crico dorsalis arytaenoideus*)	*M. cricoarytaenoideus**, *M. dilator glottidis*^, *M. thyroarytenoideus*$^{\#}$	Superficial intrinsic muscle (dilator)
M. constrictor glottidis intercricoideus	*M. constrictor glottidis**, Part A of *M. constrictor glottidis*^	Parts 2 and 3 of *M. cricoprocricoideus*	*M. crico interarytaenoideus caudolateralis, caudomedialis et intermedialis**, ^	Deep intrinsic muscle (constrictor): Lateral division
M. constrictor glottidis crico-arytenoideus	*M. constrictor glottidis**, Part A of *M. constrictor glottidis*^	Part 1 of *M. cricoprocricoideus*	*M. cricoarytenoprocricoideus*$^{\#}$	Deep intrinsic muscle (constrictor): Middle division
M. interarytenoideus rostralis	*M. constrictor glottidis**, Part B of *M. constrictor glottidis*^	*M. cricoarytaenoideus rostralis*	*M. crico interarytaenoideus rostralis**, ^; *M. cricoarytenoideus rostralis*$^{\#}$	Not mentioned
M. interarytenoideus caudalis	*M. constrictor glottidis**, Part C of *M. constrictor glottidis*^	*M. arytaenoprocricoideus*	*M. arytaeno interarytaenoideus**, ^; *M. arytenoprocricoideus*$^{\#}$	Deep intrinsic muscle (constrictor): Medial division
M. interarytenoideus superficialis	Not mentioned	Not mentioned	Not mentioned	Not mentioned
Not present	*M. thyreohyoideus superior**, *M. cricohyoideus superior*^; (Homberger this chapter: *M. cricopapillaris superficialis*)	*M. cricodermoideus; M. hyovalvularis*	*M. hyovalvularis*	Dorsal part of rostral extrinsic muscle attaching to dorsal mass
Not present	*M. cricohyoideus posterior*^ (Homberger this chapter: *M. cricopapillaris profundus*)	Tracheodermoid fibers; *M. claviculovalvularis*	*M. tracheovalvularis*^	Dorsal mass of caudolateral extrinsic muscle
M. cricohyoideus ventralis	*M. thyreohyoideus**, *M. thyrohyoideus*$^{+}$, *M. cricohyoideus*$^{\#, \$}$	*M. cricohyoideus (dorsalis)*	*M. cricohyoideus ventralis et dorsalis**, *M. thyrohyoideus ventralis et lateralis*$^{\#}$	Rostral extrinsic muscle (*M. basibranchialis laryngeus)*
M. cricohyoideus dorsalis	Not mentioned	Not mentioned	*Not mentioned*	Not mentioned
M. tracheolateralis lateralis	*M. tracheolateralis**$^{, +, \#, \$}$	*M. cricotrachealis lateralis dorsalis*	*M. trachealis lateralis pars cricoidea**, *M. tracheolateralis*$^{\#}$	Caudolateral extrinsic muscle (*M. sternotracheolaryngeus)*
M. tracheolateralis medialis	*M. tracheohyoideus**$^{, \#, \$}$	*M. cricotrachealis lateralis ventralis*	*M. trachealis lateralis pars hyoidea**, *M. tracheolateralis*$^{\#}$	Caudomedial extrinsic muscle (*M. sternotracheolaryngeus)*
M. tracheolateralis urohyalis	Not mentioned	Not mentioned	Not mentioned	Not mentioned

papillae above the floor of the esophagus and lower the rostral end of the cricoid into the laryngeal chamber (Fig. 2.20a). Conversely, a simultaneous contraction of the ventral cricohyoid muscle and lateral tracheolateral muscle lowers the laryngeal papillae toward the esophageal floor and lowers the caudal end of the cricoid into the laryngeal chamber (Fig. 2.20b).

These movements, in combination with movements and configurational changes by the hyoid skeleton (see Sect. 2.5), increase the diversity of configurational changes of the laryngeal chamber and oral cavity (Fig. 2.21). Additional movements of the paraglossum relative to the basihyal (not shown here, but see Homberger 1986) increase the number of possible configurations of the floor of the oral cavity by themselves, but also through their linkage to the raising of the laryngeal papillae through the superficial dorsal cricohyoid muscle (Fig. 2.20a).

2.6.1.2 A Complex Kinetic Larynx with Simple Extrinsic Laryngeal Muscles: Passerine Birds

The kinetic larynx of passerine birds is characterized by a complex (ventral and dorsal) cricoid, special intrinsic muscles, and simple extrinsic laryngeal muscles (Figs. 2.22 and 2.23; see Table 2.1; Bock 1972, 1978a, b, 1985b, c; Zweers and Berkhoudt 1987; Heidweiller and Zweers 1990).

The skeletal cricoid consists of two parts: A large ventral cricoid forms the floor and sides of the laryngeal cavity (Fig. 2.22). A longitudinal section reveals a series of cartilaginous concretions (see Bock 1978b), which suggests that the ventral cricoid may have evolved from a fusion of tracheal half rings similar to those in the parrot larynx. The trachea with full tracheal rings extends caudally from these half rings. A pair of dorsal cricoids articulates rostro-laterally with the caudo-dorsal ends of the ventral cricoid and caudo-medially with the unpaired procricoid (Fig. 2.22a). The procricoid articulates caudally with the converging ends of the ventral cricoid in the crico-procricoid joint and rostrally with the paired arytenoids, which form the ceiling of the laryngeal chamber and articulate each with a caudally pointing arytenoid process. Hence, the passerine larynx comprises eight articulating skeletal elements in contrast to the four skeletal elements in the psittacine larynx (see Sect. 2.6.1.1).

The functional morphology of the intrinsic dilator and constrictor laryngeal muscles of passerine birds appear to be similar to those described for parrots based on interpretations of the studies by Bock (1972, 1978b, c, 1985a, b), Zweers and Berkhoudt (1987), and Heidweiller and Zweers (1990) despite differences in the laryngeal skeleton (see Table 2.1). This interpretation will need to be corroborated because early descriptions are either not detailed enough or differ in some details in the two studies of three *Corvus* species by Bock (1978b) and Zweers and Berkhoudt (1987) so as to leave open the question whether the reported differences are due to actual specific differences or to observer error.

The passerine intrinsic laryngeal musculature, however, comprises a unique antagonistic pair of intrinsic muscles, which insert on the connective tissue pad that underlies the laryngeal papillae and anchor them to the caudal end of the dorsal cricoids (Fig. 2.23a, b; Bock 1978b; Table 2.1). The superficial cricopapillary muscle raises the dorsal cricoids and laryngeal papillae above the floor of the esophagus, and the deep cricopapillary muscle depresses them. These actions significantly modify not only the size and shape of the oral cavity but also the size and shape of the laryngeal chamber. However, the descriptions of these unique muscles by Zweers and Berkhoudt (1987) and Heidweiller and Zweers (1990) are incomplete and need to be redone in future.

The extrinsic laryngeal muscles of passerine birds are simpler than those of parrots and comprise a protracting cricohyoid muscle and a retracting tracheolateral muscle (Fig. 2.23c; Table 2.1). They do not contribute to the modifications of the size and shape of the laryngeal chamber, but can affect the shape of the oral cavity.

2.6.2 Akinetic Larynges

The akinetic larynx of many birds is characterized by a relatively simple laryngeal skeleton and simple extrinsic laryngeal muscles and is exemplified by the larynx in chickens (White and Chubb 1968;

White 1975), pigeons (Zweers et al. 1981; Zweers 1982a; Bhattacharyya 1998), herons and egrets (Fig. 2.24; Homberger 1999), and probably also in cranes (see Fisher and Goodman 1955) (Table 2.1.).

The laryngeal skeleton comprises a single trough-shaped cricoid, whose caudo-dorsal corners meet in the midline and embrace the cricoid to form the crico-procricoid joint (Fig. 2.24a, b). The holes in the floor of the cricoid of pigeons (see Zweers 1982a) indicate that this part of the cricoid may have evolved from a fusion of tracheal half rings (see Sect. 2.6.1.1.). The paired arytenoids articulate with the cranial end of the procricoid and form the ceiling of the laryngeal chamber. The trachea extends caudally from the caudal end of the cricoid.

The functional morphology of the intrinsic laryngeal dilator and constrictor muscles in akinetic larynges is very similar to those in the kinetic larynges of parrots (see Sect. 2.6.1.1) and passerine birds (see Sect. 2.6.1.2) (see White and Chubb 1968; White 1975; Zweers et al. 1981; Zweers 1982a; Bhattacharyya 1998; Fig. 2.24c). However, the mechanism of raising the laryngeal papillae above the floor of the esophagus varies among the three avian groups with akinetic larynges that have been studied and differs from that in parrots and passerine birds. In pigeons, a hyovalvular muscle passes dorsally over the intrinsic muscles and raises the laryngeal papillae, whereas the poorly defined tracheovalvular muscles probably depresses them (Zweers 1982a). In chickens, a dorsal portion of the laryngeal protractor muscle inserts on a tissue mass underlying the laryngeal papillae and raises them above the floor of the esophagus, whereas a caudolateral portion of the laryngeal retractor muscle attaching to the tissue mass depresses them (White 1975). In egrets, the ends of the arytenoid processes project caudally beyond the cricoid (Fig. 2.24), underlie the laryngeal papillae, and are supported by erectile tissue. Muscle slips originate from the flexible arytenoid processes and insert on the erectile tissue and may be part of a mechanism of raising the laryngeal papillae above the floor of the esophagus (unpublished personal observation).

The extrinsic musculature of the akinetic larynx is very similar to that in passerine birds (Table 2.1, Fig. 2.24c). In the egret, the larynx is protracted along the urohyal by the cricohyoid muscle, which originates from the apical end of the basihyal and inserts by various slips on the caudal end of the cricoid (Figs. 2.14c and 2.24c). It is retracted by the tracheolateral muscle, which originates from the pectoral fascia at the base of the neck and inserts cranially, ventrally, and caudo-dorsally on the cricoid by various slips. Because the insertions of the protractor and retractor muscles are broadly distributed over the outer surface of the cricoid, these muscles have little, if any, capacity to tilt the larynx relative to the trachea.

The descriptions of the constrictor musculature of the glottis and the muscles moving the laryngeal papillae of akinetic larynges (White and Chubb 1968; White 1975; Zweers et al. 1981; Zweers 1982a; Bhattacharyya 1998) are not detailed enough to ascertain the similarities and differences among the different avian groups and will need to be clarified. It can nevertheless be established that the large cricoid and the simple extrinsic laryngeal musculature of the akinetic larynx in several species preclude significant modifications of the size and shape of the laryngeal chamber, but that probably convergent mechanisms exist to raise and lower the laryngeal papillae during deglutition (see Table 2.2; Sect. 2.7.3.1).

2.7 Integrated Biological Roles of the Avian Jaw, Lingual and Laryngeal Apparatus

Biological roles are distinguished from functional-morphological properties by their integration into the behavior of an individual in its natural environment (Bock and von Wahlert 1965). An understanding of biological roles of morphological systems is the necessary foundation for a reconstruction of the evolutionary history based on the selective value of particular configurations of these systems. It also requires a high degree of integration of ecological, behavioral, physiological, functional and morphological observations and analyses. Therefore, eco-morphological studies of the avian jaw, lingual, or laryngeal apparatus are rare (e.g., Bowman 1961; Richards and Bock 1973; Homberger 1990, 2003).

Table 2.2 Summary of evolutionary interpretations of various configurations of the lingual and laryngeal apparatus

Configuration of organ	Behaviors	Avian group examples	Evolutionary interpretation
Loss of lingual body	Cranial-inertial feeding; simple vocalizations	Ratites, ibises, pelicans, cormorants	Multiply derived, independent and convergent conditions
Tight hyoid suspension with functional hyoid sheath	Intraoral food processing; lingual feeding; complex vocalizations	Psittaciformes, granivorous songbirds	Ancestral condition
Relaxed hyoid suspension with functional hyoid sheath	Lingual and cranial-inertial feeding; simple vocalization	Anseriformes, Galliformes	Derived condition
Loose hyoid suspension with nonfunctional hyoid sheath	Cranial-inertial feeding; simple vocalization	Herons and egrets (possibly ratites, ibises, pelicans, cormorants, hornbills)	Multiply derived, independent and convergent conditions
Kinetic larynx with tracheal half rings	Lingual feeding; complex articulated vocalization	Psittaciformes	Ancestral condition
Akinetic larynx with a trough-shaped cricoid from fused tracheal half rings	Any feeding mechanisms; simple, non-articulated vocalizations	Galliformes, Columbiformes, Ardeidae	Multiply derived, independent and convergent conditions
Kinetic larynx with a trough-shaped ventral cricoid from fused tracheal half rings and a dorsal cricoid	Lingual feeding; complex articulated vocalizations	Passerine birds	Independently derived condition

This section will try to provide a framework for future studies by providing examples of how the movements of the jaw, lingual and laryngeal apparatus are coordinated to perform particular behaviors. Because only very few functional-morphological studies of all three apparatus in a single species exist, this section also serves to highlight the need for additional integrated studies to gain a deeper understanding of the diversity, as well as general commonalities, of these apparatus. Especially glaring is the paucity of studies on the functional morphology of the avian larynx, whose crucial functions in feeding, respiration and vocalization mechanisms are usually glossed over, if mentioned at all.

2.7.1 Coordination of the Jaw and Lingual Apparatus: Evaporative Cooling Mechanisms

Most birds start to pant when the ambient temperature rises to stressful levels in order to maintain their normal body temperature, but some birds supplement this heat dissipation by gular fluttering. The difference between these groups seems to reside in the size of the body and in the structure of the hyoid suspension apparatus. An important aspect of all evaporative cooling mechanisms is the presence of submucosal venous plexuses and other vascular specializations in the tongue, palate, and nasal cavity of birds to expose the blood to the epithelial surface that is being cooled by evaporation (Gaunt 1980; Baumel et al. 1983; Midtgård 1984, 1986; Arad et al. 1989; Porter and Witmer 2016).

2.7.1.1 Gular Fluttering

In many birds (cormorants, pelicans, boobies, gannets, anhingas, frigate birds, herons, ibises, owls, roadrunners, mousebirds, caprimulgids, sandgrouse, turkey vultures, and gallinaceous, anseriform, and charadriiform species), excessive heat is dissipated by panting (i.e., increasing the breathing rate), which increases the evaporative cooling from the moist surfaces of the nasal and oral cavities and is supplemented by the characteristic behavior of gular fluttering, during which the beak is opened, the head is raised, and the gular region is stretched and set into oscillations (Howell and Bartholomew 1962; Bartholomew 1966; Lasiewski and Bartholomew 1966; Bartholomew et al. 1968; Thomas and Robin 1977; Arad et al. 1989; Yap 2011; Fam 2013; Chow 2014; Ming 2014; Campbell 2014; Amar-Singh 2015, 2016; Avila 2015; Leo-Smith and Joubert 2015; Marts 2015, 2016; Turnbull 2015; McKechnie et al. 2016). Among these birds, strigiform (owls), caprimulgid (nightjars), anseriform, and gallinaceous birds flutter their gular region synchronously with their panting (Lasiewski and Bartholomew 1966; Bartholomew et al. 1968; Lasiewski 1972; Calder and King 1974), but cormorants, pelicans, and mousebirds (Coliidae) flutter their gular region independently from their panting (Lasiewski and Snyder 1969; Lasiewski 1972; Calder and King 1974). Gular fluttering to increase evaporative cooling, instead of increased panting, requires less energy due to the small mass of the lingual apparatus relative to that of the body and avoids alkalosis (Bartholomew et al. 1968; Calder and King 1974; Baumel et al. 1983).

With rising ambient temperatures, the amplitude of the gular flutter and the expanse of the gular region increase in order to enlarge the surface of the moist mucous membranes of the oral cavity for evaporative cooling, but the frequency of the oscillation does not change, even if weights are experimentally added to the gular region (Lasiewski and Bartholomew 1966; Bartholomew et al. 1968; Lasiewski and Snyder 1969; Lasiewski 1972; Evans 1984). The underlying mechanism of gular fluttering has been variously described as movements by the lingual apparatus driven by the muscles moving the hyoid horns (Lasiewski and Bartholomew 1966; Bartholomew et al. 1968; Lasiewski and Snyder 1969; Calder and King 1974). However, the constancy of the oscillation frequency indicates that the gular region flutters at its resonant frequency, which would be the most energy-efficient rate because it would not require muscle work and would explain the observations that birds can flutter the gular region for extended times without tiring (Schmidt-Nielsen 1964; Lasiewski and Bartholomew 1966; Bartholomew et al. 1968; Lasiewski and Snyder 1969; Lasiewski 1972; Calder and King 1974). At least in pelicans, cormorants, and herons, this resonant oscillation would be supported by the loose hyoid suspension apparatus that allows an expansion and stretching of the gular region (see Sect. 2.5.1.2) and could be set in motion by an initial stretch initiated by the hyoid muscles. The construction of the hyoid suspension and lingual apparatus of other birds that flutter their gular region is insufficiently known.

2.7.1.2 Columbiform Gular Fluttering with Esophageal Pulsation

Heat-stressed pigeons release heat by panting, fluttering their gular region, and synchronously inflating and deflating (i.e., "pulsating") their upper esophagus with air. The excursions of gular fluttering and esophageal pulsation increase with the ambient temperature and, thereby, supplement the moist surface areas of the oral cavity by those of the upper esophagus, which is supplied by a vascular plexus (Gaunt 1980; Baumel et al. 1983). The capacity of the esophagus to be filled with air is characteristic for pigeons and doves and is also used for vocalizations (see Sect. 2.7.4.5) and mating rituals. However, for the esophagus to expand, air needs to be forced into it when the mouth and nares are closed, which would interfere with panting and gular fluttering. Hence, the precise mechanism underlying this behavior deserves further study.

2.7.1.3 Panting without Gular Fluttering

At elevated ambient temperatures, passeriform birds, frogmouths (Podargidae), tropic birds (*Phaeton* spp.), and New World vultures (Cathartidae) only pant, but do not engage in gular fluttering (Howell and Bartholomew 1962; Lasiewski 1972; Calder and King 1974). In open-beak panting parrots, the beak and tongue move in synchrony with the thorax (Weathers and Caccamise 1975; Weathers and

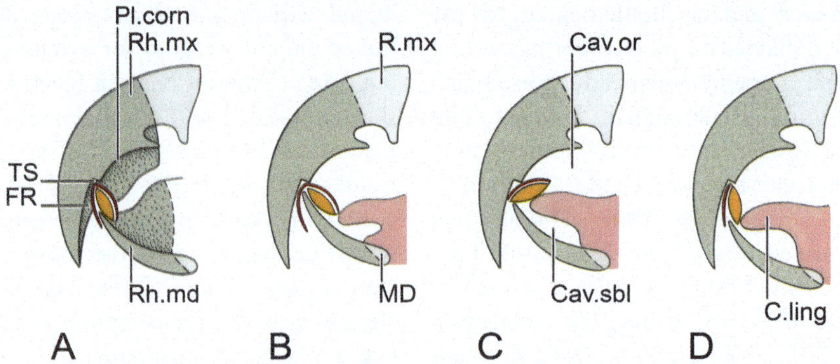

Fig. 2.25 Graphic model of the seed-shelling behavior of parrots with a psittacid bill to show the complex interactions between the beak and the tongue; for simplification, the upper beak is shown in a static position even though it is actually also moving; the tongue is omitted in (**a**) to reveal the corneous palate and sublingual floor. C.ling = lingual body; Cav.or = oral cavity; Cav.sbl = sublingual cavity; FR = filing ridges on internal surface of the projecting upper bill tip; MD = mandible; Pl.corn = corneous palate; R.mx = upper beak; Rh.mx = maxillary rhamphotheca; Rh.md = mandibular rhamphotheca; TS = transverse step [Adapted from Homberger (1980)]

Schoenbaechler 1976; Gallup et al. 2009). The absence of gular fluttering in these birds is probably due to a variety of causes. Passeriform birds are generally small and may, therefore, suffer less from heat loads and may be able to release sufficient heat through their legs and beaks in addition to panting (see also Midtgård 1981, 1986, 1988; Evans 1984; Symonds and Tattersall 2010; Greenberg et al. 2011; Campbell 2014; Campbell-Tennant et al. 2015). Seed-eating passerine birds and parrots are probably unable to flutter their gular region because of their massive tongues. These species may have yet-to-be discovered heat-releasing mechanisms without a disruption of their acid-base balance due to hyperventilation (see also Calder and King 1974). The morphology of the lingual and hyoid suspension apparatus and the gular region of the other birds that do not flutter their gular region while panting is insufficiently known.

2.7.2 Coordination of the Jaw and Lingual Apparatus: Intraoral Food Processing

Because birds lack a masticatory dentition, intraoral mechanical processing is rare among birds and best exemplified by the seed-shelling behavior, which has evolved several times independently in parrots and cockatoos and in seed-eating passerine birds with different jaw and lingual structures, configurations, and movements. The fine coordination of precise beak and tongue movements necessary for intraoral food processing requires a tight hyoid suspension apparatus as a fulcrum.

2.7.2.1 Seed-Shelling Behavior of Parrots and Cockatoos

Whereas cockatoos with a calyptorhynchid beak break up complex woody fruits and shell seeds by using the tip of the maxillary rhamphotheca as instrumental part of their beak working against the mandibular rhamphotheca, parrots and cockatoos with a psittacid beak always remove the seed shell and seed coat of seeds before swallowing them and do so in a typical manner (Homberger 1980, 1990, 1994b, 2001, 2003, 2006, 2016). They brace a seed against the transverse step and filing ridges of the corneous palate on the projecting upper bill tip and hold it in place with the tip of their tongue, while the cutting edge of the mandibular rhamphotheca cuts or cracks the seed shell (Fig. 2.25a). The outer half of the seed shell is then separated from the kernel with the cutting edge of the mandibular rhamphotheca (Fig. 2.25b), while the tongue rotates the seed (Fig. 2.25c) so that the half of the seed shell facing the oral cavity now faces the corneous palate for removal by the cutting edge of the mandibular rhamphotheca (Fig. 2.25d). This operation requires a fine coordination of raising and lowering the upper and lower beak (see Sect. 2.4.1); finely tuned lateral movements of the cutting edge of the mandible (see Sect. 2.4.3); deft back-and-forth, up-and-down and lateral movements of the hyoid apparatus; as well as up-and-down and slight lateral movements of the lingual body, which is supported by the paraglossum; and modifications of the lingual tip from spoon-shaped to flat (see Sect. 2.5.2), all of which are guided by the large number of mechanoreceptors typical for parrots (see Sect. 2.3).

This complex mechanism and the needed morphological structures and configurations evolved at least twice independently from an ancestral condition with a calyptorhynchid beak, which evolved originally as an adaptation to feeding on wood-boring and gallicole insect larvae by tearing up fibrous wood (Homberger 1990, 1994b, 2001, 2003, 2006).

2.7.2.2 Seed-Shelling Behavior of Granivorous Passerine Birds

Like parrots and cockatoos, seed-eating passerine birds possess a corneous palate and remove the shells of seeds before swallowing them, but have evolved at least two different mechanisms of seed-shelling, which are different from those in parrots and cockatoos. Thraupid-geospizine Galápagos finches, fringillid finches, and estrildid parrot finches (*Erythrura* spp.) of the subgenus *Acalanthe* shell dicotyledonous seeds by placing them with the tongue between two longitudinal ridges on the corneous palate and by cutting the shell open with back-and-forth movements of one of the lateral tomia of the mandibular rhamphotheca (Bowman 1961; Ziswiler 1965, 1969; Ziswiler et al. 1972; Nuijens and Bout 1998). In contrast, emberizid (including cardinaline), ploceid, and estrildid passerine birds dehusk mainly gramineous seeds by crushing them between a ridge on the corneous palate and the tomia of the mandibular rhamphotheca with up-down movements of the mandible (Ziswiler 1965, 1969; Ziswiler et al. 1972; Nuijens and Bout 1998).

The cutting and crushing seed-shelling methods have evolved several times independently, as is indicated by the convergently evolved, non-derivable modifications of the lingual tips and ridge patterns on the corneous palate in apparent adaptation to seed shelling (see Sect. 2.3). Lateral movements of the mandible during seed shelling were observed in the thraupid-geospizine Galápagos finches (Bowman 1961); the fringillid Chaffinch, *Fringilla coelebs* (Kear 1962); fringillid finches in general (Ziswiler 1965); the fringillid seed-crushing greenfinch, *Carduelis chloris* (Nuijens and Zweers 1997); the fringillid seed shell-cutting yellow-fronted canary, *Serinus mozambicus* (Nuijens and Bout 1998); and the emberizid seed shell-crushing northern cardinal, *Cardinalis cardinalis* (Homberger and Cozic unpubl. observations) and can be surmised for all seed-shelling mechanisms in passerine birds, which need to move their mandibles into the right place relative to the upper beak to open seeds. The presence or absence of a postorbital ligament in granivorous passerine finches is not generally correlated with the type of seed-shelling mechanism (Nuijens and Bout 1998; Nuijens et al. 2000), but seems to be correlated with the capacity for lateral mandibular movements (see Nuijens and Zweers 1997) (see Sect. 2.4.3).

2.7.3 Coordination of the Jaw, Lingual, and Laryngeal Apparatus: Feeding and Drinking Mechanisms

The process of feeding and drinking involves the transportation of a food item or of an aliquot of liquid through the oral cavity into the esophagus. In all instances, the movement of the laryngeal mound as a piston to push food items or liquids into the esophagus and the closure of the glottis need to be synchronized with the movements of the lingual and jaw apparatus. Depending on the relative size of the food item and the type of the hyoid suspension apparatus, two main feeding

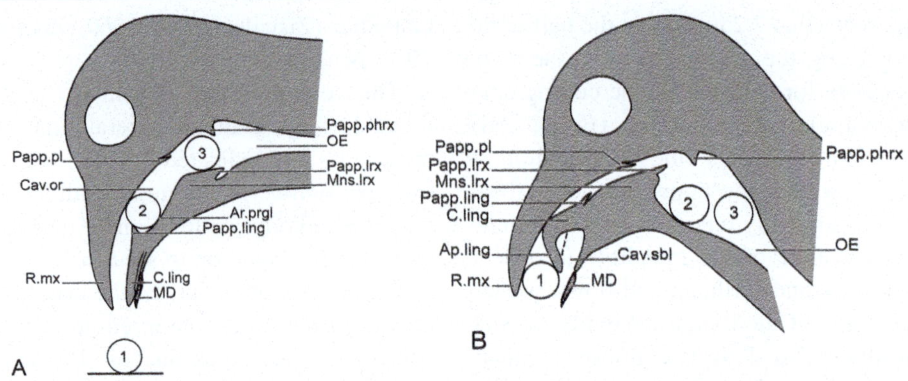

Fig. 2.26 Graphic model of the lingual deglutition mechanism in a pigeon (*Columba livia*). (**a**) Pushing food items into the pharynx and esophagus by the retracting tongue and laryngeal mound. (**b**) Retention of food items in the oral cavity by the laryngeal and pharyngeal papillae. Ap.ling = lingual tip; Ar.prgl = preglottal area; C.ling = lingual body; Cav.or = oral cavity; Cav.sbl = sublingual cavity; MD = mandible; Mns.lrx = laryngeal mound; OE = esophagus; Papp.ling = lingual papillae; Papp.lrx = laryngeal papillae; Papp.pl = palatine papillae; Papp.phrx = pharyngeal papillae; R.mx = upper beak [Adapted from Zweers (1982b)]

modes, namely lingual and cranial-inertial feeding mechanisms, can be distinguished with various intermediate and bimodal modes. In addition, specialized feeding mechanisms that are akin to drinking mechanisms take in food that is liquid or suspended in water. By varying the synchronization and extent of their basic movements, the jaw, lingual and laryngeal apparatus can adapt to the intake of a great variety of foods and liquids (see also Kooloos and Zweers 1991; Zweers 1992).

2.7.3.1 Lingual Feeding Mechanisms

Many birds swallow relatively small food items and saliva by back-and-forth movements of the tongue supported by finely coordinated opening and closing movements of the beak. The raising and flattening of the rows of lingual and laryngeal papillae on the floor of the oral cavity play an active role in this "slide-and-glue" mechanism, while the palatal and pharyngeal papillae on the ceiling of the oral cavity play an equally crucial, but passive role.

As described for pigeons (Fig. 2.26; Zweers 1982a, b; Zweers et al. 1994), an item is picked up between the tip of the tongue and upper beak, stuck on the moist lingual tip, and transported backward by the retracting tongue with raised lingual wings and lingual papillae (Fig. 2.26a; item #2). The item is held back by the palatal papillae when the tongue is protracted again by gliding underneath the item with flattened lingual wings and lingual papillae to pick up another item. During the next lingual retraction with a new item stuck to the lingual tip (Fig. 2.26b), the first items #2 and #3 are pushed into the esophagus by the raised laryngeal papillae and retained there by the pharyngeal papillae when the laryngeal papillae are flattened and the laryngeal mound is gliding underneath the items as the tongue is protracted again to pick up the next item (Fig. 2.26b). Chickens presumably use the same mechanism to swallow small food items (see White 1969; Bels and Baussart 2006).

The mechanism of raising the laryngeal papillae is central to the process of lingual feeding, but remains to be analyzed in birds with an akinetic larynx, such as pigeons and chickens (see Sect. 2.6.2). Birds with a kinetic larynx, such as parrots and passerine birds, however, raise their laryngeal papillae by raising the caudal part of their laryngeal skeleton and, thereby, affect the shape of both the oral cavity and the laryngeal chamber (see Sect. 2.6.1). Parrots, for example, raise the laryngeal papillae to push a food item into the pharynx and esophagus with the laryngeal mound by using the dorsal cricohyoid muscles and the medial and urohyal tracheolateral muscles to tilt the cricoid so as to raise its caudal end and simultaneously lower its rostral end (Fig. 2.27a). To lower and flatten the laryngeal papillae to glide the laryngeal mound underneath the food item

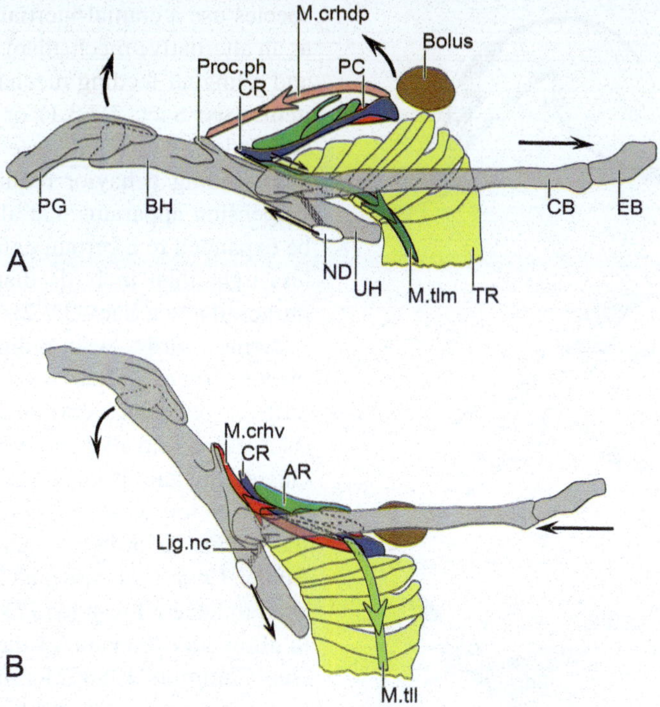

Fig. 2.27 Graphic model of the movements of the hyoid, larynx, and trachea during lingual deglutition in an African Grey Parrot (*Psittacus erithacus*). (**a**) Pushing a food item into the pharynx by retracting the hyoid with a raised caudal end of the cricoid and, thus, raised laryngeal mound. (**b**) Retention of the food item in the pharynx when protracting the hyoid with a depressed caudal end of the cricoid and, thus, flattened laryngeal mound. AR = arytenoid; BH = basihyal; Bolus = food item; CB = ceratobranchial; CR = cricoid; EB = epibranchial; Lig. nc = nodulo-ceratobranchial ligament; M.crhdp = *M. cricohyoideus dorsalis profundus*; M.crhv = *M. cricohyoideus ventralis*; M.tll = *M. tracheolateralis lateralis*; M.tlm = *M. tracheolateralis medialis*; ND = *Nodulus*; PC = procricoid; PG = paraglossum; Proc.ph = parahyal process; TR = trachea; UH = urohyal

while it is held in place by the pharyngeal papillae, parrots tilt the cricoid so as to lower its caudal end with the lateral tracheolateral and ventral cricohyoid muscles (Fig. 2.27b).

Based on the laryngeal morphology, the swallowing mechanism of passerine birds can be hypothesized to involve an analogous mechanism as the swallowing mechanism of parrots by raising and lowering the dorsal cricoids as a way to raise and flatten the laryngeal papillae to transport and retain food items (see Sect. 2.6.1.2).

2.7.3.2 Cranial-Inertial Feeding Mechanisms

Some birds swallow relatively large prey or food items by using a cranial-inertial mechanism (also called “ballistic transport” or “catch-and-throw mechanism”) to transfer the food items from the tip of the beak into the pharynx and esophagus (Fig. 2.28). If the beak is pointing downward or is held more or less horizontal, the food item is released by the beak, while the head jerks backward, and then caught again immediately with the beak, while accelerating the head forward so that the food item is now positioned closer to the pharynx. If the beak is pointing upward, gravity is also used to let the prey item fall into the pharynx.

This transport mechanism does not involve the tongue, except when grasping or holding a food item, even in species with a well-developed tongue, such as chickens and ducks (Berkhoudt 1985; Kooloos 1986; Kooloos and Zweers 1991; Zweers et al. 1994; van den Heuvel and Berkhoudt 1998; Bels and Baussart 2006; Skieresz-Szewczyk and Jackowiak 2016). These

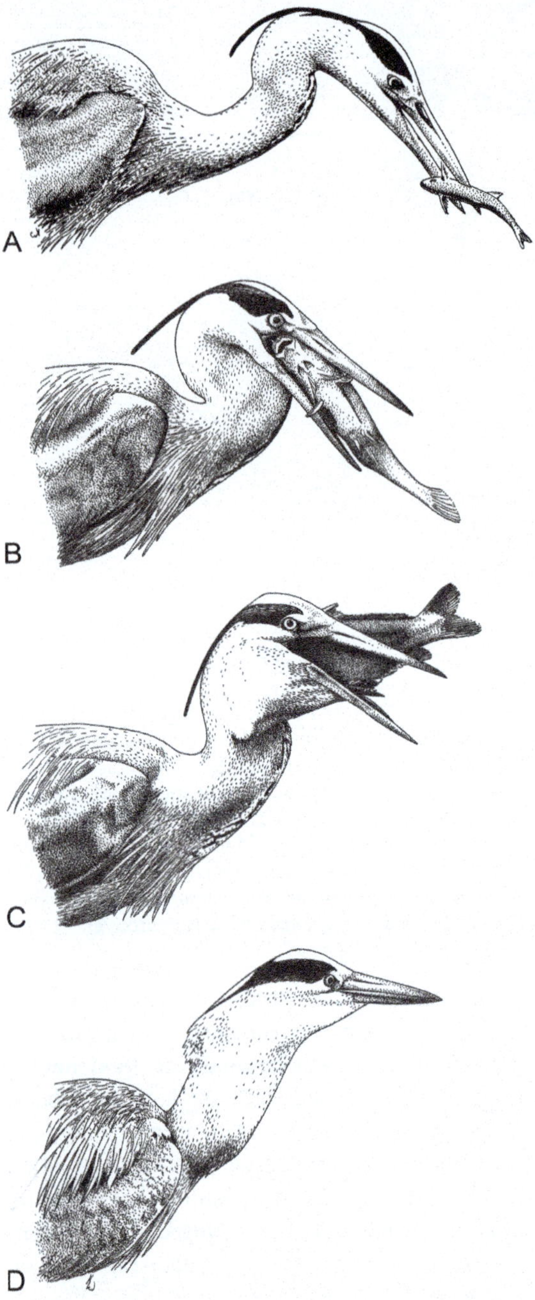

Fig. 2.28 Cranial-inertial swallowing mechanism by a Grey Heron (*Ardea cinerea*). (**a**) Beak with a speared fish. (**b**) A fish in the beak after it was shaken off and re-caught with the beak. (**c**) A fish in the expanded oral and gular cavity after it was flung into the air and caught by the open beak. (**d**) Expanded throat with fish in the esophagus (Drawn by Karen Westphal from photographs in Creutz 1981)

species use a cranial-inertial feeding mechanism as an alternative mechanism for larger food items and a lingual feeding mechanism for small food items (see Sect. 2.7.3.1) or food items that are suspended in a liquid (see Sect. 2.7.3.3). This dual feeding behavior requires a relaxed hyoid suspension apparatus that allows the pharynx to be expanded to a certain extent and is, therefore, less tight than in birds that process their food intraorally (see Sect. 2.7.2).

Some birds are obligate cranial-inertial feeders because their tongue is reduced, as in ratites, pelicans, cormorants, and hornbills (Gussekloo and Zweers 1997; Tomlinson 2000; Gussekloo and Bout 2005; Baussart and Bels 2011), or too weak to transport relatively large food items, as in herons and toucans (Fig. 2.28; Bühler 1995; Baussart et al. 2009). These species need to have a loose hyoid suspension apparatus to allow an expansion of the pharynx and throat (see Cummins 1986; Homberger 1999), whose specific construction remains to be studied for most birds besides egrets.

The obligate cranial-inertial feeding mechanism allowed the evolutionary loss of the lingual body in some birds because it is no longer involved in lingual transport, but the remainder of the hyoid apparatus, including the laryngeal mound, is still active in pushing food items into the esophagus (see also Tomlinson 2000) unless gravity is used to let large prey fall into the gullet and esophagus of upward pointing heads.

2.7.3.3 Specialized Feeding Mechanisms

Liquid foods (e.g., nectar) and food items that are suspended in water require specialized feeding mechanisms, such as nectar feeding by hummingbirds (Ewald and Williams 1982; Rico-Guevara and Rubega 2011; Rico-Guevara 2014; Rico-Guevara et al. 2015), lorikeets (Homberger 1980), and a sandpiper (Burle et al. 2013); filter feeding by ducks (Zweers et al. 1977, 1994; Kooloos et al. 1989; van der Leeuw et al. 2003; Skieresz-Szewczyk and Jackowiak 2016) and flamingos (Zweers et al. 1995); or suspension feeding in phalaropes (Rubega and Obst 1993). Except for the detailed motion analysis of the beak, tongue, and hyoid skeleton of ducks by Zweers et al. (1977), the studies of these feeding mechanisms are limited to the movements of the beak and the lingual body and tip. Integrated studies that include the laryngeal apparatus are still needed to understand these specialized feeding methods fully.

2.7.3.4 Drinking Mechanisms

In general, two main drinking mechanisms have been recognized in birds (see Zweers 1992). The most common mandibular-lingual scooping mechanism (i.e., "tip-up drinking" of Zweers 1992) has been described for ratites, chickens, passeriform birds (crows, estrildid finches), mallards, and cockatoos (Immelmann and Immelmann 1967; Homberger 1980, 1983; Heidweiller and Zweers 1990; Kooloos and Zweers 1989, 1991; Zweers 1992; Heidweiller et al. 1992; Gussekloo and Bout 2005). The less common immersion drinking (i.e., "tip-down drinking" of Zweers 1992) has been described for pigeons, mousebirds (Colidae), and some Australian estrildid finches (Immelmann and Immelmann 1967; Zweers 1982c; Peat and Gaunt 1984; Heidweiller and Zweers 1990).

The actual diversity of avian drinking mechanisms is likely to be much greater and will need to be investigated by studying the detailed movements of the beak, tongue, and larynx. For example, four different drinking mechanisms, none of them fitting exactly the two main drinking mechanisms, have been documented among Psittaciformes species (Homberger 1980, 1983). And the classic controversy regarding the drinking mechanism of sandgrouse (Pteroclidae), which immerse their bills as in "tip-down drinking", but also raise their heads as in "tip-up drinking", is still not understood in sufficient details (see Immelmann and Immelmann 1967; de Juana 1997).

2.7.4 Coordination of the Jaw, Lingual, and Laryngotracheal Apparatus: Respiration and Vocalization

While the jaw, lingual, and laryngeal apparatus need mainly to be properly aligned during the processes of respiration and ventilation to allow the free flow of air into and out of the lungs and pulmonary air sacs (Fig. 2.2; see Sect. 2.3) and generally do not move during regular breathing (Riede et al. 2013), while the larynx performs crucial gating functions separating the airways from the alimentary tract, their influence on airflow becomes more apparent during vocalization, which can differ in the degree of complexity. Complex vocalizations vary in their frequencies, amplitudes, rhythm, timbre, and articulation with various vowel-like and consonant-like sounds, as in the melodious tonal songs of songbirds or in the mimicked human speech of trained songbirds, parrots, and cockatoos. These complex vocalizations are not correlated with the structural complexity of the sound-generating syrinx (Suthers 2001), but they are correlated with the presence of a kinetic larynx (Homberger 1997; Table 2.2; see Sect. 2.6.1) and a tight hyoid suspension (Table 2.2; Sect. 2.5.2.1), which allow greater control of lingual and laryngeal movements. Falconiform raptors and hummingbirds also generate complex vocalizations, but the functional morphology of their jaw, lingual, and laryngeal apparatus has not been studied yet in sufficient details. Simple vocalizations lack the elements of melody and articulation, as in the cooing of pigeons (Fletcher et al. 2004; Riede et al. 2004), the crowing of roosters (Collias 1987), and the vocal expressions of herons (Bayer 1984). These simple vocalizations are correlated with an akinetic larynx (see Sect. 2.6.2) and a relaxed or loose hyoid suspension (Table 2.2; see Sects. 2.5.1.2 and 2.7.3.2). However, the vocal apparatus in birds is highly diverse (see Sect. 2.7.4; Riede et al. 2016) and is likely to have evolved multiple times independently from the respiratory apparatus of reptilian ancestors.

An emerging aspect of the avian vocal apparatus is the physical linkage between the sound-generating syrinx, the suprasyringeal sound-resonating tracheo-laryngo-oral complex, and the non-pulmonary subcutaneous cervical air sacs. Such a linkage is in principle given by the actual connections and physical proximity of these organs and tissues and provides a conceptual approach to understand how some birds (see Sect. 2.7.4.2) can automatically tune their sound-resonating organs to the sound generated by the syrinx so as to produce melodious songs rich in articulated sounds with harmonics and overtones (see also Suthers 2004; Fletcher et al. 2006; Goller and Cooper 2008). Another emerging aspect is the linkage between cranial kinesis and a mechanism protecting the hair cells in the auditory system during the production of loud high-frequency sounds with a wide-open beak (see Claes et al. 2016; see also Sect. 2.7.4.2). In birds, the tympanic membrane inserts along the otic process of the quadrate, so that it is automatically stretched and tensed when the quadrate rotates forward to raise the upper beak (Starck 1995; Claes et al. 2016).

2.7.4.1 Vocalization Mechanisms of Parrots

Parrots can learn to produce articulated speech that is understandable by humans (e.g., Thorpe 1959; Nottebohm 1976; Rauch 1978; Ilyichev and Silayeva 1992; Cruickshank et al. 1993; Bottoni et al. 2009; Singh et al. 2016), and several studies have demonstrated that the properties of the sounds they generate in the syrinx are affected by movements of their beak and tongue (Fig. 2.29; Nottebohm 1976; Warren and Pepperberg 1993; Patterson and Pepperberg 1994, 1998; Warren et al. 1996; Patterson et al. 1997; Beckers et al. 2004; Ohms et al. 2012). Some studies discussed the possibility that sounds generated by the syrinx are affected by the position of the laryngeal mound within the oral cavity (Scanlan 1988 *fide* Warren et al. 1996), the shape of the glottis (Warren et al. 1996; Patterson et al. 1997), the length of the passage from the nares to the choana (Nottebohm 1976), or the length of the trachea (Scanlan 1988 *fide* Warren et al. 1996). Homberger (1979, 1997, 1999) argued that the complex kinetic larynx of parrots is capable of modulating the sounds generated by the syrinx (see also Thorpe 1959).

The data presented earlier (see Sect. 2.6.1.1) indicate that the glottal lips can be bent to various degrees and shapes by differentiated actions of the complex constrictor muscle so that the glottal lips could contribute to the creation of certain consonant- and vowel-like sounds like the lips in humans. In addition, the resonating properties of the oral cavity and laryngeal chamber are controlled in subtle ways through modifications of their size and shape by the complex extrinsic laryngeal musculature. These actions by the laryngeal musculature are superimposed on movements by the lingual apparatus (see Sects. 2.5.1.1 and 2.5.2) and the jaw apparatus (see Sects. 2.4.1 and 2.4.3), which can significantly affect the shape and size of the oral cavity. The properties of the sounds generated by the syrinx are likely to be influenced also by the length of the trachea (unpublished personal observations) and by the length of the naso-choanal passage, which shortens when the upper beak is raised (see Figs. 2.2 and 2.8a, b), as well as by changes in the degree of inflation of the single large non-pulmonary subcutaneous cervical air sac and the tension of the skin of the neck (see Bignon 1889; Walsh and Mays 1984; Cozic and Homberger unpubl. observations).

2.7.4.2 Vocalization Mechanisms of Songbirds

Songbirds are characterized by their complex, melodious and tonal vocalizations. In captivity, many species can learn to produce articulated utterances that can be understood by humans (e.g., Thorpe 1959; Ilyichev and Silayeva 1992). Recent studies have come to a consensus that the suprasyringeal tract and its associated organs are responsible for modifying the harmonics and overtones of the tonal sounds generated by the syrinx in order to create formants analogous to what is known for human speech (Suthers and Goller 1997; Titze 1994). Ideas concerning the possible role of the larynx in generating complex vocalizations are long-standing (see, e.g., Shufeldt 1890; Thorpe 1959; Bock 1978b), but have not been considered in later studies of songbird vocalization.

In order to pinpoint the suprasyringeal structure responsible for complex vocalizations in songbirds, a large number of physiological-experimental and, more recently, imaging studies have tried to correlate the movements of the beak to particular sounds generated by the syrinx in

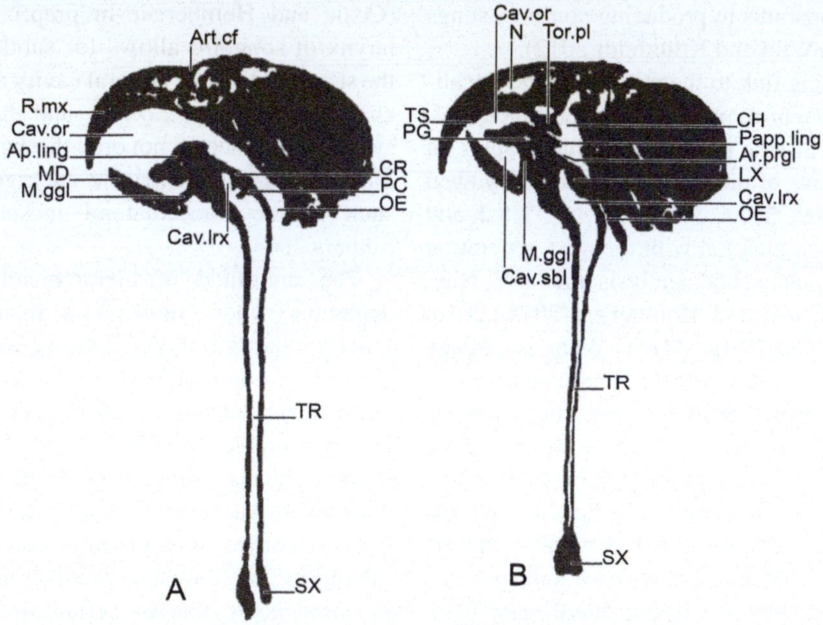

Fig. 2.29 Different positions of the jaw, lingual and laryngotracheal apparatus in an African Grey Parrot (*Psittacus erithacus*) as shown in longitudinal sections of CT data; the differences in the tracheal and syringeal lumina may be due to a slightly different level of the longitudinal sections since the trachea and bronchi cannot change their lumina to the degree shown in these images. (**a**) Open-beak position with elevated upper beak, lowered mandible, lowered and retracted hyoid skeleton, flexed paraglossum and lingual body, bulbous lingual tip, flattened laryngeal mound with the caudal end of the cricoid projecting into the laryngeal chamber, and elongated trachea. (**b**) Almost closed-beak position with lowered upper beak, raised mandible, protracted and raised hyoid skeleton, extended paraglossum and lingual body, spoon-shaped lingual tip, and shortened trachea. Ap.ling = lingual tip; Ar.prgl = preglottal area; Art.cf = craniofacial joint; Cav.lrx = laryngeal chamber; Cav.or = oral cavity; Cav.sbl = sublingual cavity; CH = choana; CR = cricoid; LX = larynx; M.ggl = *M. genioglossus*; MD = mandible; N = naris; OE = esophagus; PC = procricoid; PG = paraglossum; Papp.ling = lingual papillae; R.mx = upper beak; SX = syrinx; Tor.pl = palatal torus or cushion; TR = trachea; TS = transverse step [Adapted from Patterson et al. (1997)]

songbirds (e.g., Westneat et al. 1993; Riede et al. 2006, 2013; Schmidt and Wild 2014). In general, higher frequencies and amplitudes were correlated with open beaks, and lower frequencies and amplitudes were correlated with almost closed beaks (Westneat et al. 1993; Moriyama and Okanoya 1996; Suthers and Goller 1997; Hoese et al. 2000; Williams 2001; Goller et al. 2004; Goller and Cooper 2004; Fletcher et al. 2006; Riede et al. 2006; Suthers and Zollinger 2008; Ohms et al. 2010; Suthers et al. 2016), although some confusing inconsistencies were also observed (e.g., Suthers and Goller 1997; Nelson et al. 2005; Riede et al. 2006, 2013; Schmidt and Wild 2014). Some studies have suggested that the factor controlling the suprasyringeal sound quality resides in the size of the oral cavity (i.e., "OEC" or "oropharyngeal-esophageal cavity" of Fletcher et al. 2006; Riede et al. 2006, 2013; Riede and Suthers 2009; Ohms et al. 2010), the lateral spreading of the hyoid horns (Riede et al. 2006, 2013, 2015), or the movements of the lingual apparatus (Riede et al. 2006; Suthers et al. 2016). The role of the trachea in affecting sounds has been questioned (Suthers and Goller 1997; Goller and Cooper 2004, 2008). Other studies mentioned that the glottis (Suthers and Goller 1997; Riede et al. 2006) or the movements of the larynx relative to the lingual apparatus ("hyoid apparatus" of Riede et al. 2006, 2013) may be involved in producing complex songs, and neurophysiological studies have provided some support for the role of the oral cavity and the jaw, lingual and laryngeal apparatus in producing complex songs (Wild 1993; Wild and Krützfeldt 2012).

The missing link to the studies on the vocalization of songbirds is the kinds of functional-morphological data presented in this chapter on the movements of the jaw, lingual and laryngeal apparatus (see Sects. 2.4.2, 2.5.1.1, 2.5.2 and 2.6.1.2) in combination with the next generation of imaging and sound analysis methods (e.g., Fig. 2.30; Cozic and Homberger 2015, 2016; Homberger and Cozic 2015). What is already apparent is that the vocalization apparatus of songbirds is much more complex than what has been assumed in the past as new structures are discovered, such as the non-pulmonary subcutaneous cervical and parapatagial air sacs, and the role of the integument and cervical vertebral column in vocalizations is realized and specified (Blevins et al. 2014; Cozic and Homberger 2015, 2016; Homberger and Cozic 2015).

Hence, the emerging view on songbird vocalization is that complex and partly linked interactions of all suprasyringeal structures of the vocal tract are used by songbirds to modify and articulate their song to produce content-rich vocal communications (Cozic and Homberger in prep.). For example, the lingual apparatus interacts with the cervical vertebral column in widening or constricting the cervical region by entraining the larynx and trachea away from or toward the vertebral column. This widening or constricting of the cervical region interacts with the subcutaneous cervical and parapatagial air sacs as they passively inflate and deflate (Cozic and Homberger 2015, 2016). The movements of the lingual and laryngeal apparatus relative to the palate modify not only the size but also the shape of the oral cavity (see Sect. 2.6.1.2; see also Suthers et al. 2016; cf. Riede et al. 2006, 2013) and are linked to the spreading of the hyoid horns during the retraction of the lingual apparatus (see Sect. 2.5.1.1; see also Riede et al. 2006, 2013, 2016). The movements of the beak and jaw apparatus are responsible for the seeming, but not actual, lowering of the lingual apparatus when the beak opens widely, and the length changes of the trachea are linked to movements of the jaw, lingual, and laryngeal apparatus (Cozic and Homberger in prep.). The kinetic larynx of songbird allows for subtle changes in the shape and size of the oral cavity and laryngeal chamber (see Sect. 2.6.1.2), and the larynx and syrinx are connected not only through the trachea but also through extrinsic laryngeal muscles, such as the tracheolateral muscle (see also Suthers 2004).

The multitude of hierarchically organized apparatus, whose movements relative to each other are partly linked, allows an almost infinite number of configurations of the suprasyringeal vocal tract, thereby influencing not only the easily measurable amplitude and frequency of sounds but also generating the vowel-like and consonant-like sounds of articulated song that can be recorded and reproduced as onomatopoeic utterances by humans. Hence, the described inconsistencies between certain movements and positions of the vocal apparatus of birds on the one hand and the measured frequencies and amplitudes of the produced sound on the other hand (see, e.g., Suthers and Goller 1997; Nelson et al. 2005; Riede et al. 2006, 2013; Schmidt and Wild 2014; Suthers et al. 2016) may actually reflect the production of the more difficult-to-measure aspects of sound quality, such as timbre and vowel- and consonant-like sounds.

2.7.4.3 Vocalization Mechanisms of Chickens

Whereas chickens have a limited repertoire of soft vocalizations, roosters produce a characteristically loud call that seem to contain vowel- and consonant-like sounds (e.g., "kikerikii" or "cock-a-doodle-doo"). While crowing, the cervical vertebral column is retroflexed, and the larynx is pulled down toward the base of the neck (White 1968; Gaunt and Gaunt 1977; Claes et al. 2016), but no morphological or physiological details are known. Since chickens possess an akinetic larynx, their limited capability for modulation of their calls may derive from a yet-to-discovered mechanism (see also McLelland 1989). Claes et al. (2016) observed that rooster open their beaks more widely by raising their upper beak farther, with the effect of producing louder vocalizations and, at the

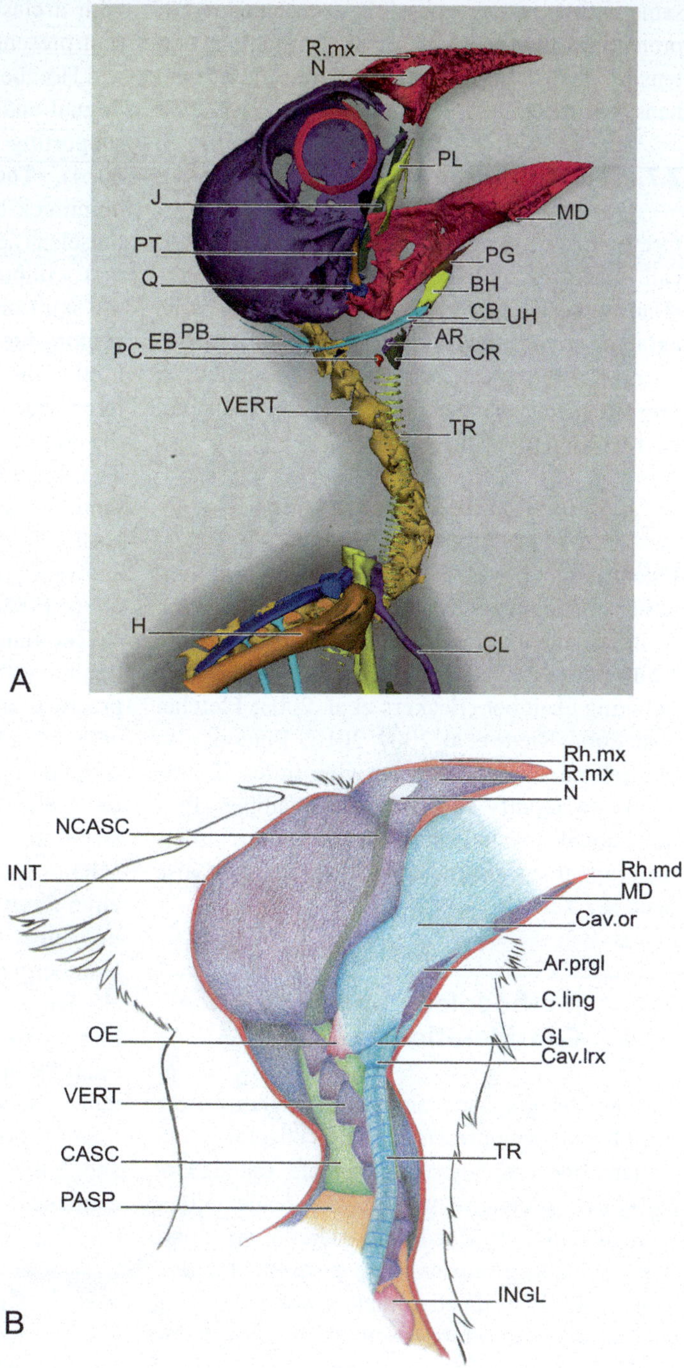

Fig. 2.30 Model of an "open-beak/tight-neck/high-frequency/large-amplitude" posture of a singing Northern Cardinal (*Cardinalis cardinalis*) with a relaxed hyoid suspension. (**a**) Animated 3D model aligned with a single frame of an X-ray video; the *right halves* of the upper jaw apparatus and mandible were removed to reveal the complete hyoid apparatus. (**b**) Graphic model of an adjusted longitudinal section of (**a**) with manually added soft tissue structures. AR = arytenoid; Ar.prgl = preglottal area; BH = basihyal; C.ling = lingual body; CASC = non-pulmonary subcutaneous cervical air sac of the cervico-cephalic air sac system; Cav.lrx = laryngeal chamber; Cav.or = oral cavity; CB = ceratobranchial; CL = clavicle (furcula); CR = cricoid; EB = epibranchial; GL = glottis; H = humerus; INGL = crop; INT = integument (skin); J = jugal bone; MD = mandible; N = naris; NCASC = passageway between naris and subcutaneous cervical air sac; OE = esophagus; PASP = parapatagial air sac of the pulmonary air sac system; PB = pharyngobranchial; PC = procricoid; PG = paraglossum; PL = palatine bone; PT = pterygoid bone; Q = quadrate bone; R.mx = upper beak or maxilla; Rh.md = mandibular rhamphotheca; Rh.mx = maxillary rhamphotheca; TR = trachea; UH = urohyal; VERT = vertebral column (Cozic and Homberger in prep.)

same time as a possible mechanism of protecting the hearing organ, stretching and tensing their tympanic membranes more than hens (see Sect. 2.7.4).

2.7.4.4 Vocalization Mechanisms of Geese and Ducks

Geese are able to modulate the frequency of their call through movements of their beaks (Hausberger et al. 1991). Since they possess an akinetic larynx, they may demonstrate that frequency changes can be effected through beak movements or changes in the shape of their oral cavity or length of their trachea.

2.7.4.5 Vocalization Mechanisms of Pigeons and Doves

Pigeons and doves are characterized by their rather monotonous vocalizations they produce while closing their beak and nostrils and inflating their upper esophagus as a sound-resonating and radiating chamber (Beckers et al. 2003; Fletcher et al. 2004; Riede et al. 2004, 2016). Pigeons also possess an akinetic larynx with little, if any, ability to modify the laryngeal chamber. It is not known how pigeons swallow air to inflate their upper esophagus before cooing (see also Sect. 2.7.1.2).

2.8 Methodology and Theoretical Considerations

Various imaging and analytical methods were used for the data presented in this chapter.

The line drawings are accurate tracings of dissected specimens, which were made with a camera lucida attached to a stereomicroscope. The skeletal specimens were positioned orthographically in a 3D coordinate space for tracing so that they could be used to create graphic models, which were constructed on the basis of functional-anatomical data gained from dissections and on the basis of biomechanical and physiological principles (see Homberger 1986, 1988b; Osborn and Homberger 2015). Whereas the jaw apparatus consists of skeletal elements acting as levers and can be analyzed with a classic free-body force diagram method (not presented here, but see Bock 1974; Osborn and Homberger 2015), this is not the case for the lingual and laryngeal apparatus, which are self-supporting tensegritous structures (see Scarr 2014). The lingual apparatus is held in a hummock-like sling of connective tissue and muscles (i.e., the avian hyoid suspension apparatus) without any direct articular connections to the skull or mandible, and the skeletal elements responsible for the closure and opening of the glottis (the arytenoids) are suspended by connective tissue and muscles above a shell created by the cricoid or tracheal half rings (see Sect. 2.6.1). The avian lingual and laryngeal apparatus are, therefore, not amenable to being analyzed by a classic free-body force diagram and need to be analyzed as integrated systems (see Homberger 1986, 1988b).

The functional models presented here are difficult to test by physiological experimentation for practical and ethical reasons, but are fully testable by advanced 3D imaging and animation techniques that can reveal the actual movements of skeletal elements in living birds. These techniques create animated 3D models based on MRI or X-ray CT data of actual individuals and align these models to frames of X-ray videos of live birds (see Fig. 2.30; Blevins et al. 2014; Homberger and Cozic 2015; Cozic and Homberger 2015, 2016). These 3D techniques, like all imaging techniques, depend on a thorough knowledge of the morphology of the system under study and, especially, of the ligamentous and articular linkages that constrain and link the movements of individual skeletal elements.

2.9 Comparison of the Jaw, Lingual and Laryngeal Apparatus in Birds and Mammals

The morphology of the head of birds and mammals differs to such a degree that direct comparisons are rarely made, and homologizations are not sensible since the two vertebrate classes are only very distantly related; the use of similar terms for analogous structures is simply a matter of convenience and efficiency. The fundamental differences between these two vertebrate classes are visible even at the level of the neuropeptide control of the feeding behavior (Tachibana and Tsutsui 2016). However, a direct comparison of the particular structures of the head after a review of the lesser-known avian condition (see Sects. 2.3, 2.4, 2.5, 2.6, and 2.7) will highlight how very different constructions can achieve similar goals in nutrition, respiration, thermoregulation, and vocalization and how physical constraints from tissue properties channel similar solutions.

The kinetic skulls of birds are lightweight with several shock-absorbing and force-redirecting articulations surrounding the sutureless cranium (Bock 1999a). With the exception of the protractor of the quadrate and the depressor of the mandible, the jaw muscles are two-joint muscles that can contract isometrically and can economize on fiber length and number (Bock 1999b). This economical arrangement of the jaw musculature and the absence of teeth allow an elongation of the jaws into beaks covered in hard-cornified sheaths for probing, hammering, and spearing. The akinetic skull of mammals, in contrast, is built of sturdy bones interconnected by sutures with significant, though minimal, flexibility and shock-absorbing capacity in order to adjust to the strains generated by the one-joint jaw muscles (Bock 1999b; Rafferty and Herring 1999; Herring and Teng 2000), whose fibers need to be sufficiently long to allow the stretching necessary to open the jaws and need to be sufficiently numerous to develop the necessary bite forces (Bock 1999b). This design and the need to retain a properly aligned masticatory heterodontic dentition limit the elongation of the face and jaws of most mammals, except in the homodontic toothed whales (Odontoceti) and in edentulous myrmecophagous mammals.

The head of birds is characterized by nasal cavities in the upper beak, a variable complexity of the turbinates (conchae), and an absence of sinus cavities. Birds lighten their heads by pneumatizing the bones of the cranium and beaks and by non-pulmonary cervico-cephalic air sacs that extend into the neck in many birds (see Sects. 2.7.4.1 and 2.7.4.2). The head of mammals, in contrast, is characterized by large sinus and nasal cavities and generally complex turbinates (Fig. 2.1b) as a means to control its weight while enlarging the face for an appropriate size of the jaws with a masticatory dentition and musculature.

The mandibular rami of birds are always fused at the tip, but, being edentulous, may contain flexion zones to allow a bowing to widen the gape (see Sect. 2.4.3). The mandible of many mammals, in contrast, possess a mandibular symphysis or syndesmosis for slight movements between the paired mandibular rami (for a review, see Scott et al. 2012). The individual tooth-bearing rami, however, are rigid to maintain the integrity of the dentition.

The jaws of birds are edentulous, but some birds have acquired cutting or crushing rhamphotheca with which they break down food items intraorally (see Sect. 2.7.2). Furthermore, the absence of teeth allowed the rostra of the jaws to become conical or, in curved beaks, at least with an apically converging profile to support the avian flight mode that depends on aerodynamically streamlined body contours to minimize drag (Homberger 2002). The jaws in mammals, in contrast, with few exceptions, as in baleen whales (Mysticeti) and many myrmecophagous mammals, bear teeth for the prehension and mechanical breakdown of food or prey.

In the oral cavity of birds, a subdivision into the pharynx is not evident, although a pharynx is often mentioned to indicate the back of the oral cavity. The oral cavity in mammals, in contrast, is subdivided into distinct chambers, such as the oral cavity proper, the pharynx, and, in humans, the supralaryngeal chamber.

The free tongue (lingual body) of birds is supported by a skeletal element, the paraglossum, which constrains the movements of the tongue by linkages through ligaments and joints (see Sect. 2.5). The lingual body and laryngeal mound form a functional unit that transports food items into the esophagus, with the laryngeal mound acting as a piston (see Sect. 2.7.3.1), except in birds that have widened their gape and gular region and are swallowing large food and prey items whole (see Sect. 2.7.3.2). The free part of the tongue of mammals, in contrast, is a muscular hydrostat devoid of hard skeletal elements and is, therefore, capable of a great variety of contortiive movements (see Sect. 2.2). In adult mammals, the free part of the tongue places food between the tooth rows for mastication and pushes the food bolus into the pharynx, where the pharyngeal constrictor muscles squeeze the food bolus into the esophagus (Schwenk 2000; Hiiemae 2000). In suckling young, the flexible tongue plays a central role in extracting milk from mammary glands (German and Crompton 2000).

The hyoid apparatus of birds is attached to and suspended from the mandible by fasciae and extrinsic muscles and only indirectly connected to the skull by tendon-sheath-like hyoid sheaths (see Sect. 2.5.1). As a result, the movements of the hyoid apparatus are mainly translational during lingual feeding, as demonstrated in pigeons (Zweers 1982c) and turkeys (Tomlinson 2000), and independent from the movements of the upper jaw apparatus. This construction of the suspension apparatus allows the hyoid to be positioned at various distances from the skull and mandible to adapt the size of the gular region and entrance to the esophagus to the maximal size of the food or prey that is swallowed (see Sect. 2.7.3.2). However, once the hyoid suspension is built below the mandible into the throat fascia with a nonfunctional hyoid sheath, as in the emu, the movements of the hyoid during deglutition are less constrained and can become more circular as inferred from the description by Tomlinson (2000). The reduced hyoid skeleton of mammals, in contrast, is directly connected to the base of the skull by a ligament or a chain of ossicles, which restricts the entrance to the esophagus at the *Isthmus faucium* (see Sect. 2.2), thereby constraining the size of food boluses that can be swallowed and necessitating a masticatory dentition to mechanically break down food items to sizes and consistencies appropriate for deglutition (Homberger 1994a, 1999; Schwenk 2000; Hiiemae 2000; Homberger and Walker 2004). As a result of the fixed hyoid attachment to the skull, the movements of the hyoid bone are roughly circular in a lateral view (see Hiiemae 2000).

A reduction of the free part of the tongue (lingual body) was made possible in some birds (see Sects. 2.3 and 2.5.1) because it is supported by a separate skeletal element that can be lost without affecting the laryngeal support provided by the hyoid apparatus (see Sect. 2.5.2) and because the possibility of widening the entrance to the esophagus through a loose hyoid suspension allows large food items to be swallowed without lingual transport (see Sect. 2.7.3.2). In mammals, in contrast, a reduction of the free part of the tongue has never occurred, presumably because of its central role in suckling, mastication, and deglutition.

An elongation of the tongue can be achieved in birds in two ways: (1) by elongating the paraglossum, which supports the lingual body, or (2) by elongating the hyoid horns, which can be pulled out of their hyoid sheaths when protracting the tongue (see Sect. 2.5.1). In mammals, in contrast, only the free fleshy tongue can be elongated by muscular-hydraulic mechanisms (Kier and Smith 1985; Naples 1999).

The bodies of larger salivary glands in birds are part of the mechanical construction of the lingual apparatus (see Sect. 2.3) and the ceiling of the oral cavity (see Sect. 2.3) and often support rows of papillae with specific roles in feeding and swallowing (see Sect. 2.7.3.1). In contrast, the larger salivary glands of mammals, such as the parotid and mandibular glands, are located external to the oral cavity and play no direct mechanical role in the movements of the jaw, lingual and laryngeal apparatus.

Mechanoreception within the oral cavity of birds is based on a great variety of lamellated mechanoreceptors not only in the soft-cornified parts of the mouth cavity and tongue but also in the hard-cornified parts of the rhamphothecae and corneous palate, but little is known about nociceptors. Oral touch and pain reception in mammals is based on nerve endings and a variety of lamellated corpuscles in the oral soft-cornified tissues and the dental pulp.

The larynx in birds has no direct muscular or ligamentous attachment to the skull and is, therefore, dependent on the hyoid skeleton for its suspension and movements within the oral cavity (see Sect. 2.6). The larynx of mammals, in contrast, is attached by paired articulations to the hyoid bone and suspended to the base of the skull by the pharyngeal connective tissue and muscles.

The laryngeal skeleton of birds varies significantly among groups, consists of fully or at least partly ossified elements (the cricoid, procricoid, and the paired arytenoids), and is devoid of vocal folds (analogous vocal membranes are located in the syrinx). The laryngeal skeleton in mammals is fairly uniform, consists of different cartilaginous elements (the epiglottis and the thyroid, cricoid and paired arytenoid cartilages), and comprises sound-generating vocal folds. The fundamental structural differences in the mammalian and avian larynges and their accessory structures (suspension apparatus, musculature), which cannot be derived from each other, suggest that they are not homologous (i.e., not derived from a common laryngeal structure in a common ancestor).

The nasal cavities of birds are located in their upper beak and usually relatively small with simpler surface-enlarging conchae to clean and moisten the air, which continues through a very long trachea and the caudal abdominal air sacs, where it may also be cleaned and moistened before reaching the lungs. The nasal cavities of mammals, in contrast, are large with a system of complex surface-enlarging turbinalia to warm, clean, and moisten the air before it reaches the lungs through a relatively short trachea.

The choana of birds is usually elongated because of the absence of a secondary palate, and the avian larynx lacks an epiglottis. Hence, the hyoid apparatus needs to raise the laryngeal mound toward the caudal end of the choana and the lingual body toward the rostral end of the choana in order to establish continuous airways from the nostrils to the larynx and lungs (Fig. 2.2). The choana of mammals, in contrast, is short because of the long secondary and soft palate so that, except in human adults, the larynx not only neatly fits in this opening but can also be hooked into the soft palate *via* the epiglottis to establish continuous airways from the nostrils to the lungs.

Evaporative cooling in birds involves panting and, in many larger birds, fluttering of the gular region at its resonating frequency as the main mechanisms (see Sect. 2.7.1.1) besides heat loss through the rhamphotheca of the beak (Tattersall et al. 2009; Greenberg et al. 2011; Campbell-Tennant et al. 2015) and the unfeathered skin of the legs and feet (Midtgård 1981, 1986, 1988; Arad et al. 1989). Evaporative cooling mechanisms in mammals comprise sweating through sweat glands in the skin or panting with open mouth and, in some species that possess large limp tongues, a tongue that hangs out of the mouth and flaps at its resonant frequency (Crawford 1962) to supplement evaporation in the large and complex nasal cavities with surface-enlarging turbinalia. In addition, various vascular specializations serve to release heat from particular organs, such as the brain or testicles. Both mammals and birds have evolved evaporative cooling methods that take advantage of the fact that certain organs can oscillate at their resonating frequency and, hence, use minimal energy and can be performed for extended durations.

A comparison of the entire vocal apparatus of mammals and birds reveals a set of nonhomologous, analogous structures and constructions (i.e., different structures with similar functions in unrelated organisms). Even though the sound-generating organ is the syrinx in birds and the larynx in mammals, the syringeal muscles and vocal membranes and the laryngeal muscles and vocal folds are remarkably similar in their physical and physiological properties (Elemans et al. 2008; Riede and Goller 2010a, b, 2014), but they are, as expected, innervated differently (Suthers 2004). The sound-modifying resonating chambers distal to the sound-generating organs also differ fundamentally in birds and mammals. Birds, and in particular parrots and songbirds with their superior capacity for producing complex vocalizations, have a large array of modifiable resonating structures at their disposal, such as the beak, the oral cavity with a moveable tongue and a movable laryngeal mound, a kinetic larynx with a modifiable laryngeal chamber, a length-adaptable trachea, and inflatable non-pulmonary and pulmonary subcutaneous air sacs (see Sects. 2.7.4.1 and 2.7.4.2). The mammalian sound-resonating structures, in contrast, comprise mainly the oral cavity whose shape can be modified by the tongue, lips and supralaryngeal cavity in humans, but may include other air-filled structures, such as the nasal cavity; palatine, pharyngeal and laryngeal diverticles and pouches; and enlarged oral cavities through retracted larynges (Frey et al. 2007; Frey and Gebler 2010).

Nevertheless, a comparison of the vocalizations of parrots and songbirds on the one hand and of humans on the other hand established another analogy, namely, that of the glottal lips of these birds (see Sect. 2.6.1.1) with the lips of humans in being capable of producing consonant- and vowel-like sounds. In adult humans, because of the permanent descent of the larynx upon maturity, an additional supralaryngeal chamber is formed between the larynx and the root of the tongue, and the shape and resonating properties of the relatively small oral cavity can be changed by subtle changes in the position of the tongue and shape and tension of the lips (Titze 1994). This complex configuration of the supralaryngeal sound-resonating cavities makes the difference between infants and adults in their capacity to produce vowellike and consonant-like sounds that can be combined to create articulated speech (Laitman and Reidenberg 1993a, but cf. Fitch 2006, 2016). The difference in the capacity to produce articulated speech between quadrupedal mammals with a descended larynx (e.g., male deer, large cats) and adult humans lies in the humans' cheeks that enclose the oral cavity and the lips that can change the shape and size of the mouth opening.

2.10 Evolutionary History and Phylogenetic Inferences of the Lingual and Laryngeal Apparatus of Birds

At this point, an evolutionary and phylogenetic interpretation of the various configurations of the avian lingual and laryngeal apparatus will have to remain preliminary because of the sketchiness of the available morphological data. Nevertheless, the data presented here suggest some new evolutionary hypotheses, indicate that certain structural configurations have arisen multiple times independently, and demonstrate the principle of mosaic evolution.

For example, a tight hyoid suspension needs to be considered an ancestral condition in comparison to a loose hyoid condition because it comprises a functional hyoid sheath, even though it is found in Psittaciformes and granivorous songbirds, which have evolved specialized intraoral feeding mechanisms and are generally considered to be more recently evolved bird groups (Table 2.2). In contrast, a loose hyoid suspension, often combined with a loss of the paraglossum and lingual body, needs to be considered derived, even though it is found in larger birds that are often considered to be close to the origin of modern birds, because its hyoid sheath has become built into the throat wall (cervical fascia) and has lost its functionality (Table 2.2). Once functionality has been lost, it cannot be reacquired in its original form. A loose hyoid suspension with or without the loss of the paraglossum was evolved multiple times independently in adaptation to the swallowing of large food or prey items and has allowed the evolution of gular fluttering as an efficient evaporative cooling mechanism in large birds, which have a less effective surface to volume ratio.

The configuration of the kinetic laryngeal skeleton of Psittaciformes (see Sect. 2.6.1.1) may be similar to an ancestral condition of the avian larynx in comparison to the larynges of most birds because its side and ventral walls are still built by distinct tracheal half rings, even though parrots are considered to have arisen later in the history of birds (Table 2.2). The akinetic larynx (see Sect. 2.6.2) with its trough-shaped cricoid evolved from a fusion of tracheal half rings with the original cricoid and can, therefore, be considered a derived condition. The kinetic larynx of passerine birds (see Sect. 2.6.1.2) is probably an independently derived condition in which only the rostral part of the original cricoid fused with the original tracheal half rings to form a trough-shaped ventral cricoid, while the caudal part of the original cricoid remained free to articulate as the dorsal cricoid with the trough-shaped ventral cricoid.

Kinetic larynges occur in birds that swallow small food items after intraoral processing, are characterized by a tight hyoid suspension (see Sect. 2.5.2.1 and 2.6.1), and can raise the caudal end of their laryngeal mound and laryngeal papillae during deglutition to push food items into their esophagus (see Sect. 2.7.3.1). One can hypothesize, therefore, that kinetic larynges evolved primarily as part of the feeding apparatus. Once evolved, however, the kinetic larynges allowed the production of a variety of vowel- and consonant-like sounds for complex vocalizations by being able to modify the laryngeal chamber (see Sects. 2.7.4.1 and 2.7.4.2; see also Homberger 1999). Perhaps counterintuitively, the structurally less complex akinetic larynges are derived from kinetic larynges, just like akinetic skulls are derived from kinetic skulls in vertebrates. Once the complex structures of diarthrodial joints and the muscles moving them are lost in an akinetic condition, they are unlikely to re-evolve a kinetic condition because several joints and their muscles would have to be recreated simultaneously and in perfect structural and functional coordination (see also Bock 1999a, b). The evolution of an akinetic larynx appears to be connected to a reduced need for a highly mobile laryngeal mound in birds that have a relaxed or loose hyoid suspension and are able to swallow larger food items by cranial-inertial means. The selective advantage of this kind of feeding mechanism may have overridden the concomitant inability to evolve complex vocalizations.

The presence of the ancestral condition of the hyoid suspension and laryngeal apparatus in an avian group (the Psittaciformes) that is generally considered to have evolved more recently (see Sects. 2.5.1.1 and 2.6.1.1; Table 2.2) and is characterized by a high degree of encephalization and a capacity for vocal learning (see Jarvis et al. 2014; Zhang et al. 2014) points to a case of mosaic evolution (*sensu* de Beer 1954a, b). Another case of mosaic evolution is found in passerine birds, which are also considered to have evolved more recently for the same reasons (see Jarvis et al. 2014; Zhang et al. 2014) and possess the ancestral condition of the hyoid suspension apparatus, but a derived condition of the laryngeal apparatus (see Sects. 2.5.1.1 and 2.6.1.2; Table 2.2). Perhaps even more remarkable is the case of mosaic evolution of the derived conditions of the hyoid suspension and laryngeal apparatus in avian groups, such as cormorants and pelicans, which have generally been considered to be close to the stem of modern birds (see Jarvis et al. 2014; Zhang et al. 2014). Of course, these limited observations will need to be corroborated and complemented by further studies on a larger group of birds in order to be able to draw general conclusions and conceive a more complete picture of the evolutionary history of birds. However, the vastly different laryngeal apparatus of the Psittaciformes and Passeriformes, which are not derivable from each other, already now cast serious doubts on the validity of the close phylogenetic relationship between the Psittaciformes and the passerine birds as proposed by Jarvis et al. (2014) and Zhang et al. (2014) and may even question the relative ancestry of avian orders in general (see Homberger 2002).

Hence, functional morphology, by being able to identify convergent and mosaic evolution, can thereby test hypotheses of homologies as well as provide data to test the plausibility of phylogenetic hypotheses that are based on molecular-genetic methods, which cannot easily distinguish homoplastic from homologous characters.

2.11 Future Work

Mammals are less diverse in their lingual and laryngeal apparatus than birds probably because of their dependence on nursing and, for most of them, mastication. Birds were not constrained by such a strict selective agent in their evolution and have evolved a variety of constructions of their jaw, lingual and laryngeal apparatus. Furthermore, as presented here, hierarchically interdependent apparatus may vary independently in birds and may result in an even greater variety of the feeding and respiratory apparatus based on a "mix-and-match" of components resulting in the phenomenon of mosaic evolution. It is, therefore, difficult to provide a general description of the avian feeding apparatus. Hence, there is an urgent need not only for detailed and comprehensive anatomical descriptions of representative species of ecologically distinct orders, but also for anatomical descriptions to be complemented by biomechanical analyses that integrate physical concepts and principles, such as tensegrity (Scarr 2014) and hydraulics (Kier and Smith 1985; Gutmann 1988; Vogel 2013). Such integrated data will enable the use of functional-morphological data to test phylogenies that are based on molecular data and *vice versa*.

This review has tried to show that a proper understanding of the functional morphology of the tongue and larynx can be achieved only within the context of the functional morphology of the head and jaw apparatus. Of course, a truly complete understanding would have to include the organs of the neck and thorax as well and even the rest of the body, which would, however, go beyond the scope of this review.

References

Abumandour MMA, El-Bakary NER. Morphological features of the tongue and laryngeal entrance in two predatory birds with similar feeding preferences: common kestrel (*Falco tinnunculus*) and Hume's tawny owl (*Strix butleri*). Anat Sci Int. 2016;1–12 doi:10.1007/s12565-016-0339-9.

Amar-Singh HSS. Little Ringed Plover gular flutter and age. 2015. YouTube. https://www.youtube.com/watch?v=kylxA7DhLSU. Accessed 24 Oct 2016.

Amar-Singh HSS. Common Snipe gular flutter. 2016. YouTube. https://www.youtube.com/watch?v=SJE1R1mAkFA. Accessed 24 Oct 2016.

Arad Z, Midtgård U, Bernstein MH. Thermoregulation in turkey vultures: vascular anatomy, arteriovenous heat exchange, and behavior. Condor. 1989;91(3):505–14.

Avila K. Double-crested Cormorant gular fluttering. 2015. YouTube. https://www.youtube.com/watch?v=GH6Mz3faGBw. Accessed 14 Oct 2016.

Balete DS, Rickart EA, Rosell-Ambal GB, Jansa S, Heany LR. Descriptions of two new species of *Rhynchomys* Thomas (Rodentia: Muridae: Murinae) from Luzon Island, Philippines. J Mammal. 2007;88(2):287–301.

Bang BG, Wenzel BM. Nasal cavity and olfactory system. In: King AS, McLelland J, editors. Form and function in birds, vol. 3. London: Academic Press; 1985. p. 195–225.

Banks WJ. Applied veterinary histology. 3rd ed. St. Louis: Mosby Year Book; 1993.

Bartholomew AA. The role of behavior in the temperature regulation of the Masked Booby. Condor. 1966;68(8):523–35.

Bartholomew GA, Lasiewski RC, Crawford EC. Patterns of panting and gular flutter in cormorants, pelicans, owls, and doves. Condor. 1968;70:31–4.

Baumel JJ, Witmer LM. Osteologia. In: Baumel JJ, King AS, Breazile JE, Evans HE, Vanden-Berge J, editors. Handbook of avian anatomy: Nomina Anatomica Avium. 2nd ed. Cambridge, MA: Publications of the Nuttall Ornithological Club No. 123; 1993. p. 45–132.

Baumel JJ, Dalley AF, Quinn TH. The collar plexus of subcutaneous thermoregulatory veins in the pigeon, *Columba livia*; its association with esophageal pulsation and gular flutter. Zoomorphology. 1983;102:215–39.

Baumel JJ, King AS, Beazile JE, Evans HE, Vanden Berge J. Handbook of avian anatomy: Nomina Anatomica Avium. 2nd ed. Cambridge, MA: Publications of the Nuttall Ornithological Club No. 123; 1993.

Baussart S, Bels V. Tropical hornbills (*Aceros cassidix*, *Aceros undulatus*, and *Buceros hydrocorax*) use ballistic transport to feed with their large beaks. J Exp Zool. 2011;315:72–83.

Baussart S, Korsoun L, Libourel P-A, Bels V. Ballistic food transport in toucans. J Exp Zool. 2009;311A:465–74.

Bayer RD. Vocalizations of Great Blue Herons at Yaquina Estuary, Oregon. Colonial Waterbirds. 1984;7:35–44.

Beckers GJL, Suthers RA, ten Cate C. Mechanisms of frequency and amplitude modulation in ring dove song. J Exp Biol. 2003;206(11):1833–43.

Beckers GJL, Nelson BS, Suthers RA. Vocal-tract filtering by lingual action in a parrot. Curr Biol. 2004;14(17):1592–7.

Bels V, Baussart S. Feeding behavior and mechanisms in domestic birds. In: Bels V, editor. Feeding in domestic vertebrates: from structure to function. Oxford: CABI Publishing; 2006. p. 33–49.

Berkhoudt H. The epidermal structure of the bill tip organ in ducks. Neth J Zool. 1976;26(4):561–6.

Berkhoudt H. Taste buds in the bill of the Mallard (*Anas platyrhynchos* L.): their morphology, distribution and

functional significance. Neth J Zool. 1977;27(3):310–31.
Berkhoudt H. The morphology and distribution of cutaneous mechanoreceptors (Herbst and Grandry corpuscles) in bill and tongue of the Mallard (*Anas platyrhynchos* L.). Neth J Zool. 1980;30(1):1–34.
Berkhoudt H. Structure and function of avian taste receptors. In: King AS, McLelland J, editors. Form and function in birds, vol. 3. London: Academic; 1985. p. 463–95.
Bhattacharyya BN. Functional morphology of the jaw muscles of some Indian insect-eating birds. Gegenbaurs Morphol Jahrb. 1982;128(2):208–54.
Bhattacharyya BN. Functional morphology of the tongue muscles of some Indian insect-eating birds. Gegenbaurs Morphol Jahrb. 1985;131(1):93–123.
Bhattacharyya BN. Comparative morphology of the avian jaw apparatus. In: Ouellet H, editor. Acta XIX Congressus Internationalis Ornithologici, National Museum of Natural Sciences, vol. II. Ottawa: University of Ottawa Press; 1988. p. 2418–26.
Bhattacharyya BN. Functional morphology of the jaw muscles of two species of Imperial Pigeons, *Ducula aenea nicobarica* and *Ducula badia insignis*. Gegenbaurs Morphol Jahrb. 1989;135(4):573–618.
Bhattacharyya BN. The role of the M. pterygoideus in closure of the beaks in certain columbid birds: a functional morphological analysis reflecting diversity in feeding. Proc Zool Soc. 1997;50(2):171–80.
Bhattacharyya BN. Functional morphology of the feeding apparatus of the Common Green Pigeon, *Treron phoenicoptera* (Latham). Proc Zool Soc. 1998;51(1):1–44.
Bhattacharyya BN. Avian jaw function: adaptation of the seven-muscle system and a review. Proc Zool Soc. 2013;66:75. doi:10.1007/s12595-012-0056-x.
Bhullar B-AS, Hanson M, Fabbri M, Pritchard A, Bever GS, Hoffman E. How to make a bird skull: major transitions in the evolution of the avian cranium, paedomorphosis, and the beak as a surrogate hand. Integr Comp Biol. 2016;56(3):389–403.
Bignon F. Contribution à l'étude de la pneumaticité chez les oiseaux. Mém Soc Zool France Paris. 1889;2:260–320 and plates X–XIII.
Blevins CE, Ge J, Suthers RA, Homberger DG. An animated 3D model of the synchronous movements of the suprasyringeal structures and organs in the neck of a vocalizing songbird, the Northern Cardinal (*Cardinalis cardinalis*). FASEB J. 2014;28(Suppl):918.8.
Bock WJ. Kinetics of the avian skull. J Morphol. 1964;114:1–42.
Bock WJ. Morphology of the tongue apparatus of *Ciridops anna* (Drepanididae). Ibis. 1972;114(1):61–78.
Bock WJ. The avian skeletomuscular system. In: Farner DS, King JR, Parkes KC, editors. Avian biology, vol. IV. New York: Academic Press; 1974. p. 109–258.
Bock WJ. Tongue morphology and affinities of the Hawaiian honeycreeper *Melamprosops phaeosoma*. Ibis. 1978a;120(4):467–79.
Bock WJ. Morphology of the larynx of *Corvus brachyrhynchos* (Passeriformes: Corvidae). Wilson Bull. 1978b;90(4):553–65.
Bock WJ. Is *Diglossa* (?Thraupinae) monophyletic? Neotrop Ornithol. 1985a;36:319–32.
Bock WJ. Relationships of the sugarbird (*Promerops*; Passeriformes, ?Meliphagidae). In: Schuchmann K-L, editor. Proceedings of the international symposium on African vertebrates: systematics, phylogeny, and evolutionary ecology. Bonn: Zoologisches Forschungsinstitut und Museum Alexander Koenig; 1985b. p. 349–74.
Bock WJ. The skeletomuscular system of the feeding apparatus of the Noisy Scrub-bird, *Atrichornis clamosus* (Passeriformes: Atrichornithidae). Rec Aust Mus. 1985c;37:193–210.
Bock WJ. Functional and evolutionary morphology of woodpeckers. Ostrich. 1999a;70(1):23–31.
Bock WJ. Cranial kinesis revisited. Zool Anz. 1999b;238(1–2):27–39.
Bock WJ, Morony J. The preglossale of *Passer* (Aves: Passeriformes)—a skeletal neomorph. J Morphol. 1978;155(1):99–110.
Bock WJ, von Wahlert G. Adaptation and the form-function complex. Evolution. 1965;19(3):269–99.
Bock WJ, Balda RP, Vander Wall SB. Morphology of the sublingual pouch and tongue musculature in Clark's Nutcracker. Auk. 1973;90(3):491–519.
Böker H. Die anatomische Konstruktion zur Erweiterung des Unterschnabels bei den Pelikanen. Anat Anz. 1938;87:294–303.
Bottoni L, Masin S, Lenti-Boero D. Vowel-like sound structure in an African Grey Parrot (*Psittacus erithacus*) vocal production. Open Behav Sci J. 2009;3:1–16. doi:10.2174/1874230000903010001.
Bowman RI. Morphological differentiation and adaptation in the Galápagos Finches, Berkeley: University of California Publications in Zoology, vol. 58; 1961. p. 1–326.
Bragulla HH, Homberger DG. Structure and functions of keratin proteins in simple, stratified, keratinized and cornified epithelia. J Anat. 2009;214(4):516–59.
Bühler P. Functional anatomy of the avian jaw apparatus. In: King AS, McLelland J, editors. Form and function in birds, vol. 2. London: Academic Press; 1981. p. 439–68.
Bühler P. Grösse, Form, und Färbung des Tukanschnabels—Grundlage für den evolutiven Erfolg der Ramphastiden? J Ornithol. 1995;136(2):187–93.
Burle MH, Rico-Guevara A, Rubega MA, Lank D. A hummingbird tongue in a shorebird head: Tuamoto sandpipers are nectar-feeders. Int Comp Biol. 2013;53(Suppl 1):e25.
Calder WA, King JR. Thermal and caloric relations of birds. In: Farner DS, King JR, Parkes KC, editors. Avian biology, vol. IV. New York: Academic Press; 1974. p. 259–413.
Campbell G. Effects of temperature on gular fluttering and evaporative water loss in four sympatric cormorants in

southern Africa [Ph.D. thesis]. Cape Town: University of Cape Town; 2014. https://open.uct.ac.za/bitstream/handle/11427/12815/thesis_sci_2014_campbell_g.pdf?sequence= 1. Accessed 24 Oct 2016.

Campbell-Tennant DJE, Gardner JL, Kearney MR, Symonds MRE. Climate-related spatial and temporal variation in bill morphology over the past century in Australian parrots. J Biogeogr. 2015;42(6):1163–75. doi:10.1111/jbi.12499.

Chow L. Spotted wood-owl and gular fluttering. 2014. http://www.besgroup.org/2014/04/21/spotted- wood-owl-and-gular-fluttering/. Accessed 24 Oct 2016.

Claes R, Muyshondt PGG, Van Hoorebeke L, Dhaene J, Dirckx JJJ, Aerts P. The effect of craniokinesis on the middle ear of domestic chickens (*Gallus gallus domesticus*). J Anat. 2016; doi:10.1111/joa.12566.

Collias NE. The vocal repertoire of the Red Junglefowl: a spectrographic classification and the code of communication. Condor. 1987;89:510–24.

Cozic AM, Homberger DG. The paired cervico-cephalic air sacs and their role in the vocalizations of songbirds. FASEB J. 2015;29(April, Suppl):867.9.

Cozic AM, Homberger DG. The role of cervical air sacs in the vocalization of songbirds. Anat Rec. 2016;299 (Special Feature, 41):258–9.

Crawford EC. Mechanical aspects of panting in dogs. J Appl Physiol. 1962;17:249–51.

Creutz G. Der Graureiher: *Ardea cinerea*. Die Neue Brehm-Bücherei, A. Lutherstadt Wittenberg: Ziemsen Verlag; 1981.

Crole MR, Soley JT. Comparative distribution and arrangement of Herbst corpuscles in the oropharynx of the ostrich (*Struthio camelus*) and emu (*Dromaius novaehollandiae*). Anat Rec. 2014;297:1338–48.

Crole MR, Soley JT. Contrasting morphological evidence for the presence of taste buds in *Dromaius novaehollandiae* and *Struthio camelus* (Palaeognathae, Aves). Zoomorphology. 2015;134 (3):499–507.

Cruickshank AJ, Gautier J-P, Chappuis C. Vocal mimicry in wild African Grey Parrots *Psittacus erithacus*. Ibis. 1993;135(3):293–9.

Cummins CL. The morphology of the hyoid apparatus and gular region of the Snowy Egret, *Egretta thula* (Molina), (Aves: Ardeidae) [M.S. thesis]. Baton Rouge: Louisiana State University; 1986.

Cunningham SJ, Alley MR, Castro I, Potter MA, Cunningham M, Pyne MJ. Bill morphology of ibises suggests a remote-tactile sensory system for prey detection. Auk. 2010;127(2):308–16.

Danner RM, Gulson-Castillo ER, James HF, Dzielski SA, Frank DC, Sibbald ET, Winkler DW. Habitat-specific divergence of air conditioning structures in bird bills. Auk. 2017;134(1):65–75. doi:10.1642/AUK-16-107.1.

Dawson MM, Metzger KA, Baier DB, Brainerd EL. Kinematics of the quadrate bone during feeding in mallard ducks. J Exp Biol. 2011;214(12):2036–46.

De Beer G. Archaeopteryx lithographica. London: Trustees of the British Museum (Natural History); 1954a.

De Beer GR. Archaeopteryx and evolution. Oxford Meeting of the British Association for the Advancement of Science, No. 42. 1954b Sep. p. 1–11.

Doran GA. Review of the evolution and phylogeny of the mammalian tongue. Acta Anat. 1975;91(1):118–29.

Doran GA, Baggett H. A structural and functional classification of mammalian tongues. J Mammal. 1971;52 (2):427–9.

de Juana E. Family Pteroclidae (Sandgrouse). In: del Hoyo J, Elliott A, Sargatal J, editors. Handbook of the birds of the world, Sandgrouse to Cuckoos, vol. 4. Barcelona: Lynx Edicions; 1997. p. 30–57.

Dyce KM, Sack WO, Wensing CJG. Textbook of veterinary anatomy. Philadelphia, PA: WB Saunders Company; 1987.

Dzerzhinsky FYA. Biomechanical analysis of the avian jaw apparatus [in Russian]. Moscow, Russia: Moscow State University Press; 1972.

Elemans CPH, Mead AF, Rome LC, Goller F. Superfast vocal muscles control song production in songbirds. PLoS One. 2008;3(7):e2581. doi:10.1371/journal.pone.0002581.

Erdoğan S, Iwasaki S. Function-related morphological characteristics and specialized structures of the avian tongue. Ann Anat. 2014;196(2–3):75–87.

Erdoğan S, Pèrez W, Alan A. Anatomical and scanning electron microscopic investigations of the tongue and laryngeal entrance of the Long-legged Buzzard (*Buteo rufinus*, Cretschmar, 1829). Microsc Res Tech. 2012a;75(9):1245–52.

Erdoğan S, Sağsöz H, Akbalik ME. Anatomical and histological structure of the tongue and histochemical characteristics of the lingual salivary glands in the chukar partridge (*Alectoris chukar*, Gray 1830). Br Poult Sci. 2012b;53(3):307–15.

Esselstyn JA, Achmadi AS, Rowe KC. Evolutionary novelty in a rat with no molars. Biol Lett. 2012;8(6):990–3.

Evans HE, Martin GR. Organa sensuum [Organa sensoria]. In: Baumel JJ, King AS, Breazile JE, Evans HE, Vanden Berge J, editors. Handbook of avian anatomy: Nomina Anatomica Avium. 2nd ed. Cambridge, MA: Publications of the Nuttall Ornithological Club No. 123; 1993. p. 585–611.

Evans RM. Development of thermoregulation in young white pelicans. Can J Zool. 1984;62(56):808–13.

Ewald PW, Williams WA. Function of the bill and tongue in nectar uptake by hummingbirds. Auk. 1982;99 (3):573–6.

Fam DN. New Zealand Birds: Australasian Gannet adult panting. 2013. YouTube. https://www.youtube.com/watch?v=HpSCeVXb_D4. Accessed 24 Oct 2016.

Fisher HI, Goodman DC. The myology of the whooping crane, *Grus americana*. Illinois Biological Monographs, vol. 24, No. 2. Urbana: University of Illinois Press; 1955. p. 1–127.

Fitch T. Production of vocalizations in mammals. In: Brown K, editor. Encyclopedia of language and linguistics. 2nd ed. Oxford: Elsevier; 2006. p. 115–21.

Fitch WT, de Boer B, Mathur N, Ghazanfar AA. Monkey vocal tracts are speech-ready. Sci Adv. 2016;2(12): e1600723. doi:10.1126/sciadv.1600723.

Fletcher NH, Riede T, Beckers GJL, Suthers RA. Vocal tract filtering and the "coo" of doves. J Acoust Soc Am. 2004;116(6):3750–6.

Fletcher NH, Riede T, Suthers RA. Model for vocalization by a bird with distensible vocal cavity and open beak. J Acoust Soc Am. 2006;119(2):1005–111.

Foelix RF. Vergleichend-morphologische Untersuchungen an den Speicheldrüsen körnerfressender Singvögel. Zool Jahrb Anat. 1970;87:523–87.

Frey R, Gebler A. Mechanisms and evolution of roaring-like vocalization in mammals. In: Brudzynski SM, editor. Handbook of mammalian vocalization—an integrative neuroscience approach. London: Academic Press; 2010. p. 439–50.

Frey R, Gebler A, Fritsch G, Nygrén K, Weissengruber GE. Nordic rattles: the hoearse vocalization and the inflatable laryngeal air sac of reindeers (*Rangifer tarandus*). J Anat. 2007;210(2):131–59.

Gallup AC, Miller ML, Clark AB. Yawning and thermoregulation in budgerigars, *Melopsittacus undulatus*. Anim Behav. 2009;77:109–13.

Gaunt SLL. Thermoregulation in doves (Columbidae): a novel esophageal heat exchanger. Science. 1980;210:445–7.

Gaunt AS, Gaunt SLL. Mechanics of the syrinx in *Gallus gallus*. II. Electromyographic studies of *ad libitum* vocalizations. J Morphol. 1977;152(1):1–20.

Genbrugge A, Herrel A, Boone M, van Hoorebeke L, Podos J, Dirckx J, Aerts P, Adriaens D. The head of the finch: the anatomy of the feeding system in two species of finches (*Geospiza fortis* and *Padda oryzivora*). J Anat. 2011;219(6):676–95.

Genbrugge A, Adriaens D, de Kegel B, Brabant L, van Hoorebeke L, Podos J, Dirckx J, Aerts P, Herrel A. Structural tissue organization in the beak of Java and Darwin's finches. J Anat. 2012;21(5):383–93.

German RZ, Crompton AW. The ontogeny of feeding in mammals. In: Schwenk K, editor. Feeding: form, function, and evolution in tetrapod vertebrates. San Diego, CA: Academic Press; 2000. p. 449–57.

Gerritsen AFC, Meiboom A. The role of touch in prey density estimation by *Calidris alba*. Neth J Zool. 1986;36:530–62.

Gerritsen AFC, Sevenster JG. Foraging behavior and bill anatomy in sandpipers. In: Duncker H-R, Fleischer G, editors. Vertebrate morphology, Fortschritte der Zoologie, vol. 30; 1985. p. 237–9.

Ghetie V, Atanasiu I. Die Myologie des Zungenbeinaufhängeapparates und der Zunge, beim Hühner- und Wassergeflügel. Rev Biol Bucharest. 1962;7:85–94.

Goller F, Cooper BG. Peripheral motor dynamics of song production in the Zebra Finch. Ann N Y Acad Sci. 2004;1016:130–52.

Goller F, Cooper BG. Peripheral mechanisms of sensorimotor integration during singing. In: Zeigler HP, Marler P, editors. Neuroscience of birdsong. New York: Cambridge University Press; 2008. p. 99–114.

Goller F, Mallinckrodt MJ, Torti SD. Beak gape dynamics during song in Zebra finch. J Neurobiol. 2004;59 (3):289–303.

Goodman DC, Fisher HI. Functional anatomy of the feeding apparatus in waterfowl (Aves: Anatidae). Carbondale, IL: Southern Illinois University Press; 1962.

Gottschaldt K-M. Structure and function of avian somatosensory receptors. In: King AS, McLelland J, editors. Form and function in birds, vol. 3. London: Academic Press; 1985. p. 375–461.

Greenberg R, Danner R, Olsen B, Luther D. High summer temperature explains bill size variation in salt marsh sparrows. Ecography. 2011;35(2):146–52. doi:10.1111/j.1600-0587.2011.07002.x.

Güntert M, Ziswiler V. Konvergenzen in der Struktur von Zunge und Verdauungstrakt nektarfressender Papageien. Rev Suisse Zool. 1972;79:1016–26.

Gussekloo SWS. Feeding structures in birds. In: Bels V, editor. Feeding in domestic vertebrates: from structure to function. Oxford: CABI Publishing; 2006. p. 14–32.

Gussekloo SWS, Bout RG. The kinematics of feeding and drinking in paleognathous birds in relation to cranial morphology. J Exp Biol. 2005;208 (17):3395–407.

Gussekloo SWS, Zweers GA. Feeding adaptations in the greater rhea (*Rhea americana*; Ratitae). J Morphol. 1997;232(3):262.

Gussekloo SWS, Vosselman MG, Bout RG. Three-dimensional kinematics of skeletal elements in avian prokinetic and rhynchokinetic skulls determined by Roentgen stereophotogrammetry. J Exp Biol. 2001;204(10):1735–44.

Gutmann WF. The hydraulic principle. Am Zool. 1988;28:257–66.

Haggard P, de Boer L. Oral somatosensory awareness. Neurosci Biobehav Rev. 2014;47:4690484. doi:10.1016/j.neubiorev.2014.09.015.

Harrison DFN. The anatomy and physiology of the mammalian larynx. Cambridge, UK: Cambridge University Press; 1995

Hausberger M, Black JM, Richard J-P. Bill opening and sound spectrum in barnacle goose loud calls: individuals with 'wide mouths' have higher pitched voices. Anim Behav. 1991;42(2):319–22.

Heidweiller J, Zweers GA. Drinking mechanisms in the zebra finch and the bengalese finch. Condor. 1990;92 (1):1–28.

Heidweiller J, van Loon JA, Zweers GA. Flexibility of the drinking mechanism in adult chickens (*Gallus gallus*) (Aves). Zoomorphology. 1992;111:141–59.

Herring SW, Teng S. Strain in the braincase and its sutures during function. Am J Phys Anthropol. 2000;112:575–93.

Hiiemae KM. Feeding in mammals. In: Schwenk K, editor. Feeding: form, function, and evolution in tetrapod vertebrates. San Diego, CA: Academic; 2000. p. 411–48.

Hoese WJ, Podos J, Boetticher NC, Nowicki S. Vocal tract function in birdsong production: experimental manipulation of beak movements. J Exp Biol. 2000;203(12):1845–55.

Homberger DG. Functional morphology of the larynx in the parrot *Psittacus erithacus*. Am Zool. 1979;19(3):988.

Homberger DG. Funktionell-morphologische Untersuchungen zur Radiation der Ernährungs- und Trinkmethoden der Papageien (Psittaci). Bonner zoologische Monographien, vol. 13. Bonn: Zoologisches Forschungsinstitut und Museum Alexander Koenig; 1980. p. 1–192.

Homberger DG. Nonadaptive evolution of avian drinking methods. Am Zool. 1983;23(4):894.

Homberger DG. The lingual apparatus of the African Grey Parrot, *Psittacus erithacus* Linné (Aves: Psittacidae): description and theoretical mechanical analysis. Ornithol Monogr. 1986;39:1–232.

Homberger DG. Comparative morphology of the avian tongue. In: Ouellet H, editor. Acta XIX Congressus Internationalis Ornithologici, vol. II. Ottawa: University of Ottawa Press; 1988a. p. 2427–35.

Homberger DG. Models and tests in morphology: the significance of description and integration. Am Zool. 1988b;28(1):217–29.

Homberger DG. Filing ridges and transversal step of the maxillary rhamphotheca in Australian cockatoos (Psittaciformes: Cacatuidae): a homoplastic structural character evolved in adaptation to seed shelling. In: van den Elzen R, Schuchmann K-L, Schmidt-Koenig K, editors. Proceedings of the international 100th DO-G meeting: current topics in avian biology. Bonn: German Ornithological Society. 1990. p. 43–48.

Homberger DG. The hyoid suspension apparatus as a structural constraint of feeding mechanisms in birds and mammals. J Morphol. 1994a;220(3):355.

Homberger DG. Oekomorphologie der rotschwänzigen Rabenkakadu-Arten (*Calyptorhynchus* spp.) in Australien: Beispiel einer multispektiven Biodiversitätsstudie als Grundlage für die Rekonstruktion der Evolutionsgeschichte einer Artengruppe. In: Gutmann WF, Mollenhauer D, Peters DS, editors. Senckenberg-Buch 70: Morphologie und Evolution. Frankfurt a.M: Kramer; 1994b. p. 425–34.

Homberger DG. The role of the larynx in articulated vocalization of birds. Am Zool. 1997;37(5):136A.

Homberger DG. The avian tongue and larynx: multiple functions in nutrition and vocalization. In: Adams N, Slotow R, editors. Proceedings of the 22nd International Ornithological Congress. Durban, South Africa: University of Natal; 1999. p. 94–113.

Homberger DG. The case of the cockatoo bill, horse hoof, rhinoceros horn, whale baleen, and turkey beard: the integument as a model system to explore the concepts of homology and non-homology. In: Dutta HM, Datta Munshi JS, editors. Vertebrate functional morphology: horizon of research in the 21st century. Enfield, NH: Science Publishers Inc.; 2001. p. 317–43.

Homberger DG. The aerodynamically streamlined body shape of birds: implications for the evolution of birds, feathers, and avian flight. In: Zhou Z, Zhang F, editors. Proceedings of the 5th symposium of the Society of Avian Paleontology and Evolution, Beijing, 1–4 June 2000. Science Press, Beijing, 1–4 June 2002. p. 227–252.

Homberger DG. The comparative biomechanics of a prey-predator relationship: the adaptive morphologies of the feeding apparatus of Australian Black-Cockatoos and their foods as a basis for the reconstruction of the evolutionary history of the Psittaciformes. In: Bels VL, Gasc J-P, Casinos A, editors. Vertebrate biomechanics and evolution. Oxford, UK: BIOS Scientific Publishers; 2003. p. 203–28.

Homberger DG. The classification and the status of wild populations of parrots. In: Luescher A, editor. Manual of parrot behavior. Ames, IA: Blackwell Publishing; 2006. p. 3–12.

Homberger DG. Comparative beak morphology of two subspecies of Australian Red-tailed Black-Cockatoos: small changes with significant functional effects as a model for macroevolutionary processes. Anat Rec. 2016;24:131–2.

Homberger DG, Brush AH. Functional morphological and biochemical correlations of the keratinized structures of the African Grey Parrot (*Psittacus erithacus* L.). Zoomorphology. 1986;106:103–14.

Homberger DG, Cozic AM. New insights in the functional anatomy of the neck and its organs in songbirds. Integr Comp Biol. 2015;55(Suppl 1):e82.

Homberger DG, Meyers RA. The morphology of the lingual apparatus of the domestic chicken, *Gallus gallus*, with special attention to the structure of the fasciae. Am J Anat. 1989;186(3):217–57.

Homberger DG, Walker WF. Vertebrate dissection. 9th ed. Florence: Brooks/Cole; 2004.

Homberger DG, Ziswiler V. Funktionell-morphologische Untersuchungen am Schnabel von Papageien. Rev Suisse Zool. 1972;79:1038–48.

Howell TR, Bartholomew GA. Temperature regulation in the red-tailed tropic bird and the red-footed booby. Condor. 1962;64:6–18.

Ilyichev V, Silayeva O. Talking birds. Moscow: Nauka Publishers; 1992.

Immelmann K, Immelmann G. Verhaltensökologische Studien an afrikanischen und australischen Estrildiden. Zool Jb Syst. 1967;94:609–86.

Jackowiak H, Godynicki S. Light and scanning electron microscopic study of the tongue in the white tailed eagle (*Haliaeetus albicilla*, Accipitridae, Aves). Ann Anat. 2005;187(3):251–9.

Jackowiak H, Ludwig M. Light and scanning electron microscopic study of the structure of the ostrich (*Struthio camelus*) tongue. Zool Sci. 2008;25:188–94.

Jackowiak H, Andrzejewski W, Godynicki S. Light and scanning electron microscopic study of the tongue in the Cormorant *Phalacrocorax carbo* (Phalacrocoracidae, Aves). Zool Sci. 2006;23:161–7.

Jackowiak H, Skieresz-Szewczyk K, Godynicki S, Iwasaki S, Meyer W. Functional morphology of the tongue in the Domestic Goose (*Anser anser f. domestica*). Anat Rec. 2011;294(9):1574–84.

Jarvis E, Mirarab S, Aberer AJ, Li B, Houde P, Li C, Ho SYW, Faircloth BC, Nabholz B, Howard JT, Suh A, Weber CC, da Fonseca RR, Li J, Zhang F, Li H, Zhou L, Narula N, Liu L, Ganapathy G, Boussau B, Bayzid MS, Zavidovych V, Subramanian S, Gabaldón T, Capella-Gutiérrez S, Huerta-Cepas J, Rekepalli B, Munch K, Schierup M, Lindow B, Warren WC, Ray D, Green RE, Bruford MW, Zhan X, Dixon A, Li S, Li N, Huang Y, Derryberry EP, Bertelsen MF, Sheldon FH, Brumfield RT, Mello CV, Lovell PV, Wirthlin M, Schneider MPC, Prosdocimi F, Samaniego JA, Velazquez AMV, Alfaro-Núñez A, Campos PF, Petersen B, Sicheritz-Ponten T, Pas A, Bailey T, Scofield P, Bruce M, Lambert DM, Zhou Q, Perelman P, Driskell AC, Shapiro B, Xiong Z, Zeng Y, Liu S, Li Z, Liu B, Wu K, Xiao J, Yinqi X, Zheng Q, Zhang Y, Yang H, Wang J, Smeds L, Rheindt FE, Braun M, Fjeldsa J, Orlando L, Barker FK, Jonsson KA, Johnson W, Koepfli K-P, O'Brien S, Haussler D, Ryder OA, Rahbek C, Willerslev E, Graves GR, Glenn TC, McCormack J, Burt D, Ellegren H, Alström P, Edwards SV, Stamatakis A, Mindell DP, Cracraft J, Braun EL, Warnow T, Jun W, Gilbert MTP, Zhang G. Whole-genome analyses resolve early branches in the tree of life of modern birds. Science. 2014;346 (6215):1320–31.

Kalyakin MV, Dzerzhinsky FYA. Some aspects of trophic adaptations in bulbuls (Aves, Pycnonotidae) as seen from functional morphology of feeding apparatus. Zool Zurn. 1997;76(7):836–44. [in Russian]

Kear J. Food selection in finches with special reference to interspecific differences. Proc Zool Soc London. 1962;138(2):163–204.

Kier WM, Smith KK. Tongues, tentacles and trunks: the biomechanics of movement in muscular-hydrostats. Zool J Linnean Soc. 1985;83:307–24.

Kooloos JGM. A conveyer-belt model for pecking in the mallard (*Anas platyrhynchos* L.). Neth J Zool. 1986;36(1):47–87.

Kooloos JGM, Zweers GA. Mechanics of drinking in the mallard (*Anas platyrhynchos*, Anatidae). J Morphol. 1989;199:327–47.

Kooloos JGM, Zweers GA. Integration of pecking, filter feeding and drinking mechanisms in waterfowl. Acta Biotheor. 1991;39:107–40.

Kooloos JGM, Kraaijeveld AR, Langenbach GEJ, Zweers GA. Comparative mechanics of filter feeding in *Anas platyrhynchos*, *Anas clypeata* and *Aythya fuligula*. Zoomorphology. 1989;108(5):269–90.

Korzun LP, Érard C, Gasc J-P, Dzerzhinsky FJ. Bill and hyoid apparatus of pigeons (Columbidae) and sandgrouse (Pteroclidae): a common adaptation to vegetarian feeding? C R Biol. 2008;331(1):64–87.

Krulis V. Struktur und Verteilung von Tastrezeptoren im Schnabel-Zungenbereich von Singvögeln, im besonderen der Fringillidae. Rev Suisse Zool. 1978;85(2):385–447.

Laitman JT, Reidenberg JS. Specializations of the human upper respiratory and upper digestive systems as seen through comparative and developmental anatomy. Dysphagia. 1993a;8(4):318–25.

Laitman JT, Reidenberg JS. Comparative anatomy of the mammalian epiglottis: functional and evolutionary implications. Anat Rec. 1993b;237(Suppl 1):76.

Larson JE, Herring SW. Movement of the epiglottis in mammals. Am J Phys Anthropol. 1996;100:71–82.

Lasiewski RC. Respiratory function in birds. In: Farner DS, King JR, Parkes KC, editors. Avian biology, vol. II. New York: Academic Press; 1972. p. 287–342.

Lasiewski RC, Bartholomew GA. Evaporative cooling in the poor-will and the tawny frogmouth. Condor. 1966;68:253–62.

Lasiewski RC, Snyder GH. Responses to high temperatures in nestling double-crested and pelagic cormorants. Auk. 1969;86:529–40.

Leder SB, Burrell MI, Van Dale DJ. Epiglottis is not essential for successful swallowing in humans. Ann Otol Rhinol Laryngol. 2010;119(12):795–8.

Leiber A. Vergleichende Anatomie der Spechtzunge. Zoologica. 1907;51:1–79 and plates I–VI.

Leo-Smith B, Joubert B. Grey go-away birds & gular flutter. 2015. YouTube. https://www.youtube.com/watch?v=8mkrd83SAa0. Accessed 24 Oct 2016.

Marts K. Juvenile white ibis exhibit gular fluttering sea pines. 2015. YouTube. https://www.youtube.com/watch?v=S572YOLG9Vc. Accessed 24 Oct 2016.

Marts K. Young anhinga does gular fluttering. 2016. YouTube. https://www.youtube.com/watch?v=mbtYLUGnQ38. Accessed 24 Oct 2016.

Mason JR, Clark L. The chemical senses in birds. In: Whittow GC, editor. Sturkie's avian physiology. 5th ed. San Diego, CA: Academic; 2000. p. 39–56.

Matsuo K, Palmer JB. Anatomy and physiology of feeding and swallowing—normal and abnormal. Phys Med Rehabil Clin N Am. 2008;19(4):691–707.

McKechnie AE, Smit B, Whitfield MC, Noakes MJ, Talbot WA, Garcia M, Gerson AR, Wolf BO. Avian thermoregulation in the heat: evaporative cooling capacity in an archetypal desert specialist, Burchell's sandgrouse (*Pterocles burchelli*). J Exp Biol. 2016;219:2137–44.

McLelland J. Digestive system. In: King AS, McLelland J, editors. Form and function in birds, vol. 1. London: Academic Press; 1979. p. 69–181.

McLelland J. Larynx and trachea. In: King AS, McLelland J, editors. Form and function in birds, vol. IV. London: Academic Press; 1989. p. 69–103.

McLelland J. Apparatus digestorius [Systema alimentarium]. In: Baumel JJ, King AS, Beazile JE,

Evans HE, Vanden Berge J, editors. Handbook of avian anatomy: Nomina Anatomica Avium. 2nd ed. Cambridge, MA: Publications of the Nuttall Ornithological Club No. 123; 1993. p. 301–27.
Meyers RA, Myers RR. Mandibular bowing and mineralization in Brown Pelicans. Condor. 2005;107(2):445–9.
Midtgård U. The *rete tibiotarsale* and arteriovenous associations in the hind limbs of birds: a comparative morphological study on counter-current heat exchange systems. Acta Zool. 1981;62:67–87.
Midtgård U. Blood vessels and the occurrence of arteriovenous anastomoses in cephalic heat loss areas of mallards, *Anas platyrhynchos* (Aves). Zoomorphology. 1984;104:323–35.
Midtgård U. The peripheral circulatory system in birds: a morphological and physiological study of some adaptations to temperature regulation [Ph.D. dissertation]. Copenhagen, Denmark: University of Copenhagen; 1986.
Midtgård U. Comparative morphology of the avian circulatory system. In: Ouellet H, editor. Acta XIX Congressus Internationalis Ornithologici, vol. II. Ottawa: University of Ottawa Press; 1988. p. 2445–54.
Ming LT. Grey herons—panting and preening. 2014. YouTube. http://www.besgroup.org/2014/03/27/grey-herons-panting-and-preening/#more. Accessed 24 Oct 2016.
Moriyama K, Okanoya K. Effect of beak movement in singing Bengalese finches (*Lonchura striata var. domestica*). J Acoust Soc Am. 1996;100(4):2643.
Naples VL. The morphology and function of the hyoid region in the tree sloths, *Bradypus* and *Choloepus*. J Mammal. 1986;67(4):712–24.
Naples VL. Morphology, evolution and function of feeding in the giant anteater (*Myrmecophaga tridactyla*). J Zool (Lond). 1999;249:19–41.
Nelson BS, Beckers GJL, Suthers RA. Vocal tract filtering and sound radiation in a songbird. J Exp Biol. 2005;208(2):297–308.
Nickel R, Schummer A, Seiferle E, Siller WG, Wight PAL. Anatomy of the domestic birds. Berlin: Verlag Paul Parey; 1977.
Nickel R, Schummer A, Seiferle E, Frewein J, Wilkens H, Wille K-H. The locomotor system of the domestic mammals. Berlin: Verlag Paul Parey; 1986.
Nottebohm F. Phonation in the Orange-winged Amazon Parrot, *Amazona amazonica*. J Comp Physiol A. 1976;108:157–70.
Nuijens FW, Bout RG. The role of two jaw ligaments in the evolution of passerines. Zoology. 1998;101(1):24–33.
Nuijens FW, Zweers GA. Characters discriminating two seed husking mechanisms in finches (Fringillidae: Carduelinae) and Estrildids (Passeridae: Estrildinae). J Morphol. 1997;232:1–33.
Nuijens FW, Hoek AC, Bout RG. The role of the postorbital ligament in the zebra finch (*Taeniopygia guttata*). Neth J Zool. 2000;50(1):75–88.
Ohms VR, Snelderwaard PC, ten Cate C, Beckers GJL. Vocal tract articulation in zebra finches. PLoS One. 2010;5(7):e11923. doi:10.1371/journal.pone.0011923.
Ohms VR, Beckers GJL, ten Cate C, Suthers RA. Vocal tract articulation revisited: the case of the monk parakeet. J Exp Biol. 2012;215(1):85–92.
Olsen AM, Westneat MW. Linkage mechanisms in the vertebrate skull: structure and function of three-dimensional, parallel transmission systems. J Morphol. 2016;277(12):1570–83. doi:10.1002/jmor.20596.
Osborn ML, Homberger DG. The human shoulder suspension apparatus: a causal explanation for bilateral asymmetry and a fresh look at the evolution of human bipedality. Anat Rec. 2015;298(9):1572–88.
Paton DC, Collins BG. Bill and tongues of nectar-feeding birds: a review of morphology, function and performance, with intercontinental comparisons. Aust J Ecol. 1989;14(4):473–506.
Patterson DK, Pepperberg IM. A comparative study of human and parrot phonation: acoustic and articulatory correlates of vowels. J Acoust Soc Am. 1994;96 (2, Part 1):634–48.
Patterson DK, Pepperberg IM. Acoustic and articulatory correlates of stop consonants in a parrot and a human subject. J Acoust Soc Am. 1998;103 (4):2197–215.
Patterson DK, Pepperberg IM, Story BH, Hoffman EA. How parrots talk: insights based on CT scans, image processing and mathematical models. Proc SPIE. 1997;3033:14–24.
Peat CM, Gaunt AS. Mechanics of drinking in doves. Lawrence (KS): abstracts of Posters and Lectures of the 102nd stated meeting of the American Ornithologists' Union, No. 129. 1984.
Porter WMR, Witmer LM. Avian cephalic vascular anatomy, sites of thermal exchange, and the rete ophthalmicum. Anat Rec. 2016;299(11):1461–86.
Powell IL, Jones KL, Carpenter JH, Tully TN. Captive Hispaniolan Parrots (*Amazona ventralis*) can discriminate between experimental foods with sodium concentrations found in Amazonian mineral licks. Wilson J Ornithol. 2017;129(1):1818–185.
Rafferty KL, Herring SW. Craniofacial sutures: morphology, growth, and *in vivo* masticatory strains. J Morphol. 1999;242:167–79.
Rauch N. Struktur der Lautäusserungen eines Sprache imitierenden Graupapageis (*Psittacus erithacus* L.). Behaviour. 1978;66(1–2):65–104.
Redd TC, Dubansky BH, Osborn ML, Tully TN, Homberger DG. A registration algorithm for the identification of individual parrots based on the patterns of filing ridges on their upper bill tip. Intl J Biometrics Bioinform (IJBB). 2012;6(3):68–91. http://www.cscjournals.org/journals/IJBB/issue-manuscripts.php?v=6&i=3
Reiss KZ. Feeding in myrmecophagous mammals. In: Schwenk K, editor. Feeding: form, function, and evolution in tetrapod vertebrates. San Diego, CA: Academic; 2000. p. 459–85.
Richards LP, Bock WJ. Functional anatomy and the adaptive evolution of the feeding apparatus in the Hawaiian honeycreeper Genus *Loxops* (Drepanididae). Ornithol Monogr. 1973;15:1–173.

Rico-Guevara A. Morphology and function of the drinking apparatus in hummingbirds [Ph.D. dissertation]. Storrs, CT: University of Connecticut; 2014.

Rico-Guevara A, Rubega MA. The hummingbird tongue is a fluid trap, not a capillary tube. Proc Natl Acad Sci U S A. 2011;108(23):9356–60.

Rico-Guevara A, Fan T-H, Rubega MA. Hummingbird tongues are elastic micropumps. Proc R Soc B. 2015;282(1813):1–8.

Riede T, Goller F. Peripheral mechanisms for vocal production in birds—differences and similarities to human speech and singing. Brain Lang. 2010a;115(1):69–80.

Riede T, Goller F. Functional morphology of the sound-generating labia in the syrinx of two songbird species. J Anat. 2010b;216(1):23–36.

Riede T, Goller F. Morphological basis for the evolution of acoustic diversity in oscine songbirds. Proc R Soc B. 2014;281:20132306. doi:10.1098/rspb.2013.2306.

Riede T, Suthers RA. Vocal tract motor patterns and resonance during constant frequency song: the white-throated sparrow. J Comp Physiol A. 2009;195:183–92.

Riede T, Schilling N, Goller F. The acoustic effect of vocal tract adjustments in zebra finches. J Comp Physiol A. 2013;199:57–69.

Riede T, Forstmeier W, Kempenaers B, Goller F. The functional morphology of male courtship dispalys in the Pectoral Sandpiper (*Calidris melanotos*). Auk. 2015;132(1):65–77.

Riede T, Beckers GJL, Blevins W, Suthers RA. Inflation of the esophagus and vocal tract filtering in ring doves. J Exp Biol. 2004;207(23):4025–36.

Riede T, Suthers RA, Fletcher NH, Blevins WE. Songbirds tune their vocal tract to the fundamental frequency of their song. Proc Natl Acad Sci U S A. 2006;103(14):5543–8.

Riede T, Eliason CM, Miller EH, Goller F, Clarke JA. Coos, booms, and hoots: the evolution of closed-mouth vocal behavior in birds. Evolution. 2016;70(8):1734–46.

Rubega MA, Obst BS. Surface-tension feeding in phalaropes: discovery of a novel feeding mechanism. Auk. 1993;110(2):169–78.

Sağsöz H, Erdoğan S, Akbalik ME. Histomorphological structure of the palate and histochemical profiles of the salivary palatine glands in the Chukar partridge (*Alectoris chukar*, Gray 1830). Acta Zool. 2013;94:382–91.

Salmons S. Muscle. In: Williams PL, editor. Gray's anatomy. 38th ed. New York: Churchill Livingston; 1995. p. 737–900.

Scarr G. Biotensegrity: the structural basis of life. Pencaitland, Scotland: Handspring Publishing; 2014.

Schmidt MF, Wild JM. The respiratory-vocal system of songbirds: anatomy, physiology, and neural control. Prog Brain Res. 2014;212:297–335. doi:10.1016/B978-0-444-63488-7.00015-X.

Schmidt-Nielsen K. Desert animals: physiological problems of heat and water. London: Oxford University Press; 1964.

Schummer A, Nickel R, Sack WO. The viscera of the domestic mammals. 2nd ed. Berlin: Verlag Paul Parey; 1979.

Schwenk K. An introduction to tetrapod feeding. In: Schwenk K, editor. Feeding: form, function, and evolution in tetrapod vertebrates. San Diego, CA: Academic; 2000. p. 21–61.

Scott JE, Hogue AS, Ravosa MJ. The adaptive significance of mandibular symphyseal fusion in mammals. J Evol Biol. 2012;25(4):661–73. doi:10.1111/j.1420-9101.2012.02457.x.

Shields RT. On the development of tendon sheaths. Contrib Embryol. 1923;25(73):53–61 & 1 plate.

Singh R, Kumar A, Lehana P. Investigations of the quality of speech imitated by Alexandrine Parrot (*Psittacula eupatria*). Circuits Syst Signal Process. 2016;35(11):2292–314. doi:10.1007/s00034-016-0395-3.

Shufeldt RW. The myology of the raven (*Corvus corax sinuatus*): a guide to the study of the musculature system in birds. London: Macmillan; 1890.

Skieresz-Szewczyk K, Jackowiak H. Morphofunctional study of the tongue in the domestic duck (*Anas platyrhynchos f. domestica*, Anatidae): LM and SEM study. Zoomorphology. 2016:135–255. doi:10.1007/s00435-016-0302-2.

Soons J, Herrel A, Genbrugge A, Adriaens D, Aerts P, Dirckx J. Multi-layered bird beaks: a finite-element approach towards the role of keratin in stress dissipation. J R Soc Interface. 2012;9(73):1787–96. doi:10.1098/rsif.2011.0910.

Starck JM. Comparative anatomy of the external middle ear of palaeognathous birds. Adv Anat Embryol Cell Biol. 1995;131:1–137.

Susi FR. Keratinization in the mucosa of the ventral surface of the chicken tongue. J Anat. 1969;105(3):477–86.

Suthers RA. Peripheral vocal mechanisms in birds: are songbirds special? Neth J Zool. 2001;51(2):217–42.

Suthers RA. Vocal mechanisms in birds and bats: a comparative view. An Acad Bras Cienc. 2004;76(2):247–52.

Suthers RA, Goller F. Motor correlations of vocal diversity in songbirds. In: Nolan V, Ketterson ED, Thompson CF, editors. Current ornithology, vol. 14. New York, NY: Plenum Press; 1997. p. 235–88.

Suthers RA, Zollinger SA. From brain to song: the vocal organ and vocal tract. In: Zeigler HP, Marler P, editors. Neuroscience of birdsong. Cambridge, NY: Cambridge University; 2008. p. 78–98.

Suthers RA, Rothgerber JR, Jensen KK. Lingual articulation in songbirds. J Exp Biol. 2016;219(4):491–500.

Symonds MRE, Tattersall GJ. Geographical variation in bill size across birds species provides evidence for Allen's rule. Am Nat. 2010;176(2):188–97.

Tachibana T, Tsutsui K. Neuropeptide control of feeding behavior in birds and its difference with mammals. Front Neurosci. 2016;10:485. doi:10.3389/fnins.2016.00485.

Tattersall GJ, Andrade DV, Abe AS. Heat exchange from the toucan bill reveals a controllable vascular thermal radiator. Science. 2009;325(5939):468–70. doi:10.1126/science.1175553.

Thomas DH, Robin AP. Comparative studies of thermoregulatory and osmoregulatory behaviour and physiology of five species of sandgrouse (Aves: Pterocliidae) (sic!) in Morocco. J Zool (Lond). 1977;183:229–49.

Thorpe WH. Talking birds and the mode of action of the vocal apparatus of birds. Proc Zool Soc London. 1959;132(3):441–55.

Titze IR. Principles of voice production. Englewood Cliffs, NJ: Prentice Hall, Inc; 1994.

Tomlinson CAB. Feeding in paleognathous birds. In: Schwenk K, editor. Feeding: form, function, and evolution in tetrapod vertebrates. San Diego, CA: Academic Press; 2000. p. 395–4.

Trunov VL, Korzun LP, Dzerzhinsky FYA. Morphofunctional features of feeding adaptations in barbets (Megalaima, Capitonidae). Bull Moscow Nat Chall. 1996;101(5):39–49. [in Russian]

Turnbull R. Cormorant gular flutter. 2015. YouTube. https://www.youtube.com/watch?v=ge1U1beHwAE. Accessed 24 Oct 2016.

Vanden Berge JC, Zweers GA. Myologia. In: Baumel JJ, King AS, Breazile JE, Evans HE, Vanden-Berge J, editors. Handbook of avian anatomy: Nomina Anatomica Avium. 2nd ed. Cambridge, MA: Publications of the Nuttall Ornithological Club No. 123; 1993. p. 301–27.

van den Heuvel WF. Kinetics of the skull in the chicken (*Gallus gallus domesticus*). Neth J Zool. 1991;42 (4):561–82.

van den Heuvel JW, Berkhoudt H. Pecking in the chicken (*Gallus gallus domesticus*): motion analysis and sterotypy. Neth J Zool. 1998;48(3):273–303.

van der Leeuw AHJ, Kurk K, Snelderwaard PC, Bout RG, Berkhoudt H. Conflicting demands on the trophic system of Anseriformes and their evolutionary implications. Anim Biol. 2003;53(3):259–301.

van Gennip EMSJ. The osteology, arthrology and myology of the jaw apparatus of the pigeon (*Columba livia* L.). Neth J Zool. 1986;36(1):1–46.

van Hemert C, Handel CM, Blake JE, Swor RM, O'Hara TM. Microanatomy of passerine hard-cornified tissues: beak and claw structure of the black-capped chickadee (*Poecile atricapillus*). J Morphol. 2012;273(2):226–40.

Vogel S. Comparative biomechanics: life's physical world. 2nd ed. Princeton, NJ: Princeton University Press; 2013.

Walker WF, Homberger DG. A study of the cat, with references to human beings. 5th ed. Philadelphia, PA: Saunders College Publishing; 1993.

Walsh MT, Mays MC. Clinical manifestations of cervicocephalic air sacs in psittacines. Comp Cont Educ. 1984;6(9):783–9.

Warren DK, Pepperberg IM. Cineradiographic analysis of the mechanisms of vowel production in an African Grey Parrot, *Psittacus erithacus*. Am Zool. 1993;33 (5):107A.

Warren DK, Patterson DK, Pepperberg IM. Mechanisms of American English vowel production in a Grey Parrot (*Psittacus erithacus*). Auk. 1996;113(1):41–58.

Weathers WW, Caccamise DF. Temperature regulation and water requirements of the Monk Parakeet, *Myiopsitta monachus*. Oecologia. 1975;18:329–42.

Weathers WW, Schoenbaechler DC. Regulation of body temperature in the budgerygah, *Melopsittacus undulatus*. Aust J Zool. 1976;24:39–47.

Westneat MW, Long JH, Hoese W, Nowicki S. Kinematics of birdsong: functional correlation of cranial movements and acoustic features in sparrows. J Exp Biol. 1993;182(1):147–71.

White SS. Movements of the larynx during crowing in the domestic cock. J Anat. 1968;103(2):390–2.

White SS. Mechanisms involved in deglutition in *Gallus domesticus*. J Anat. 1969;104(1):177.

White SS. Larynx. In: King AS, editor. Aves respiratory system, p. 1891–1997. In: Getty R, editor. Sisson's and Grossman's the anatomy of the domestic animals, 5th ed. Philadelphia: W.B. Saunders; 1975. p. 1883–1918.

White SS, Chubb JC. The muscles and movements of the larynx of *Gallus domesticus*. J Anat. 1968;102 (3):575.

Wild JM. Descending projections of the songbird nucleus robustus archistriatalis. J Comp Neurol. 1993;338 (2):225–41.

Wild JM, Krützfeldt NOE. Trigeminal and telencephalic projections to jaw and other upper vocal tract premotor neurons in songbirds: sensorimotor circuitry for beak movements during singing. J Comp Neurol. 2012;520(3):590–605.

Williams H. Choreography of song, dance and beak movements in the zebra finch (*Taeniopygia guttata*). J Exp Biol. 2001;204(20):3497–506.

Yap F. Savanna nightjar—gular fluttering. 2011. YouTube. http://www.besgroup.org/2011/09/01/savanna-nightjar-gular-fluttering/. Accessed 24 Oct 2016.

Zhang G, Li C, Li B, Larkin DM, Lee C, Storz JF, Antunes A, Greenwold MJ, Meredith RW, Ödeen A, Cui J, Zhou Q, Xu L, Pan H, Wang Z, Jin L, Zhang P, Hu H, Yang W, Hu J, Xiao J, Yang Z, Liu Y, Xie Q, Yu H, Lian J, Wen P, Zhang F, Li H, Zeng Y, Xiong Z, Liu S, Zhou L, Huang Z, An N, Wang J, Zheng Q, Xiong Y, Wang G, Wang B, Wang J, Fan Y, da Fonseca RR, Alfaro-Núñez A, Schubert M, Orlando L, Mourier T, Howard JT, Ganapathy G, Pfenning AR, Whitney O, Rivas MV, Hara E, Smith J, Farré M, Narayan J, Slavov G, Romanov MV, Borges R, Machado JP, Khan I, Springer MS, Gatesy J, Hoffmann FG, Opazo JC, Håsted O, Sawyer RH, Kim H, Kim K-W, Kim HJ, Cho S, Li N, Huang Y, Bruford MW, Zhan X, Dixon A, Bertelsen MF, Derryberry EP, Warren W, Wilson RK, Li S, Ray DA, Green RE, O'Brien SJ, Griffin D, Johnson WE, Haussler D, Ryder OA, Willerslev E, Graves GR, Alström P, Fjeldså J, Mindell DP, Edwards SV, Braun EL, Rahbek C, Burt DW, Houde P, Zhang Y, Yang H, Wang J, Jarvis ED, Gilbert MTP, Wang J. Comparative genomics reveals insights into avian genome evolution and adaptation. Science. 2014;346 (6215):1311–20.

Ziswiler V. Zur Kenntnis des Samenöffnens und der Struktur des hörnernen Gaumens bei körnerfressenden Oscines. J Ornithol. 1965;106(1):1–48.

Ziswiler V. Adaptive Radiation innerhalb der Prachtfinkengattung *Erythrura* Swainson. Rev Suisse Zool. 1969;76:1095–105.

Ziswiler V. Zungenfunktionen und Zungenversteifung bei granivoren Singvögeln. Rev Suisse Zool. 1979;86 (4):823–31.

Ziswiler V, Trnka V. Tastkörperchen im Schlundbereich der Vögel. Rev Suisse Zool. 1972;79(Suppl):307–18.

Ziswiler V, Güttinger HR, Bregulla H. Monographie der Gattung *Erythrura* Swainson, 1837 (Aves, Passeres, Estrildidae). Bonn Zool Monogr. 1972;2:1–158.

Zubkova EN, Korzun LP. Morphofunctional aspects of the trophic specialization of the frugivorous green broadbill *Calyptomena viridis* (Passeriformes, Eurylaimidae): a comparative analysis. Biol Bull. 2014;41(9):788–800.

Zusi RL. Patterns of diversity in the avian skull. In: Hanken J, Hall BK, editors. The skull, Patterns of structural and systematic diversity, vol. 2. Chicago: University of Chicago Press; 1993. p. 259–437.

Zweers GA. Structure, movement, and myography of the feeding apparatus of the mallard (*Anas platyrhynchos* L.): a study in functional anatomy. Neth J Zool. 1974;24(4):323–467.

Zweers G. The feeding system of the pigeon (*Columba livia* L.). Adv Anat Embryol Cell Biol. 1982a;73:1–108.

Zweers GA. Pecking of the pigeon (*Columba livia* L.). Behavior. 1982b;81(2–4):173–230.

Zweers GA. Drinking of the pigeon (*Columba livia* L.). Behaviour. 1982c;80(3–4):274–317.

Zweers GA. Behavioural mechanisms of avian drinking. Neth J Zool. 1992;42(1):60–84.

Zweers GA, Berkhoudt H. Larynx and pharynx of crows (*Corvus corone* L. and *C. monedula* L., Passeriformes: Corvidae). Neth J Zool. 1987;37(3–4):365–93.

Zweeers GA, Berkhoudt H, Vanden Berge JC. Behavioral mechanisms of avian feeding. In: Bels VL, Chardon M, Vandewalla P, editors. Comparative and environmental physiology, vol 18: Biomechanics of feeding in vertebrates. Berlin: Springer; 1994. p. 241–79.

Zweers GA, Gerritsen AFC, van Kraanenburg-Voogd HJ. Mechanics of feeding of the mallard (*Anas platyrhynchos* L.; Aves, Anseriformes): the lingual apparatus and the suction-pressure pump mechanism of straining. In: Hecht MK, Szalay FS, editors. Contributions to vertebrate evolution, vol. 3. Basel: S. Karger; 1977. p. 1–109.

Zweers GA, van Pelt HC, Beckers A. Morphology and mechanics of the larynx of the pigeon (*Columba livia* L.): a drill-chuck system (Aves). Zoomorphology. 1981;99(1):37–96.

Zweers GA, de Jong F, Berkhoudt H, Vanden Berge JC. Filter feeding in flamingos (*Phoenicopterus ruber*). Condor. 1995;97(2):297–324.

Pulmonary Transformations of Vertebrates 3

C.G. Farmer

Abstract
The structure of the lung subserves its function, which is primarily gas exchange, and selection for expanded capacities for gas exchange is self-evident in the great diversity of pulmonary morphologies observed in different vertebrate lineages. However, expansion of aerobic capacities does not explain all of this diversity, leaving the functional underpinnings of some of the most fascinating transformations of the vertebrate lung unknown. One of these transformations is the evolution of highly branched conducting airways, particularly those of birds and mammals. Birds have an extraordinarily complex circuit of airways through which air flows in the same direction during both inspiration and expiration, unidirectional flow. Mammals also have an elaborate system of conducting airways; however, the tubes arborize rather than form a circuit, and airflow is tidal along the branches of the bronchial tree. The discovery of unidirectional airflow in crocodilians and lizards indicates that several inveterate hypotheses for the selective drivers of this trait cannot be correct. Neither endothermy nor athleticism drove the evolution of unidirectional flow. These discoveries open an uncharted area for research into selective underpinning of unidirectional airflow.

Keywords
Unidirectional flow • Bronchial tree • Respiration • Evolution • Vertebrate • Lung • Bird • Mammal • Reptile

Contents

3.1 Introduction

False facts are highly injurious to the progress of science, for they often endure long; but false views, if supported by some evidence, do little harm, for everyone takes a salutary pleasure in proving their falseness: and when this is done, one path towards

C.G. Farmer (✉)
Trinity College Dublin, University of Dublin, Dublin 2, Ireland

Department of Biology, University of Utah, Salt Lake City, UT, USA
e-mail: cg.frmr@gmail.com

J.N. Maina (ed.), *The Biology of the Avian Respiratory System*, DOI 10.1007/978-3-319-44153-5_3

error is closed and the road to truth is often at the same time opened. (Darwin 1871)

Many of the great transformations in the vertebrate story have been revealed through the study of the fossil record and by functional and developmental analyses of living vertebrates. Unraveling how the trunk, the head and neck, and appendages such as fins, wings, and limbs changed, as walking tetrapods evolved from swimming fish and flying birds evolved from running dinosaurs, has been possible using these approaches. Unlike the skeletal system, many of the major transformations of the vertebrate lung remain shrouded in mystery because it leaves little trace in the fossil record. There is general agreement that the vertebrate lung is an ancient organ, but there is little consensus on just how old it is, how and why it originated, or the importance of various selective factors driving its transformations as vertebrates radiated. Yet comparisons of lung structure in extant lineages reveal that this organ has undergone extraordinary modifications. The central function of the lung is gas exchange between air and blood, and so one might expect little structural diversity that is not directly related to differences in the capacity of various lineages for gas exchange, yet the lung is one of the most diverse organs of vertebrates, and this diversity is not necessarily correlated with differences in aerobic capacity. For example, the volume of the lungs with respect to body weight varies tremendously. In European chameleons, the lung is approximately 125 ml for 100 g body weight, occupying an enormous portion (about 54%) of the body, whereas in small mammals (mouse, rat, and rabbit), the lung volume is 6.4 ml for 100 g body weight and occupies just 6.3% of the body (Perry and Duncker 1978). Some vertebrates have completely lost this organ. The ancestors of plethodontid salamanders had lungs, but this lineage subsequently came to rely instead on gas exchange across the skin and buccopharyngeal region (Whitford and Hutchison 1965). Reduction or loss of one lung has evolved repeatedly in many, but not all, limbless tetrapods (Farmer 2011 and references therein; Van Wallach 1998). In many snakes, Gymnophiona, and most limbless lizards, there is a reduction in the left lung, but in amphisbaenids there is a smaller right lung than left (Butler 1895; Van Wallach 1998). In some animals, the lung has taken the opposite path from the reduction seen in limbless tetrapods and has expanded beyond the thoracoabdominal cavity to enter the bones of the axial and even the appendicular skeleton. One of the most striking examples of this phenomenon, known as pneumatization, is seen in the butterfly fish (*Pantodon buchholzi*) (Nysten 1962) (Fig. 3.1), but it is also common in birds.

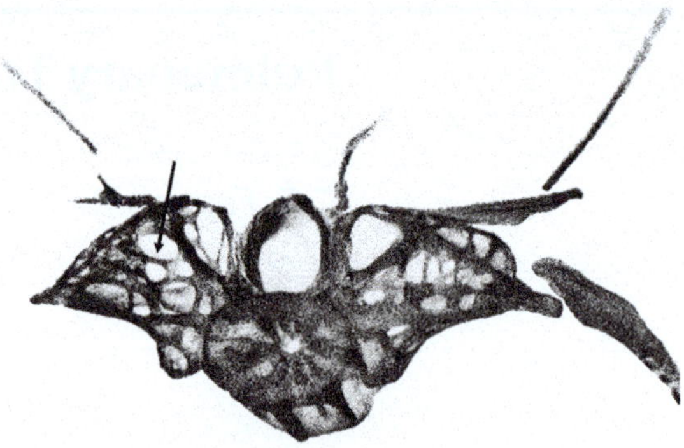

Fig. 3.1 Vertebra from the butterfly fish, *Pantodon buchholzi*, showing extensive pneumatization. *Arrow* points to a pneumatic space (from Nysten 1962)

The biomechanics of the musculoskeletal systems that ventilate the lung also differ. For example, a buccal pump is used to push air into the lungs of fishes and amphibians, while amniotes use costal inspiration (Brainerd and Owerkowicz 2006). However, a similar positive pressure gular pump supplements costal ventilation in many lizards (Brainerd and Owerkowicz 2006). Crocodilians supplement costal ventilation by employing their gastralia (Farmer and Carrier 2000) and their diaphragmaticus muscle-liver complex (Gans and Clark 1976), birds use dorsoventral movements of the sternum (Brackenbury 1987; Duncker 1971), and mammals supplement costal ventilation with the diaphragm.

The internal organization of the lung varies as much between lineages as the external architecture (Duncker 1978; Milani 1894, 1897; Perry 1983). The distribution of tissues where gas exchange occurs is relatively uniform in mammals and many amphibians but highly concentrated, generally in a dorsocranial region, in other animals, such as birds and snakes, while the remainder of their respiratory system is relatively devoid of blood vessels (Maina 2006; Maina et al. 1989, 1999). The topography of the conducting airways is also highly variable. In some lineages, such as amphibians and the tuatara, the primary bronchus terminates at the lung hilum (Fig. 3.2), and there is neither an internal network of tubes nor significant septation forming distinct chambers; the lung parenchyma

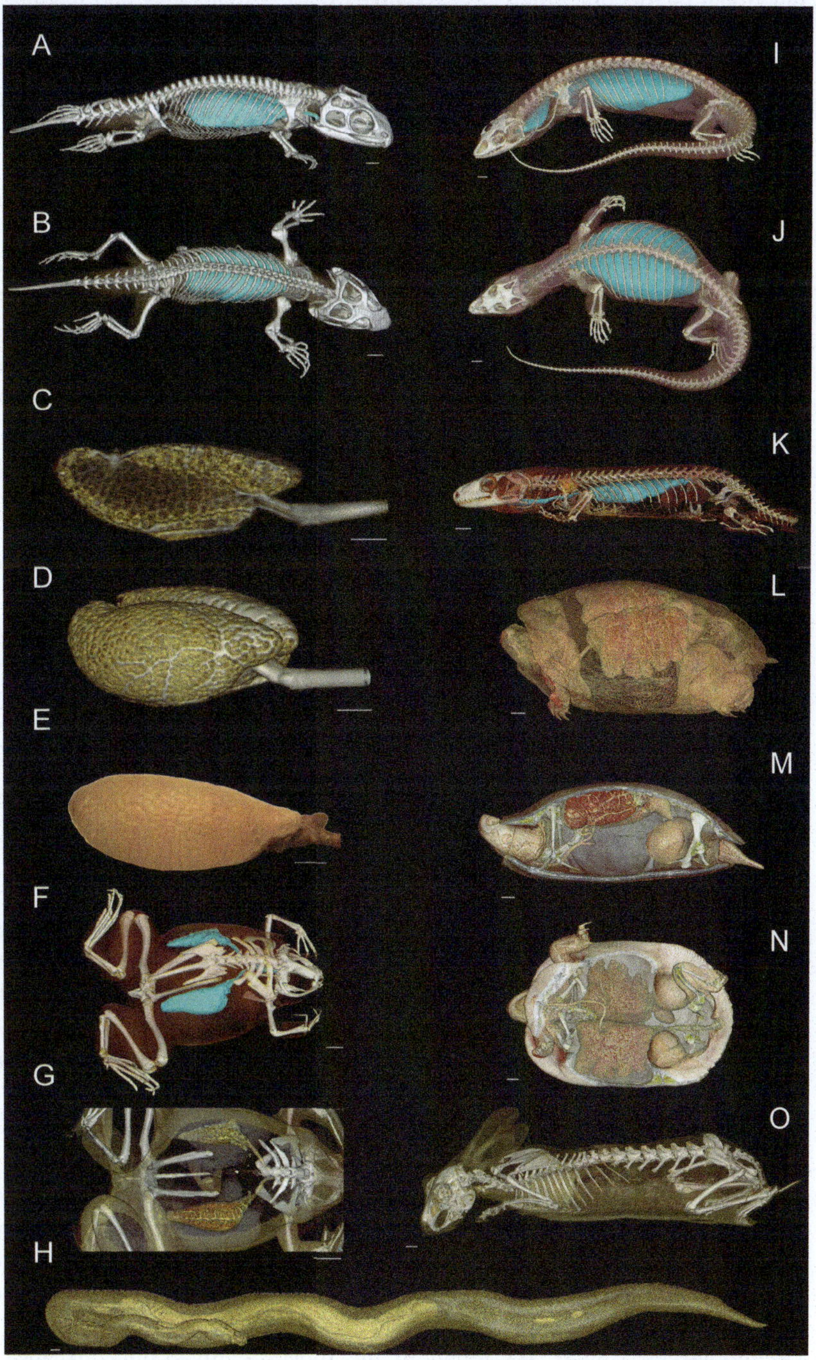

Fig. 3.2 Computed tomography data of lungs in tetrapods. (**a–e**) Tuatara. (**a**) Lateral and (**b**) dorsal views showing large, inflated lungs (*blue*); (**c**) lateral view of lung with parasagittal cut; (**d**) dorsocranial view of both lungs; (**e**) photograph in lateral view. (**f–g**) African clawed frog. (**f**) Dorsal view; (**g**) dorsal view with coronal cut to show the lung lumen. (**h**) dorsal view of a ball python. (**i–j**) Savannah monitor. (**i**) Dorsolateral and (**j**) dorsal views. (**k**) Bearded dragon in lateral view. (**l–n**) *Red-eared slider*. (**l**) Dorsolateral view; (**m**) lateral with parasagittal cut; (**n**) ventral view. (**o**) Lateral view of a rabbit. Scans were made on live unanesthetized (frog, turtle, bearded dragon, python) and anesthetized (rabbit) animals as well as specimens with artificially inflated lungs (tuatara, monitor). *Scale bar* = 1 cm

simply arises directly from the visceral pleura in wide, polygonal structures. In stark contrast, in birds and mammals there is an extensive conducting network.

Such large differences in pulmonary architecture raise cardinal questions about the evolution of this organ. Can this diversity be explained by functional factors other than gas exchange? If so, what are they? Did key environmental factors, such as changing levels of atmospheric oxygen and carbon dioxide over time, give rise to this diversity? Have the forces associated with the mechanics of breathing influenced the internal design of the lungs? Are there unique aspects to the development of the lung that generate a greater range of morphologies than are found in other organs and that underpin the exceptional diversity of this organ? Thus, the lung presents a very exciting opportunity to an evolutionary biologist aiming to understand the unfolding of this diversity, from the ancient fishes that first evolved lungs as a supplement to respiration with gills to the highly sophisticated lungs of extant birds and mammals.

3.2 When Did the Vertebrate Lung Evolve?

Biologists of the 1800s used the lung as a character that distinguished fish from amphibians. Accordingly, when the lungfish were discovered, they were initially classified as amphibians. For example, in an early description of the Australian lungfish, Gerard Krefft wrote to the Zoological Society of London (Krefft 1870) describing a "gigantic Amphibian allied to the Genus *Lepidosiren*." It is now clear that these animals should be classified as fishes and that most of the basal living bony fishes have a lung, even if the organ is no longer used for air breathing. For example, adult coelacanths have a vestigial lung, but a sizable lung, with bony plates as covering, is found in early embryos (Cupello et al. 2015). Similar plates can be detected in the fossil remains of many coelacanths (Cupello et al. 2015), leaving a hard-part correlate for the presence of lungs. Among the ray-finned fishes (Actinopterygii), lungs are found in the most primitive family, the Polypteridae (reedfish, bichirs), as well as in the Neopterygii (gar, bowfin). It was suggested by Sagemehl (Sagemehl 1885) that the single, dorsal gas bladder that is common in teleosts evolved gradually from paired, ventral lungs. However, differences in development have been proposed to indicate that lungs and gas bladders are not homologous organs (Hsia et al. 2013). More recently, patterns of expression of cassettes of genes in the gas bladder of zebrafish and in lungs (Cass et al. 2013), as well as the presence of vestigial pulmonary arteries in *Acipenser* and *Polyodon*, two primitive fishes that do not breathe air (Longo et al. 2013), shore up the homologies of these organs and corroborate the scenario that lungs were present at least as early as the first bony fishes (Liem 1989), and their descendants either inherited a lung or descended from lineages in which the lung underwent a transition to a gas bladder.

Thus, early fishes evolved an organ that will later be critical to one of the greatest transformations of vertebrates, the transition from an aquatic lifestyle to a terrestrial one, at least 100 million years before this transition took place (Boucot and Janis 1983). The fact that lungs are not an adaptation for a terrestrial existence, but initially served some other function, means that they are a remarkable example of an exaptation. Furthermore, their origin in this distant past is perplexing because the early bony fishes were marine (Boucot and Janis 1983), and an inveterate view of the selective pressure driving lung evolution is aquatic hypoxia, which is more commonly encountered in freshwater habitats, such as the Amazon River basin, than in saltwater ecosystems, which are stirred by tides and waves (Kramer et al. 1978). Thus, their ancient origin raises questions of the initial selective pressure for the origin of lungs in marine environments. Other aspects of the traditional paradigm are also in need of scrutiny. For example, although the African and South American lungfishes were once believed to be good analogues for understanding the evolution of aerial respiration, it has been argued that these are not good models for the first fishes that evolved lungs because their gill filaments are so reduced that they are obligate air breathers (Farmer 1997 and references therein). These fish drown if not allowed to breathe air! This cannot have been the initial condition for the first fishes that evolved lungs. Better analogues are the Australian lungfish, the Polypteridae, and the Neopterygii. In all of these groups, activity is a strong stimulus to breathe air, even when the fish are swimming in well-oxygenated water and could, presumably, choose to obtain the requisite oxygen with the gills (Farmer 1997). The answer may lie not in the level of oxygen available in the water as much as the level of oxygen available to the heart. The ventricles of most fishes have little or no coronary circulation and obtain oxygen from blood that is contained within the lumen of the heart. The level of oxygen in this blood is increased when the fish breathe with their lungs. The lungs, therefore, may have been favored as a mechanism to expand aerobic capacities and may have been particularly important in increasing the power of the heart (Farmer 1997).

3.3 Transformations in the Vertebrate Lung

The transition to a terrestrial lifestyle created new selective pressures on the respiratory system. In the clade that became most highly specialized for terrestriality, the amniotes, the lung is the primary organ for gas exchange and gills, skin, and other organs contribute minimally or not at all. There are a few exceptions, such as the use of cloacal bursae for aquatic respiration by some turtles (Mathie and Franklin 2006), but the majority of amniotes rely on the lungs to exchange both carbon dioxide and oxygen between blood and gases contained within the lungs. An expanded capacity for gas exchange is needed for terrestrial locomotion because the metabolic cost of transport is considerably greater for walking tetrapods compared to swimming fish. That is, for an animal to move a gram of mass a given distance, the cost is higher in terrestrial than in aquatic vertebrates. Furthermore, for an animal to be able to sustain this extra cost, it must provide the energy through aerobic metabolic pathways. Another important selective factor driving the design of the lungs is that the power required for locomotion, that is, the energetic cost for a given unit of time, is a function of locomotor mode. For example, flapping flight has a greater power requirement than either terrestrial or aquatic locomotion. Thus, selection on the lungs for expanded capacity will be greatest for these flying vertebrates (Maina 2000). Because pulmonary gas exchange occurs exclusively by diffusion, and rates of diffusion are proportional to the surface area through which the gases diffuse and inversely proportional to the distance over which diffusion takes place, it is possible to analyze morphometrically the capacity of the lung for gas exchange, the anatomical diffusion factor, which is the ratio of the surface area for diffusion divided by the harmonic mean thickness of the blood-gas barrier. Assessment of anatomical diffusion factor reveals that mode of locomotion coevolves with lung capacity (Maina 1998, 2000).

As gas-exchange surface area within the lung increased, an effective mechanism was needed to bring gases into contact with the gas-exchange parenchyma. This presumably led to the evolution of the extensive network of conducting bronchi of mammals and birds. A less extensive system of airways has evolved in crocodilians, chelonians, and some lizards. The topography of these networks influences patterns of airflow. It is therefore important to the efficacy of gas exchange, as well as in how inhaled particulate matter is deposited, and in the metabolic cost of breathing. The effects of these networks on the flow of lung gases have been studied largely in the mammalian and avian respiratory systems, both empirically and using either physical or computational models, but are poorly understood in other lineages.

3.4 The Conducting Airways of the Bronchoalveolar Lung

The amniote trachea is a fibromuscular tube supported by cartilaginous rings. In amniotes with two lungs, the trachea bifurcates to form two primary bronchi that carry air to and from each lung. In mammals, the tubes continue to arborize within the lungs, with the shape of the larger intrapulmonary bronchi being preserved by cartilaginous rings and plates. The cartilage gradually disappears as the tubes become smaller, and by the time the tubes are about 1 mm in diameter, they are no longer supported by cartilage and are termed bronchioles (Weibel 1963, 1984). Bronchioles continue to arborize until they terminate in the gas-exchange units, the alveoli, which are heavily invested with blood capillaries—in humans there are approximately 1000 pulmonary capillaries for each alveolus—and this massive investment of blood capillaries creates an enormous gas-exchange surface area. A tendency for the alveoli and bronchioles to collapse in the absence of the cartilaginous support that is present in the more proximal airways is countered by elastic tissues in the walls, or septa, of the alveoli. A honeycombed topography of the alveoli enables the septa to mechanically interact in such a way to help keep the alveoli open. Although there is some variation in the branching patterns of the conducting airways of different mammalian lineages (Fig. 3.3) (Montiero and Smith 2014), the structure of the lungs is sufficiently similar that a bronchoalveolar lung is generally believed to have been present in the common ancestor of extant mammals, providing a minimum date for the origin of the organ in the Triassic period (reviewed in Farmer 2015a).

The diameters, taper, and branching patterns of the airways play an important role in the flow of lung gases, for example, influencing whether there is laminar or turbulent flow, as well as the resistance to airflow (Weibel 1963; West 1995). Laminar flow through a tube consists of a series of concentric, cylindrical layers of flow, with the layer nearest the wall moving most slowly, due to frictional forces with the wall, and the central cylinders most rapidly. With laminar flow, the resistance is determined by the viscosity of the fluid, the length of the tube, and is inversely proportional to the radius of the tube to the fourth power, if the walls of the tube are smooth, in the well-known Poiseuille's law. The resistance is also dependent on the pressure difference driving flow and inversely related to the velocity of the flow. If the flow is turbulent, where unsteady vortices appear chaotically and interact with each other, the resistance to flow is proportional to the difference in pressure driving the flow, divided by the velocity squared. The resistance is more heavily influenced by the density of the fluid, rather than by the viscosity. Turbulent flow tends to occur with high velocities, high gas density, and large tube radii in a relationship known as the Reynolds (Re) number. Flow changes from laminar to turbulent when the Reynolds number exceeds 2000:

$$\mathrm{Re} = (\rho \times \mathrm{Ve} \times D)/\eta$$

where η = viscosity, ρ = density, Ve = linear velocity, and D = airway diameter.

In humans, airflow in the trachea and the larger airways is turbulent, but in the small-diameter airways, it is laminar. Thus, the Reynolds number is important in determining the resistance to flow in individual airways. Furthermore, the resistance throughout the airways depends on the topography, with resistance of tubes arranged in series being additive, but for tubes arranged in parallel, the reciprocal of the total resistance is the sum of the reciprocals of the resistance of each tube. Thus, many small-diameter tubes, when arranged in parallel, create a low total resistance in spite of Poiseuille's law, because flow in them tends to have a low Reynolds number and is laminar. For these reasons, the total resistance to flow in the smallest airways of the bronchoalveolar lung is very low, with the largest component of airway resistance occurring in the medium-sized bronchi (Weibel 1963, 1984). In mammals, airway resistance changes depending on the state of the lung, decreasing with increasing lung volume.

During idealized, normal ventilation, air leaves the lung by the same path it entered, tidal airflow. However, the patterns of flow are complicated, and under some conditions, biases can arise. Pendelluft ("swinging air"), or interregional airflow that arises from inhomogeneous inflation or deflation of the lungs, can result in significant exchange in volume at local bifurcations (Greenblatt et al. 2014). Pendelluft may be a relatively important factor influencing

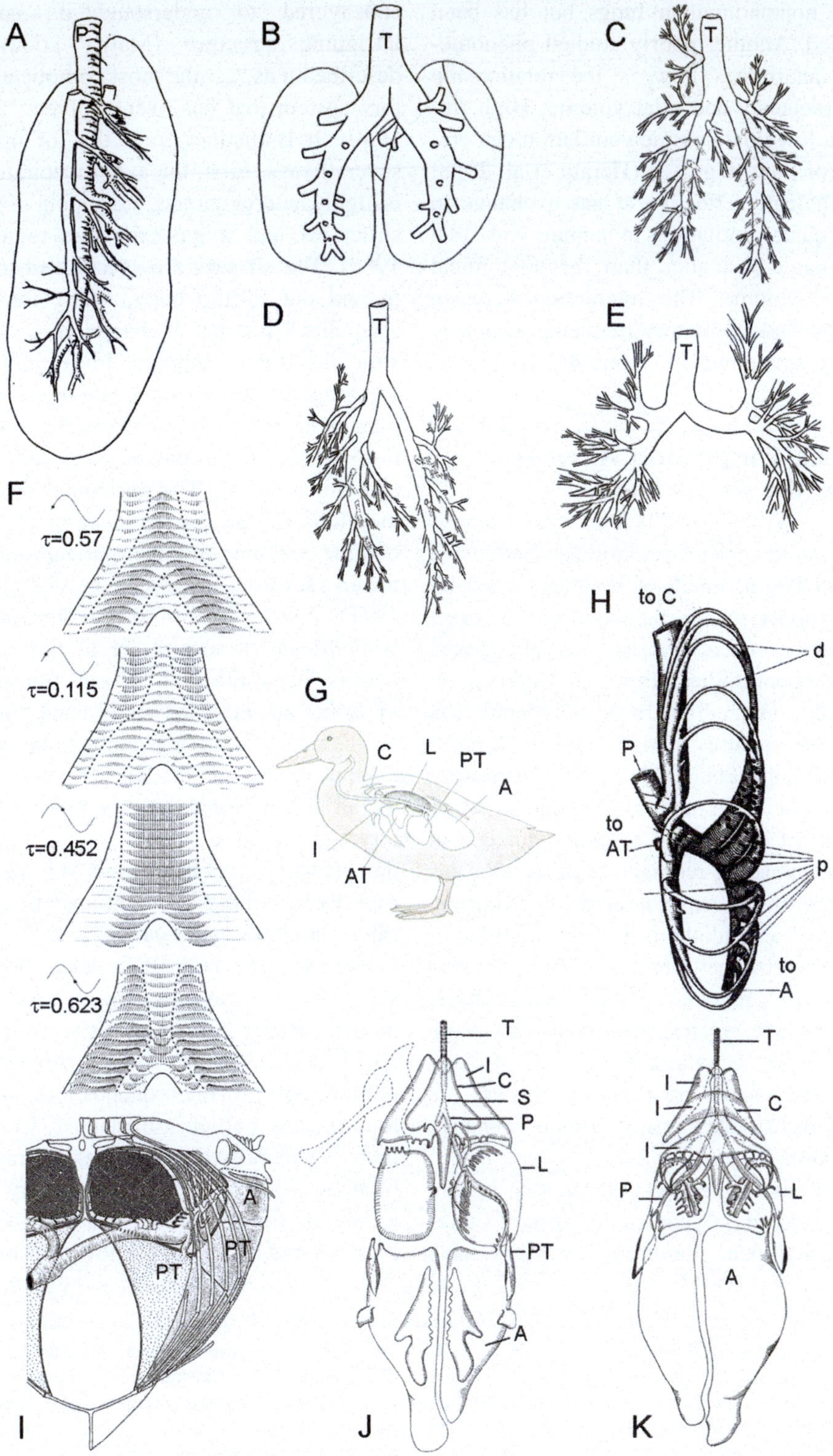

Fig. 3.3 Conducting airways and patterns of airflow in mammals and birds. Ventral view of mammalian bronchial tree: (**a**) the left lung of a dugong (after Pick 1907); (**b**) echidna (after Perry et al. 2000); and (**c**) pig, (**d**) rat, and (**e**) human (all after Montiero and Smith 2014). (**f**) Patterns of airflow with high-frequency ventilation during four phases of inspiration (positive trace) and expiration. Note the direction of flow differs in the center of the

airflow in nonmammalian lungs but has been little studied. Another poorly studied phenomenon that determines flow is the relationship between frequency and tidal volume. High frequency but low tidal volume ventilation can create biases of flow (Fig. 3.3) (Heraty et al. 2008) and is sufficiently effective at gas exchange to enable adequate ventilations in humans with tidal volumes that are smaller than the anatomical dead space volume. The interaction between tidal volume and respiratory frequency is poorly studied in nonmammalian respiratory systems.

3.5 The Conducting Airways of the Avian Lung

Birds have undergone a fascinating and extensive adaptive radiation, much of it underpinned by their ability to fly, which enabled birds to interact with their environment in a new way and opened ecological opportunities. Powered flight is an energetically demanding form of locomotion and therefore requires a high capacity for gas exchange (Maina 2000). Thus, the respiratory and cardiovascular systems, as well as other components of the oxygen cascade, changed in tandem to enable the requisite high rates of gas exchange with the environment. In addition, birds evolved an endothermic thermoregulatory strategy, whereby they produce sufficient internal heat to regulate body temperature. Endotherms have greater rates of gas exchange at rest than do ectothermic animals. Consequently, the avian respiratory system is renowned for its great capacity for gas exchange (Maina 2000).

Descriptions of the avian respiratory system date back hundreds of years to the work of Coiter (1573). Since then, numerous scientists have endeavored to understand its unique and intriguing structure (Maina 2002). Akester described it as "... the most complicated respiratory system that has ever evolved..." (Akester 1960). It is distinct from that of mammals in several ways. First, the avian respiratory system is highly heterogeneous, consisting of ventilatory structures and a gas-exchange region (Maina 1989). The air sacs are specialized to move air in and out of the body, while the lungs are specialized for gas exchange (Fig. 3.3). These structures lie in separate body compartments: the lungs in the *cavum pulmonale* and the air sacs in the *cavum subpulmonale*, with the exception of the abdominal air sacs that lie in the abdominal cavity. The horizontal septum forms the floor of the *cavum pulmonale*, while the oblique septum and the pericardium separate the air sacs from the rest of the viscera (Duncker 1971). There is a set of sacs in the cranial part of the bird and a second set in the caudal part. During inspiration, the air sacs are pulled open to bring air into the body, and the air sacs are compressed to expel air from the bird on expiration. Ostia in the horizontal septum enable the air sacs to connect to the rest of the respiratory system within the *cavum pulmonale*. Here, the primary bronchus enters the ventromedial aspect of the lung and continues to course dorsally, laterally, and caudally in a drawn-out S-shaped curve (Fig. 3.3). The abdominal air sacs arise from the distal ends of the intrapulmonary bronchi. A set of secondary bronchi, known as ventrobronchi, branch near the point of entry of the bronchus into the lung (the hilum), in a bottlebrush manner. More distally there is generally a region of the intrapulmonary bronchus that is free of secondary branches, known as the mesobronchus, after which arise more secondary airways, known as laterobronchi

Fig. 3.3 (continued) airways compared to near the walls (after Heraty et al. 2008). (**g**) Schematic of duck illustrating position of the air sacs (after Schmidt-Neilsen 1971); (**h**) dorsal view of the conducting airways in the right lung of a goose (after Brackenbury 1971); (**i**) cranial view of conducting airways, air sacs, and lungs of an idealized bird (after Duncker 1971); (**j**) dorsal and (**k**) ventral views of conducting airways, lungs, and air sacs of a chicken (after Akester 1960). *Air sacs* = *A* abdominal, *AT* anterior thoracic, *C* cervical, *I* interclavicular, *PT* posterior thoracic, *P* primary bronchus, *d* dorsobronchi, *L* lung, *p* parabronchi, and *T* trachea

and dorsobronchi (Fig. 3.3). The second tier of airways divides further to form smaller-diameter tubules, the parabronchi, which anastomose. Air capillaries radiate off the parabronchial lumina and are the site of gas exchange (Duncker 1971). Thus, the conducting airways form a circuit, analogous to the arteries, capillaries, and veins of the blood circulation, and like the flow of blood, air travels in a consistent direction through most of these airways during both phases of ventilation, unidirectional airflow.

Early studies and speculations about patterns of airflow in the avian lung attributed the fact that the flow was unidirectional to activity of physical valves (Bethe 1925; Wolf 1933); however, a lack of evidence for either sphincters or valve leaflets and the fact that the valve remains effective in dead animals eventually led to the conclusion that some sort of aerodynamic valve gives rise to unidirectional flow (Dotterweich 1936; Hazelhoff 1951) (reviewed in Butler et al. 1988). Although glass models demonstrated that topography alone can create unidirectional flow (Dotterweich 1936; Hazelhoff 1951), compared to these models, the topography of the avian conducting airways is complex, and questions remain regarding how the avian anatomy gives rise to one-way flow.

Other aspects of the avian respiratory system remain enigmatic. How, when, and why did the bird lung evolve? Did non-avian dinosaurs and pterosaurs have an avian-like respiratory system? Studies on the sister taxon of birds, the crocodilians, are providing insight into aspects of the respiratory system in avian forerunners.

3.6 Airflow in the Crocodilian Lung

> In the Crocodile ...and Alligator the bronchus enters the lung near its center, and passes somewhat obliquely into the lung until it reaches the junction of the lower and middle third; here it breaks up into eight or fifteen tubular passages. These tubular passages are studded with a great many air-sacs.... In these animals the lung for the first time gives a structure as it is found in Mammals. There are many air-sacs, which in turn communicate with a common cavity, or atrium, all of which communicate with a single terminal bronchus. A single lobule of the mammalian lung is simply enlarged to form the lung of the Crocodile; the lung of the former is only a conglomerate of that of the latter. (Miller 1893) p. 171
>
> Each bronchus is continued directly backwards into a wide canal, which dilates into an oval sac-like cavity at the posterior end of the lung, representing the mesobronchium with the posterior air-sac in birds. In the dorsal and mesial wall of the mesobronchium there are five or six apertures, which lead into as many canals, representing the entobronchia in birds. These pass, the anterior two almost directly forwards, and the others more or less obliquely, to the dorsal margin; and they lie quite superficially on the mesial face of the lung. The first is very much larger than the others, and ends in a dilatation at the anterior end of the lung. It is united with the second by transverse branches. Along the ventral margin of the lung there are four saclike chambers, which communicate, in the case of the two anterior, with the entobronchia, and, in the case of the two posterior, with the mesobronchium. Finally, there are two very large canals, external to these, which communicate with the mesobronchium by large apertures in its dorsal wall, and give off branches to the outer face of the lung, representing the ectobronchial system of birds. The orifices with which the surfaces of all these canals, except the anterior half of the mesobronchium, are thickly set, lead into depressions, which are often so deep as to become cylindrical passages, simulating the parabronchia of birds. Thus, notwithstanding all the points of difference, there is a fundamental resemblance between the respiratory organs of Birds and 'those of' Crocodiles pointing to some common form (doubtless exemplified by some of the extinct Dinosauria), of which both are modifications. p. 569 (Huxley 1882)

These scientists could not have arrived at more different conclusions regarding the structure of the crocodilian lung. Is it like an enlarged mammalian lobule or is it more like the lungs of birds? It is certainly distinct from that of birds in lacking air sacs. However, it is birdlike in that the conducting bronchi do not form a bronchial tree, but anastomose to form a circuit. In much of this network, airflow is unidirectional, flowing caudad in the airway that arises from the most proximal ostium of the primary bronchus (Fig. 3.4)

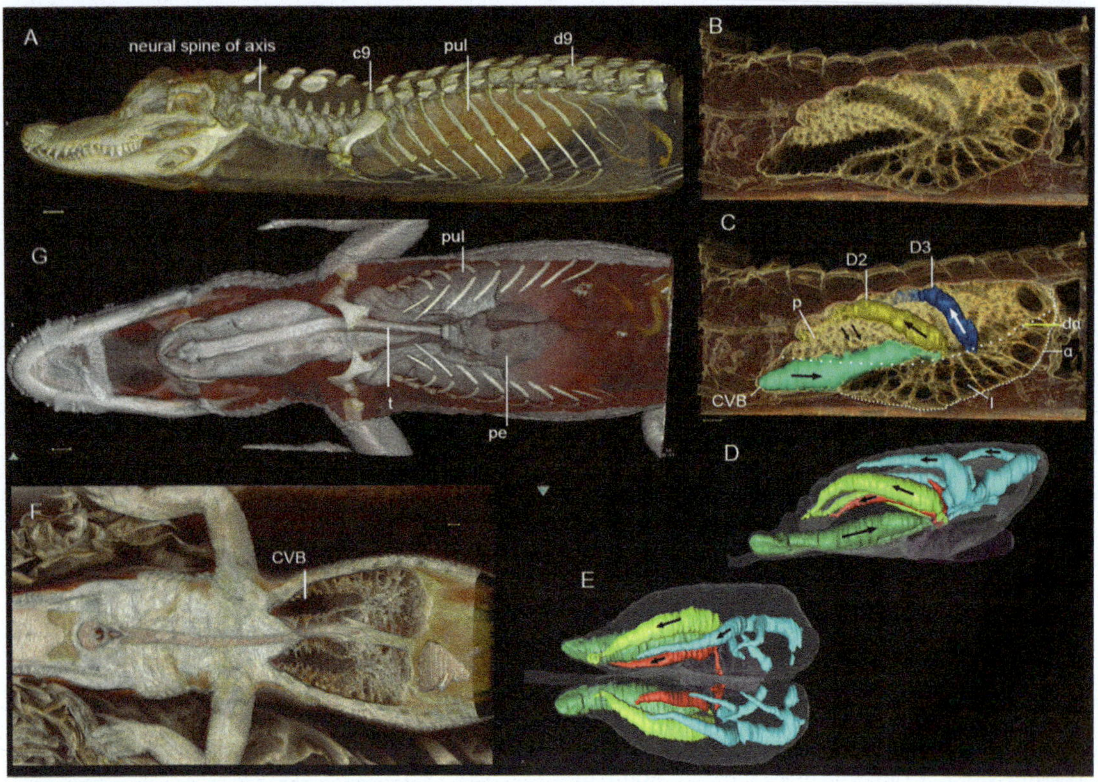

Fig. 3.4 Crocodilian anatomy and airflow. Computed tomography of American alligator in lateral (**a–d**) and ventral (**e–g**) view. **c–d** show voxels of the major airways in *color*. Air flows craniad in the *blue*, *red*, and *yellow airways* and caudad in the *green airway*, the cervical ventrobronchus (CVB). pul, lung; pe, pericardium. l, avascular locules hypothesized to be homologous to the avian air sacs; alpha, region hypothesized to be homologous to the oblique septum; da, region hypothesized to be homologous to the horizontal septum (after Farmer 2015b). scale bar, 1 cm

and flowing craniad in the other airways (Farmer 2010; Farmer and Sanders 2010). Closer inspection shows other features that are extremely bird-like. For example, the gas-exchange tissue is concentrated in the dorsal regions of the lungs, the ventral and caudal portions containing locules with few blood vessels (Fig. 3.4). It has been hypothesized that these locules are homologs of the avian air sacs (Farmer 2015b). Furthermore, the ostia for most of these locules lie on the floor of the intrapulmonary bronchus, in a location analogous to the horizontal septum of birds, which contain the ostia to the air sacs.

There are a number of implications for the discovery of unidirectional flow in crocodilians. First, because crocodilians and birds are crown-group archosaurs, unidirectional flow was probably present in pseudosuchians, pterosaurs, and non-avian dinosaurs. Second, avian style air sacs are not requisite for unidirectional flow since the simpler structures of crocodilians are sufficient to act as bellows. What, then, is the primary function of avian air sacs? These structures vary considerably between species, and investigations are wanted to shed light on the functional underpinnings of this variation. Third, the bones of crocodilians are not pneumatized. Furthermore, as pointed out previously, animals without air sacs show pneumaticity (Fig. 3.1). Therefore pneumaticity is an imperfect hard-part correlate for the presence of air sacs. Fourth, the presence of unidirectional flow in a semiaquatic, ectothermic, sit-and-wait predator suggests that it is not an adaptation for endothermy or for flight. Thus, alternative explanations should be sought for the functional benefits of unidirectional flow. Fifth, since unidirectional flow did not arise to support the high metabolic demands of flight or endothermy, it is possible unidirectional flow may be more widespread and could occur in other ectotherms, such as lizards, turtles, amphibians, and lung breathing fish.

3.7 Airflow in the Lepidosaur Lung

At least two species of lepidosaurs, with strikingly different lungs, have unidirectional airflow: the green iguana (Cieri et al. 2014) and the savannah monitor (Schachner et al. 2014). In both species, gases flow from caudal toward cranial during inspiration and expiration in large regions of the lungs. Iguana lungs are paired organs that occupy a large part of the body cavity when fully inflated (Fig. 3.5). The cranial end of the dorsocranial chamber underlies cervical vertebra 4 and overlies the heart extending to thoracic vertebra number 5 (T5). The cranial end of the ventrocaudal chamber abuts the heart, and the caudal end extends to approximately T10. A horizontally (coronally) oriented septum partitions the lung into a dorsocranial chamber and a ventrocaudal chamber. The intrapulmonary bronchus enters the lung ventromedially and, at the hilum, opens dorsally such that the ventral portion, still containing partial rings of cartilage, forms part of the floor of the horizontal septum. An ostium in the distal portion of this bronchus is the sole connection between the two chambers. The chambers are cavernous structures with a few partial septa and small niches along the walls and along the horizontal septum (Fig. 3.5). In both of the chambers, there is caudal-to-cranial flow during both phases of ventilation throughout much of the volume of the chambers. The mechanisms driving this unidirectional flow are unresolved. Convective momentum, whereby high-speed fluid streams are created when air is pulled through the primary bronchus during inspiration, may play a dominant role. These jets of air, one in the cranial chamber and one in the caudal chamber, widen as they enter the cavernous chambers, and as they move along, they entrain air on all sides of the stream. This process, known as the Coanda effect, decreases the pressure in between the jet and the walls of the lung, a very unstable situation. Eddies arising from the partitions in the walls or from the niches that are common along the horizontal septum and certain regions of the lung (Fig. 3.5) may serve to deflect the stream, where it can then lock onto a wall and keep flowing. If these internal structures serve such a function, one would expect them to be reasonably well conserved between lungs of different individuals, a testable hypothesis. Furthermore, the hypothesis that the Coanda effect plays a role in directing and creating unidirectional flow needs testing.

3.8 Future Directions

A number of areas are in need of more work. More studies are needed on the phylogenetic distribution, fluid dynamics mechanisms, effects on gas exchange, and functional significance of the patterns of airflow. Specifically, to be able to map changes through time using phylogenetic bracketing, data on patterns of airflow in amphibians, chelonians, as well as additional measurements on lepidosaurs with diverse lung morphologies are sorely needed. Insight into mechanisms for these patterns of flow will come from additional structure-function studies, as well as computational fluid dynamics models. Finally, it will be important to identify and to test key hypotheses for the effects on blood gases and on the functional significance of these patterns of flow.

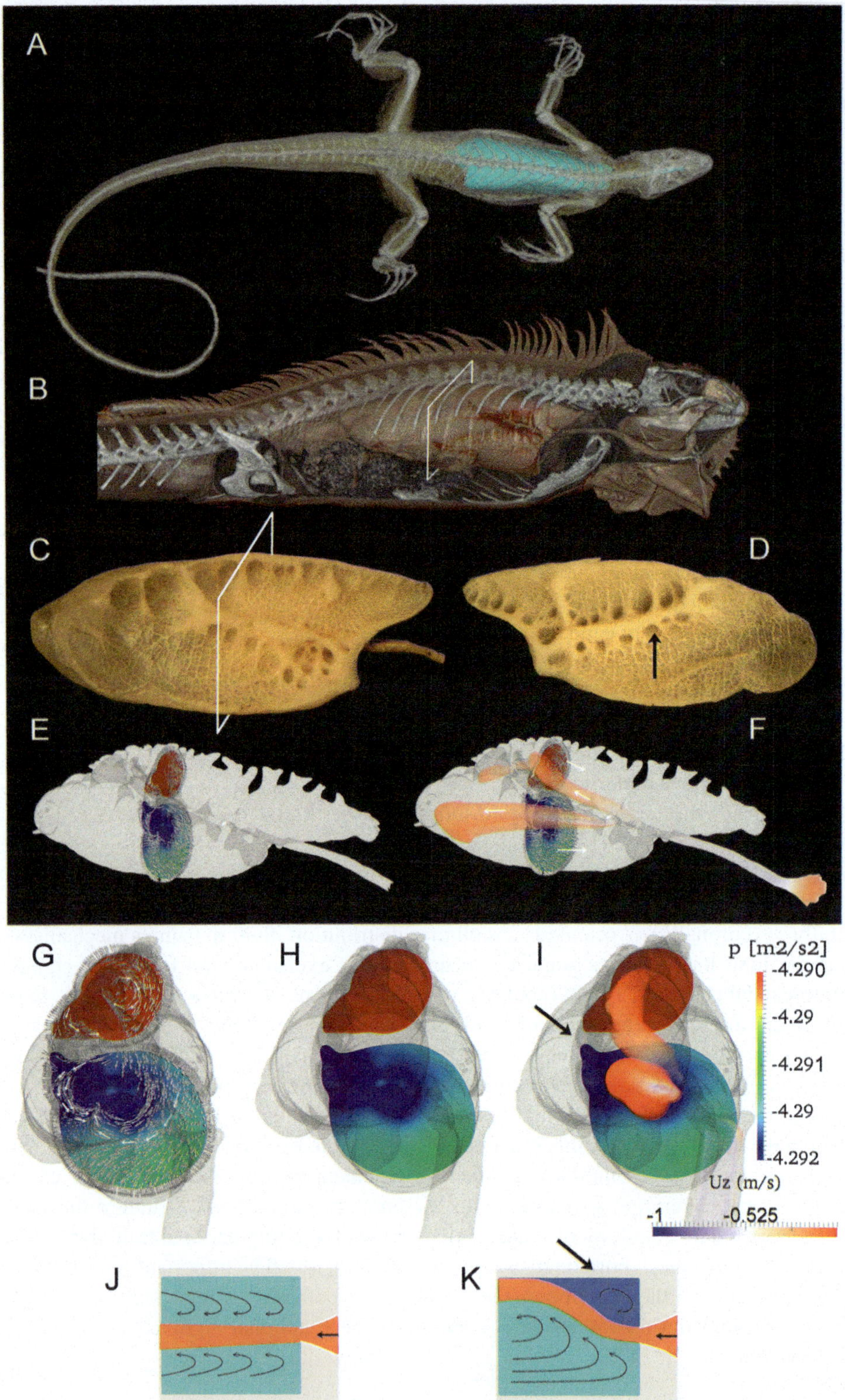

Fig. 3.5 Anatomy and patterns of airflow in iguana lungs. (**a**) Computed tomography data of a green iguana (*Iguana iguana*) in dorsal view with pulmonary voxels colored turquoise. (**b**) Lateral view of iguana showing trachea and right lung. *White square* shows region of the lung where pressure is illustrated in the computational

Acknowledgments I thank Brent Craven and Robert Cieri for modeling work and advice, and Robert Cieri, Scott Echols, Adam Huttenlocker, Jeremy Klingler, Nicola Nelson, and Brett Gartrell, for help obtaining CT data.

References

Akester A. The comparative anatomy of the respiratory pathways in the domestic fowl (Gallus domesticus), pigeon (*Columba livia*) and domestic duck (Anas platyrhyncha). J Anat. 1960;4:487–505.

Angrist SW. Fluid control devides. Sci Am. 1964;211:80–8.

Bethe A. Atmung: Allgemeines und Vergleichendes. In: Bethe A, Bergmann EG, editors. Handbuch der normals und pathologischen Physiology, vol. 2. Berlin: Springer; 1925. p. 1–36.

Boucot A, Janis C. Environment of the early Paleozoic vertebrates. Palaeogeogr Palaeoclimatol Palaeoecol. 1983;41:251–87.

Brackenbury JH. Airflow dynamics in the avian lung as determined by direct and indirect methods. Respir Physiol. 1971;13:318–29.

Brackenbury JH. Ventilation of the lung-air sac system. In: Seller TJ, editor. Bird respiration, vol. II. Boca Raton, FL: CRC Press; 1987. p. 39–69.

Brainerd EL, Owerkowicz T. Functional morphology and evolution of aspiration breathing in tetrapods. Respir Physiol Neurobiol. 2006;154:73–88.

Butler G. On the complete or partial suppression of the right lung in the Amphisbaenidae and the left lung in snakes and snake-like lizards and amphibians. Proc Zool Soc Lond. 1895;1895:691–712.

Butler JP, Banzett RB, Fredberg JJ. Inspiratory valving in avian bronchi: aerodynamic considerations. Respir Physiol. 1988;72:241–56.

Cass AN, Servetnick MD, McCune AR. Expression of a lung developmental cassette in the adult and developing zebrafish swimbladder. Evol Dev. 2013;15:119–32.

Cieri RL, Craven BA, Schachner ER, Farmer CG. New insight into the evolution of the vertebrate respiratory system and the discovery of unidirectional airflow in iguana lungs. Proc Natl Acad Sci USA. 2014;111:17218–23.

Coiter V. Anatomie Avium. In Externarum et internarum praecipalium humani corporis partium tabulae atque anatomicae exercitationes observationesque varieae. Germany: Norimbergae; 1573. p. 130–3.

Cupello C, Brito PM, Herbin M, Meunier FJ, Janvier P, Dutel H, Clément G. Allometric growth in the extant coelacanth lung during ontogenetic development. Nat Commun. 2015;6:8222.

Darwin C. The descent of man, and selection in relation to sex. London: John Murray; 1871.

Dotterweich H. Die Atmung der Vögel. Z f vergl Physiol. 1936;23:744–70.

Duncker HR. The lung air sac system of birds. A contribution to the functional anatomy of the respiratory apparatus. Ergeb Anat Entwicklungsgesch. 1971;45:1–171.

Duncker HR. General morphological principles of amniotic lungs. In: Piiper J, editor. Respiratory function in birds. Heidelberg: Springer; 1978.

Farmer C. Did lungs and the intracardiac shunt evolve to oxygenate the heart in vertebrates? Paleobiology. 1997;23:358–72.

Farmer CG. The provenance of alveolar and parabronchial lungs: insights from paleoecology and the discovery of cardiogenic, unidirectional airflow in the American alligator (*Alligator mississippiensis*). Physiol Biochem Zool. 2010;83:561–75.

Farmer CG. On the evolution of vascular patterns of tetrapods. J Morphol. 2011;272:1325–41.

Fig. 3.5 (continued) model (**e**–**i**). (**c**) Dried iguana lung that has been cut in the parasagittal plane in lateral view. (**d**) Lateral wall of dried lung rotated 180° from plane of cut about the vertical axis. *Black arrow* shows region of the ventral chamber where pressure was modeled during inspiration (**e**–**i**) and where small niches are present on both sides of the septum separating the craniodorsal chamber from the caudoventral chamber. (**e**) Lateral view of mesh generated from computed tomography data of iguana lung showing axial slice of pressure over gas density during inspiration (*scale bar* in part I). Model is from Cieri et al. (2014). *Small silver arrows* indicate direction of airflow in the slice. (**f**) Same as in (**e**) but with the inspiratory jet visualized (*velocity scale* as in part I). *White arrows* have been superimposed to indicate direction of flow for ease of visualization. (**g**) Cranial view of axial pressure over gas density during inspiration, with *small silver arrows* indicating direction of airflow and *white arrows* superimposed for ease of visualizing airflow. Note the low-pressure zone associated with the niches on the lateral walls of the horizontal septum. (**h**) Same as in (**g**) but with the arrows removed for ease of visualization of the zones of pressure. (**i**) Same as in (**h**) but with the inspiratory streams in both the dorsocranial and ventrocaudal chambers visualized, showing the cranial to caudal path of the streams (*orange jet*). (**j**) Illustration of the Coanda effect for a high-speed stream of air flowing into a wide container. The stream entrains air from both sides and becomes broader, carrying more air away from the ostium (inlet of nozzle) in the center of the chamber. This causes a drop in pressure in the zones between the stream and the walls, which in turn results in flow along the walls in the direction of the ostium. This is an unstable situation. A disturbance in the flow or the shape of the container can cause the equalizing flow returning along the walls to push the stream toward one wall, where it will lock onto the wall as it continues to flow (after Angrist 1964). In the iguana lung during inspiration, flow in most of the caudal chamber is craniad, with direction of flow of the high-speed stream caudad

Farmer CG. The evolution of unidirectional pulmonary airflow. Physiology. 2015a;30:260–72.

Farmer CG. Similarity of crocodilian and avian lungs indicates unidirectional flow is ancestral for archosaurs. Integr Comp Biol. 2015b;55:962–71.

Farmer CG, Carrier DR. Pelvic aspiration in the American alligator (*Alligator mississippiensis*). J Exp Biol. 2000;203:1679–87.

Farmer CG, Sanders K. Unidirectional airflow in the lungs of alligators. Science. 2010;327:338–40.

Gans C, Clark B. Studies on ventilation of *Caiman crocodilus* (Crocodilia: Reptilia). Respir Physiol. 1976;26:285–301.

Greenblatt EE, Butler JP, Venegas JG, Winkler T. Pendelluft in the bronchial tree. J Appl Physiol. 2014;117:979–88.

Hazelhoff EH. Structure and function of the lung of birds. Poult Sci. 1951;30:3–10.

Heraty KB, Laffey JG, Quinlan NJ. Fluid dynamics of gas exchange in high-frequency oscillatory ventilation: in vitro investigations in idealized and anatomically realistic airway bifurcation models. Ann Biomed Eng. 2008;36:1856–69.

Hsia CCW, Schmitz A, Lambertz M, Perry SF, Maina JN. Evolution of air breathing: oxygen homeostasis and the transitions from water to land and sky. Compr Physiol. 2013;3:849–915.

Huxley TH. On the respiratory organs of *Apteryx*. Proc Zool Soc Lond. 1882;1882:560–9.

Kramer DL, Lindsey CC, Moodie GEE, Stevens ED. The fishes and the aquatic environment of the central Amazon basin, with particular reference to respiratory patterns. Can J Zool. 1978;56:717–29.

Krefft G. Description of a gigantic amphibian allied to genus Lepidosiren, from Wide Bay district, Queensland. Proc Zool Soc Lond. 1870;1870:221–4.

Liem KF. Respiratory gas bladders in teleosts: functional conservatism and morphological diversity. Am Zool. 1989;29:333–52.

Longo S, Riccio M, McCune AR. Homology of lungs and gas bladders: insights from arterial vasculature. J Morphol. 2013;274:687–703.

Maina JN. The morphometry of the avian lung. In: King AS, McLelland J, editors. Form and function in birds, vol. 4. London: Academic; 1989. p. 307–68.

Maina JN. The gas exchangers: structure, function, and evolution of the respiratory processes. New York: Springer; 1998.

Maina JN. What it takes to fly: the structural and functional respiratory refinements in birds and bats. J Exp Biol. 2000;203:3045–64.

Maina JN. Some recent advances on the study and understanding of the functional design of the avian lung: morphological and morphometric perspectives. Biol Rev Camb Philos Soc. 2002;77:97–152.

Maina JN. Development, structure, and function of a novel respiratory organ, the lung-air sac system of birds: to go where no other vertebrate has gone. Biol Rev Camb Philos Soc. 2006;81:545–79.

Maina JN, King AS, Settle G. An allometric study of pulmonary morphometric parameters in birds, with mammalian comparisons. Philos Trans R Soc Lond Ser B Biol Sci. 1989;326:1–57.

Maina JN, Veltcamp CJ, Henry J. Study of the spatial organization of the gas exchange components of the snake lung-the sandboa *Eryx colubrinus* (Reptilia: Ophidia: Colubridae)-by latex casting. J Zool. 1999;247:81–90.

Mathie JN, Franklin CE. The influence of body size on the diving behaviour and physiology of the bimodally respiring turtle, Elseya albagula. J Comp Physiol B. 2006;176:739–47.

Milani A. Beiträge zur Kenntniss der Reptilienlunge. Zool Jahrb. 1894;7:545–92.

Milani A. Beiträge zur Kenntnis der Reptilienlunge II. Zool Jahrb. 1897;10:93–156.

Miller WS. The structure of the lung. J Morph. 1893; VIII:165–89.

Montiero A, Smith RL. Bronchial tree architecture in mammals of diverse body mass. Int J Morphol. 2014;32:312–6.

Nysten M. Étude anatomique des rapports de la vessie aérienne avec l'axe vertébral chez *Pantodon buchholzi* Peters. Ann Mus R l'Afrique Cent Sci Zool. 1962;8:187–220.

Perry SF. Reptilian lungs: functional anatomy and evolution. Berlin: Springer; 1983.

Perry SF, Duncker HR. Lung architecture, volume and static mechanics in five species of lizards. Respir Physiol. 1978;34:61–81.

Perry SF, Schmitz A, Andersen NA, Wallau BR, Nicol S. Descriptive study of the diaphragm and lungs in the short-nosed echidna, *Tachyglossus aculeatus* (Mammalia: monotremata). J Morphol. 2000;243:247–55.

Pick FK. Zur feineren Anatomie der Lunge von Halicore dugong. Archiv für Naturgeschichte. 1907;73:245–72.

Sagemehl M. Beitrage zur vergleichenden Anatomie der Fische-III. Das Cranium der Characiniden nebst allgemeinen Bemerkungen uber die mit dem Weber'schen Apparat versehenen Physostomenfamilien. Morphol Jahrb. 1885;1885:1–119.

Schachner ER, Cieri RL, Butler JP, Farmer CG. Unidirectional pulmonary airflow patterns in the savannah monitor lizard. Nature. 2014;506:367–70.

Schmidt-Neilsen K. How birds breathe. Sci Am. 1971;225:72–9.

Van Wallach V. The lungs of snakes. In: Gans C, Gaunt AS, editors. Biology of the Reptilia, Morphology G. Visceral Organs, vol. 19. Ithaca, NY: Society for the Study of Amphibians and Reptiles; 1998. p. 93–295.

Weibel ER. Morphometry of the human lung. Berlin: Springer; 1963.

Weibel ER. The pathway for oxygen. Cambridge: Harvard University Press; 1984.

West JB. Respiratory physiology—the essentials. Baltimor: Williams and Wilkins; 1995.

Whitford WG, Hutchison VH. Gas exchange in salamanders. Physiol Zool. 1965;38:228–42.

Wolf S. Zur kenntnis von Bau und Funktion der Reptilienlunge. Zool Jahrb Abt Anat Ontol. 1933;57:139–90.

4 Flying High: The Unique Physiology of Birds that Fly at High Altitudes

Graham R. Scott and Neal J. Dawson

Abstract

Birds that fly at high altitudes must support vigorous exercise in oxygen-thin environments. Here we discuss the characteristics that help high-fliers sustain the high rates of metabolism needed for flight at elevation. Many traits in the O_2 transport pathway distinguish birds in general from other vertebrates. These include enhanced gas-exchange effectiveness in the lungs, maintenance of O_2 delivery and oxygenation in the brain during hypoxia, augmented O_2 diffusion capacity in peripheral tissues, and a high aerobic capacity. These traits are not high-altitude adaptations, because they are also characteristic of lowland birds, but are nonetheless important for hypoxia tolerance and exercise capacity. However, unique specializations also appear to have arisen, presumably by high-altitude adaptation, at every step in the O_2 pathway of some highland species. The distinctive features of high-fliers include an enhanced hypoxic ventilatory response and/or an effective breathing pattern, large lungs, haemoglobin with a high O_2 affinity, further augmentation of O_2 diffusion capacity in the periphery, and multiple alterations in the metabolic properties of cardiac and skeletal muscle. These unique specializations improve the uptake, circulation, and efficient utilization of O_2 during high-altitude hypoxia. High-altitude birds also have larger wings than their lowland relatives, presumably to reduce the metabolic costs of staying aloft in low-density air. High-fliers are therefore unique in many ways, but the relative roles of adaptation and plasticity (acclimatization) in high-altitude flight are still unclear. Disentangling these roles will be instrumental if we are to understand the physiological basis of altitudinal range limits and how they might shift in response to climate change.

Keywords

Capillarity • Haemoglobin • Hypobaria • Hypoxia • Mitochondria • Oxidative Capacity • Respiration

Contents

4.1 Introduction

High-altitude environments pose numerous challenges to animal life. The physical

G.R. Scott (✉) • N.J. Dawson
Department of Biology, McMaster University, 1280 Main Street West, Hamilton, ON, Canada, L8S 4K1
e-mail: scottg2@mcmaster.ca

J.N. Maina (ed.), *The Biology of the Avian Respiratory System*, DOI 10.1007/978-3-319-44153-5_4

environment changes dramatically on ascent, with declines in oxygen availability, temperature, air density, and humidity. Despite these challenges, many animals live successfully in the high mountains. Birds are particularly diverse in montane regions—many live at over 4000 m above sea level and some cross high mountain ranges during their migration. One of the best studied examples is the bar-headed goose, which crosses the Himalayan Mountains during its biannual migration between South and Central Asia and reaches altitudes up to 7000 m or more (Takekawa et al. 2009; Köppen et al. 2010; Hawkes et al. 2011, 2013; Bishop et al. 2015) (Fig. 4.1). While some species are unique to high elevation, others are found across broad elevational gradients (McCracken et al. 2009a; Wilson et al. 2013; McGuire et al. 2014).

The decreases in total barometric pressure ('hypobaria') and O_2 partial pressure ('hypoxia') at high altitude are inescapable, unlike elevational declines in temperature and humidity, which can be buffered by local climatic variation. Hypobaria has unique consequences for flying animals, because the mechanical power output needed to sustain lift increases in thin air (Altshuler and Dudley 2006; Stalnov et al. 2015). This amplifies the already high metabolic rates needed for flapping flight (Chai and Dudley 1995)—which are already from 10- to 15-fold above resting levels during steady flight in a wind tunnel at sea level (Ward et al. 2002)—in an environment where the O_2 available to fuel metabolism is limited. According to Tucker, 'some birds perform the strenuous activity of flapping flight at altitudes in excess of 6100 m, an altitude at which resting, unacclimated man is in a state of incipient hypoxic collapse' (Tucker 1968). How then can O_2 supply processes meet the high O_2 demands of flight at high altitudes? What unique physiological characteristics allow the highest-flying species to sustain the most metabolically costly form of vertebrate locomotion at elevations that can barely support life in many other animals?

In order to properly address these questions, one must consider the properties of the O_2 pathway that transports O_2 from the environment to the sites of O_2 demand throughout the body. This pathway is composed of a series of cascading physiological 'steps' (Fig. 4.2): (1) ventilation of the lungs with air; (2) diffusion of O_2 across the pulmonary gas-exchange surface, from the air to the blood; (3) circulation of O_2 throughout the body in the blood; (4) diffusion of O_2 from the blood to mitochondria in tissues (the pectoralis muscle is the primary site of O_2 consumption during flight); and (5) metabolic utilization of O_2 to generate ATP by oxidative phosphorylation. Although not a strict component of the O_2 transport cascade, properties of intracellular ATP turnover will also have important consequences for matching O_2 supply and O_2 demand. Not surprisingly, the answer to how birds fly at high altitudes lies, at least partly, in the characteristics of this pathway.

This chapter provides an update to a previous review on the importance of both ancestral and derived characteristics in the O_2 transport pathway of birds that fly at high altitudes (Scott 2011). Many features of birds in general probably endowed high-fliers with numerous exaptations (also known as preadaptations), but many uniquely derived and presumably adaptive traits also appear to be important for high-altitude flight.

4.2 The Benefits of Being Avian

The hypoxia tolerance of birds has frequently been suggested to be greater than that of mammals. Although some ectothermic vertebrates are even more tolerant of hypoxia, birds possess a relatively high tolerance when considering the increase in metabolic demands associated with endothermy. Early work showed that lowland house sparrows (*Passer domesticus*) behaved normally and could even fly for short periods in a wind tunnel, at a simulated altitude of 6100 m (Tucker 1968). In contrast, domestic rodents were comatose and unable to maintain body temperature at the same simulated altitude. Comparisons of the few species for which tolerance (survival) data are available also support the suggestion that birds are more tolerant of

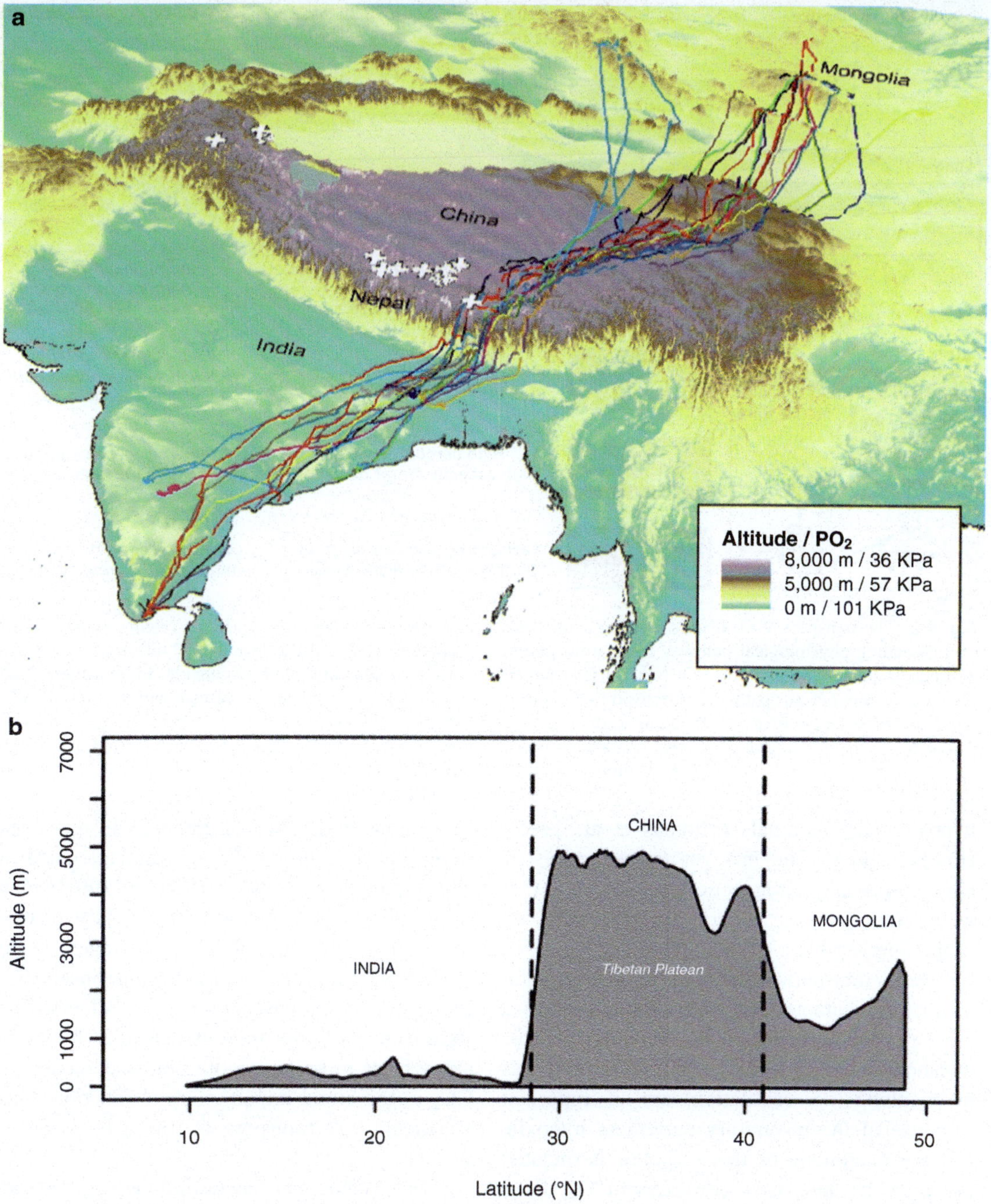

Fig. 4.1 Bar-headed geese migrate at high altitudes during their biannual migration across the Himalayas. (**a**) Satellite tracks of bar-headed geese from a Mongolian population on their southward migration across the Himalayas into India (individual geese shown as *coloured lines*). *White crosses* represent the world's highest mountains, whose peaks are all over 8000 m elevation. (**b**) The northward migration requires that bar-headed geese ascend the mountains rapidly (<1 day) at a very steep incline (minimum climb rates of 0.8–2.2 km/h) (Hawkes et al. 2011) [modified from Hawkes et al. (2013) and Scott et al. (2015)]

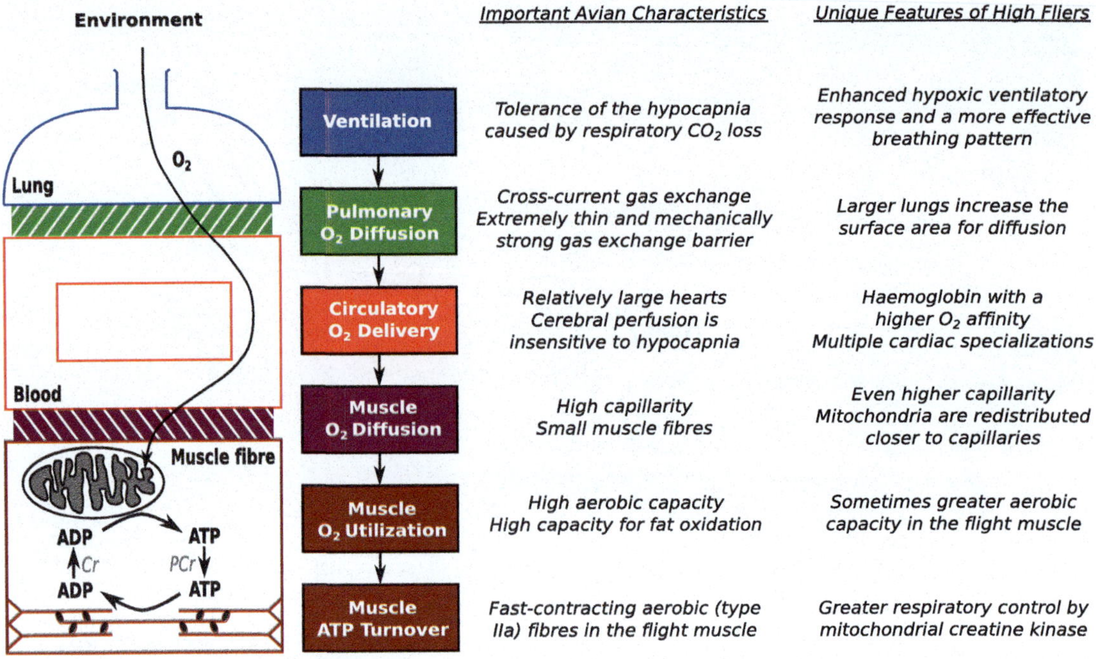

Fig. 4.2 The transport of O_2 occurs along several steps of a cascading physiological pathway from atmospheric air to the mitochondria in tissue cells (e.g. muscle fibres). The effectiveness of this pathway at transporting O_2 during hypoxia is imperative for flight at high altitudes, which depends upon several distinctive characteristics of birds in general (*left column*) and many unique features that have evolved in high-fliers (*right column*). The properties of O_2 utilization and ATP turnover in the flight muscle are also important to consider in high-fliers, such as how ATP equivalents are moved between sites of ATP supply and demand (which can occur via phosphocreatine, PCr, by virtue of the creatine kinase shuttle; see text) [modified from Scott (2011)]

hypoxia than mammals (Thomas et al. 1995). However, this issue has yet to be addressed with rigorous phylogenetic comparisons that incorporate species in both groups that are adapted to hypoxia.

The O_2 transport pathway of birds has several distinctive characteristics that should support a greater capacity for vigorous activity and aerobic metabolism during hypoxia (Fig. 4.2). Increases in breathing (i.e. ventilation) are an important response of the respiratory system to hypoxia, and the magnitude of this response is dictated primarily by the partial pressures of O_2 (PO_2) and CO_2 (PCO_2) and the pH of arterial blood (Scott and Milsom 2009). The decline in arterial PO_2 ('hypoxaemia') drives the increase in ventilation, whose secondary consequence is an amplification of CO_2 loss to the environment. This causes hypocapnia (low PCO_2 in the blood), which reflexively inhibits breathing and causes an acid-base disturbance. It has been suggested that birds have a higher tolerance of hypocapnia than mammals (Scheid 1990), which could arise from an ability to rapidly restore blood pH in the face of CO_2 challenges (Dodd et al. 2007) and from the hypocapnic insensitivity of the brain vasculature (see below). The significance of this tolerance is that it would allow birds to breathe more before depletion of CO_2 in the blood impairs normal function, thus enhancing O_2 transport to the gas-exchange surface.

The structure and function of the lungs is perhaps the best-known advantage of avian respiratory systems. The many distinctive features of bird lungs are the subject of an extensive literature that unfortunately can be dealt with only briefly here (Piiper and Scheid 1972; Scheid 1990; Maina 2006, 2015; Cieri and Farmer 2016). Air flows in one direction through the gas-exchange units of avian lungs ('parabronchioles'), and the arrangement of airway and vascular vessels creates a functionally cross-current gas exchanger. This differs substantially from the lungs of most other terrestrial vertebrates, in which gases flow in and out of terminal gas-exchange units ('alveoli' in mammals) such that capillary blood equilibrates with air having a uniform gas composition. The important consequence of this difference is that avian lungs can attain a superior efficiency for gas exchange in normoxia and moderate hypoxia, although their advantage diminishes as hypoxia becomes severe (Scheid 1990). The capacity for pulmonary O_2 diffusion is also greater in birds due to the exceptional thinness and large surface area of the exchange tissue. Nevertheless, the diffusion barrier appears to be mechanically stronger in birds than in mammals, so pulmonary blood flow and pressure can increase without causing stress failure (West 2009). Each of these distinctive features of avian lungs should improve O_2 loading into the blood during hypoxia.

The capacity for delivering O_2 throughout the body in the systemic circulation may be higher in birds than in other vertebrates. Birds have larger hearts and cardiac stroke volumes than mammals of similar body size (Grubb 1983), suggesting that birds are capable of higher cardiac outputs. If this were indeed the case, birds would have an enhanced capacity for convective delivery of O_2 in the blood during hypoxia. Cardiac output increases seven to eightfold during flight (Peters et al. 2005) and threefold or more during hypoxia at rest (Black and Tenney 1980), but maximum cardiac output has yet to be determined in birds, particularly during flight in hypoxia.

The distribution of blood flow throughout the body has consequences for hypoxia tolerance, and the mechanisms regulating this distribution are altered in birds compared to mammals. Hypoxaemia per se causes a preferential redistribution of O_2 delivery towards sensitive tissues like the heart and brain and away from more tolerant tissues (e.g. intestines). However, increases in O_2 delivery to the brain are offset in mammals at high altitudes due to the respiratory hypocapnia induced by increases in breathing. This causes a constriction of cerebral blood vessels that can completely abolish the hypoxaemic stimulation of cerebral blood flow. In contrast, the cerebral vessels of birds are insensitive to hypocapnia, such that blood flow is allowed to increase and O_2 delivery is maintained (Faraci 1991). This and possibly other distinctive features of the avian cerebral circulation (Bernstein et al. 1984) should improve brain oxygenation during hypoxia. Coupled with the inherently higher tolerance of avian neurons to low cellular O_2 levels (Ludvigsen and Folkow 2009), due at least in part to an ability to reversibly shut down the metabolic demands of cerebellar activity (Geiseler et al. 2015), the central nervous system of birds appears to be well protected from cellular damage induced by O_2 lack. Nevertheless, an intriguing question that has yet to be addressed is whether heightened blood flow increases intracranial pressure in birds, as frequently occurs in humans (Wilson et al. 2009). If so, birds may face the secondary challenge of avoiding or tolerating cerebral oedema and other neurological syndromes that can result from excessive intracranial pressure in mammals.

The capacity for O_2 to diffuse from the blood into the tissues is higher in birds compared to mammals and other vertebrates. The best evidence for this difference is the systematically higher ratio of capillary surface area to muscle-fibre surface area in the flight muscle of birds compared to the locomotory muscles of mammals (Mathieu-Costello 1990). At least two factors account for this difference: (1) the tight mesh of capillaries surrounding avian muscle fibres, due to a high degree of branching between longitudinal vessels, and (2) the smaller aerobic fibres of birds compared to similar-sized mammals (Mathieu-Costello 1990). The heart and brain also have higher densities of capillaries in birds compared to mammals (Faraci 1991). Diffusion of O_2 from the blood to the mitochondria in various tissues should therefore be higher in birds than in other vertebrates during hypoxaemia.

Although these distinctive characteristics of birds should enhance hypoxia tolerance by improving the overall capacity for O_2 transport, being avian is not in itself sufficient for flight at high altitudes. The flight muscle of birds has a very high aerobic capacity, which in most species arises by virtue of fast-contracting oxidative fibres (type IIa) that have abundant mitochondria (Mathieu-Costello 1990; Scott et al. 2009a), and the high rates of metabolism during flight are supported primarily by lipid fuels (Jenni and Jenni-Eiermann 1998; McWilliams et al. 2004; Weber 2009). Lipid oxidation is essential for supporting long-duration flight, but it amplifies the amount of O_2 required to produce a given amount of ATP when compared to carbohydrate oxidation. The metabolic demands of flight are further intensified at high altitudes by hypobaria, which requires that birds flap harder to produce lift (Chai and Dudley 1995). The implication of these factors is that high-altitude flight requires very high rates of O_2 transport when very little O_2 is available. This is clearly not possible for most lowland birds—many species cannot tolerate severe hypoxia (Black and Tenney 1980) and some fly long distances to avoid high-elevation barriers during their migration (Irwin and Irwin 2005). What then are the uniquely derived attributes that differentiate the high-fliers?

4.3 The Unique Attributes of High-Fliers

The physiology of birds that fly at high altitudes differs in many ways from that of lowland birds. Some of the best support for this conclusion comes largely from studies of the bar-headed goose (*Anser indicus*), a species that can tolerate severe hypoxia (inspired O_2 tensions ~2.8 kPa, equivalent to 12,000 m elevation) (Black and Tenney 1980), can sustain their maximal running speed on a treadmill at ~7 kPa (simulating the hypoxia at 8500 m) (Hawkes et al. 2014), and can fly at over 7000 m while migrating across the Himalayas (Fig. 4.1). Studies of bar-headed geese have revealed many important insights into the physiological basis for high-altitude flight and, when coupled with comparative phylogenetic approaches, its evolutionary origins. Our discussion of the unique attributes of high-fliers will focus largely—out of necessity—on this species but will also highlight work in other species when possible. Most of the previous work looking for inherent differences between high- and low-altitude birds compared animals in a common environment at sea level. This will be the case in the following discussion unless otherwise stated.

It is useful to begin our discussion by outlining the most influential steps in the O_2 transport pathway during exercise in hypoxia. This issue has been addressed in waterfowl using theoretical modelling to calculate the physiological control coefficient for each step in the pathway (Fig. 4.3) (Scott and Milsom 2006, 2009). This approach allows physiological traits to be altered individually so that their influence

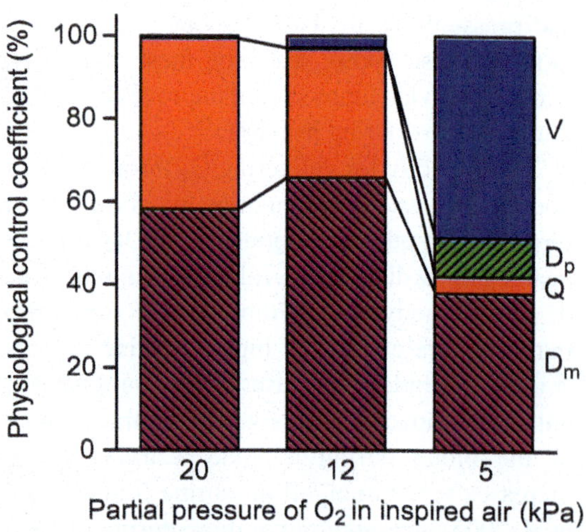

Fig. 4.3 Physiological control analysis of flux through the O_2 transport pathway in waterfowl. The influence of the respiratory system (ventilation, *V*, and the capacity for pulmonary O_2 diffusion, D_p) on O_2 transport increases and that of circulatory O_2 delivery capacity (*Q*, the product of maximum cardiac output and the O_2 carrying capacity of the blood) declines as hypoxia becomes more severe. The capacity for O_2 diffusion in the muscle (D_m) has a large influence on O_2 transport at all partial pressures of inspired O_2. Control coefficients were calculated using theoretical modelling of the respiratory system with a haemoglobin P_{50} that is typical of highland birds and are defined as the fractional change in O_2 transport rate divided by the fractional change of any given step in the O_2 transport pathway. Expressed as a percentage, the control coefficients for all steps in the pathway will sum to 100 [modified from Scott and Milsom (2006, 2009) and Scott (2011)]

on the whole O_2 pathway can be assessed without compensatory changes in other traits. A physiological trait with a larger control coefficient will have a greater influence on flux through the pathway, so an increase in the capacity of this trait will have a greater overall benefit. Interestingly, the proportion of control vested in each step was dependent on the inspired O_2 (Fig. 4.3). At sea level (inspired O_2 tensions ~20 kPa) and in moderate hypoxia (~12 kPa, equivalent to ~4500 m elevation), circulatory O_2 delivery capacity (which incorporates both maximum cardiac output and blood haemoglobin concentration) and the capacity for O_2 diffusion in the muscle retained most of the control over pathway flux (Fig. 4.3). In contrast, ventilation and the capacity for O_2 diffusion in the lungs became much more influential in severe hypoxia (~5 kPa, roughly 9000 m), while muscle diffusion remained important and circulatory O_2 delivery capacity became less so (Fig. 4.3). These results suggest that every step in the O_2 transport pathway can be influential and that the relative benefit of each step changes with altitude.

High capacities at several steps in the O_2 transport pathway have been shown to distinguish high-flying birds from their lowland relatives (Fig. 4.2), confirming our theoretical predictions. The first step of this pathway, ventilation, appears to be enhanced in some high-flying birds to improve O_2 uptake into the respiratory system. Bar-headed geese breathe more than low-altitude waterfowl when exposed to severe hypoxia (inspired O_2 tensions ~3.1–4.7 kPa) (Black and Tenney 1980; Scott and Milsom 2007), and the magnitude of their ventilatory response is greater than in any other bird species yet studied (Scott and Milsom 2009). Bar-headed geese also breathe with a more effective breathing pattern, taking much deeper breaths (i.e. higher tidal volumes) than low-altitude birds during hypoxia. Being born and raised in captivity at 3200 m elevation at Lake Qinghai in China has also been shown to further amplify the hypoxic ventilatory response compared to bar-headed geese raised in captivity at sea level (Lague et al. 2016). There are at least two mechanistic causes for the differences between bar-headed geese and lowland waterfowl: (1) ventilatory insensitivity to respiratory hypocapnia and (2) a blunting of the metabolic-depression response to hypoxia (Scott and Milsom 2007; Scott et al. 2008). These differences increase the PO_2 of the air that ventilates the pulmonary gas-exchange surface during hypoxia. Whether similar enhancements of the hypoxic ventilatory response occur in other bird species that live and/or fly at high altitudes is not yet clear. Nevertheless, bar-headed geese and numerous other highland species sampled at high altitudes have enlarged lungs (Carey and Morton 1976; Scott et al. 2011), which should enhance the second step of the O_2 transport pathway by increasing the area of the gas-exchange surface. The respiratory system of high-altitude birds therefore seems capable of loading more O_2 into the blood during hypoxia than that of lowland birds.

The circulatory delivery of O_2 throughout the body is also enhanced in high-altitude birds. The most pervasive mechanism for sustaining the circulation of O_2 in hypoxia is an alteration in the O_2-binding properties of haemoglobin in the blood. Numerous high-altitude birds, such as the bar-headed goose (Fig. 4.4), Andean goose (*Chloephaga melanoptera*) (Black and Tenney 1980), Tibetan chicken (*Gallus gallus*) (Gou et al. 2007), Ruppell's griffon (*Gyps rueppellii*) (Weber et al. 1988), Andean house wren (*Troglodytes aedon*) (Galen et al. 2015), and several populations of Andean ducks (Natarajan et al. 2015), are known to possess haemoglobins with an increased O_2 affinity. This can dramatically increase O_2 delivery and pulmonary O_2 loading in hypoxia by increasing the saturation of haemoglobin (and thus the O_2 content of the blood) at a given PO_2 (Fig. 4.4a) and can in doing so greatly improve flux through the O_2 transport pathway (Scott and Milsom 2006). In bar-headed geese, haemoglobin-O_2 binding is also more sensitive to temperature than in other birds and mammals (Meir and Milsom 2013), which could substantially increase the unloading of O_2 at the tissues if there is thermal heterogeneity between the lungs and flight muscle during flight (Maginniss et al. 1997; Scott and Milsom 2006). The genetic and structural bases for haemoglobin

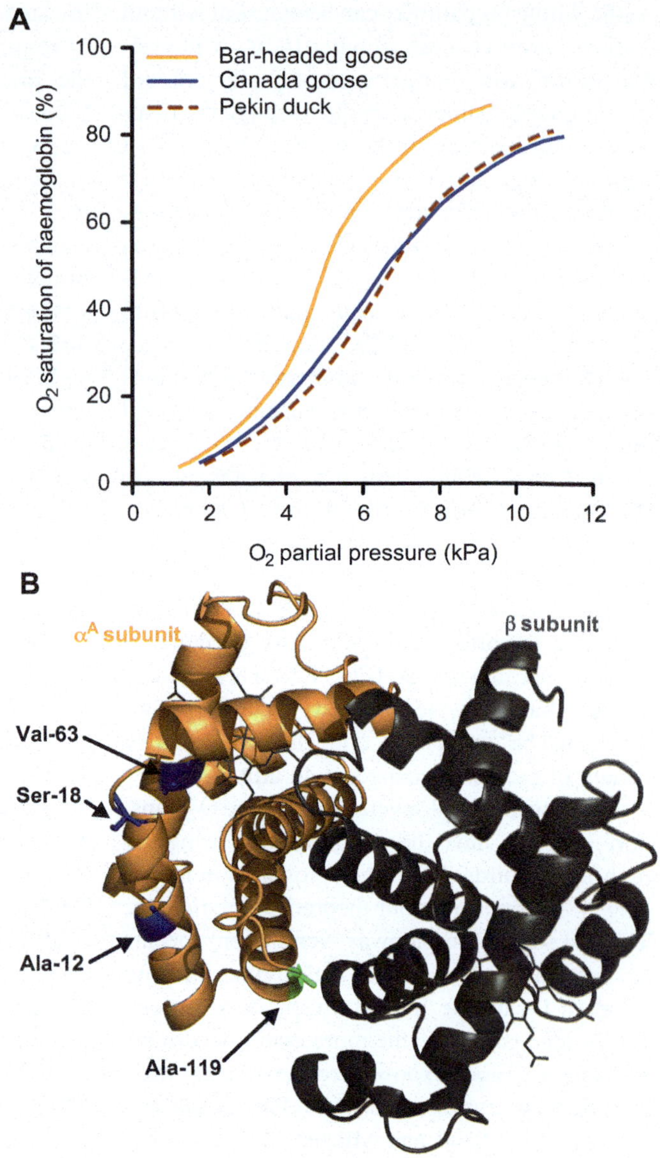

Fig. 4.4 High-altitude adaptations in the haemoglobin (Hb) of bar-headed geese. (**a**) The O_2 affinity of bar-headed goose Hb is higher than that of lowland waterfowl, as reflected by a leftward shift in the O_2 equilibrium curve of blood (measured at a pH of 7.3). Redrawn from Black and Tenney (1980). (**b**) The α^A subunit of bar-headed goose Hb contains four uniquely derived amino-acid substitutions (*blue* and *green*). Ala-119 (*green*) has a large influence on O_2 binding because it alters the interaction between α and β subunits (see text). For simplicity, only one out of two α and β subunits that compose the complete Hb tetramer is shown. This cartoon was drawn in PyMOL from the previously published structure of oxygenated Hb (Zhang et al. 1996) (Protein Data Bank ID, 1A4F) [modified from Scott (2011)]

adaptations to high altitude have been resolved in many species. For example, the bar-headed goose possesses a major (HbA) and minor (HbD) form of haemoglobin, whose α subunits contain four (α^A) (Fig. 4.4b) and two (α^D) uniquely derived amino-acid substitutions, respectively (McCracken et al. 2010). One of the substitutions in α^A (proline-119→alanine) (*green* in Fig. 4.4b) is thought to cause a large increase in O_2 affinity (Jessen et al. 1991) by altering the interaction between α and β subunits and destabilizing the deoxygenated state of the protein (Zhang et al. 1996). The convergent increases in haemoglobin-O_2 affinity that have been observed in several species can sometimes arise from parallel evolution at a genetic level (McCracken et al. 2009a) but more often involve multiple distinct paths of molecular evolution (Natarajan et al. 2015). Highland haemoglobin genotypes can even be maintained when gene flow from low altitudes is high, presumably because they are strongly favoured by natural selection (McCracken et al. 2009b).

The circulation of O_2 may also be sustained in hypoxia by specializations in the heart that safeguard cardiac output. Bar-headed geese have a higher density of capillaries in the left ventricle of the heart (Fig. 4.5a), which should help maintain the O_2 tension in cardiac myocytes and thus preserve function when hypoxaemia occurs at high altitudes (Scott et al. 2011). Cellular function could also be challenged if the production of reactive O_2 species increases at high altitudes, as what occurs in some lowland animals when declining O_2 levels at cytochrome c oxidase (COX, the enzyme that consumes O_2 in oxidative phosphorylation) shifts the electron transport chain of mitochondria towards a more reduced state (akin to a buildup of electrons) (Aon et al. 2010). However, COX from bar-headed goose hearts has a higher affinity for its substrate (cytochrome c in its reduced state) (Fig. 4.5b), which could allow the electron transport chain to operate in a less reduced state and thus minimize oxidative damage by reactive O_2 species (Scott et al. 2011). A possible cause of this difference is a single mutation in subunit 3 of the COX protein, which occurs at a site that is otherwise conserved across vertebrates (tryptophan-116→arginine) (*green* in Fig. 4.5c) and appears to alter inter-subunit interactions (Scott et al. 2011). Differences in the activity of various other metabolic enzymes in the heart distinguish high- and low-altitude populations of wild-caught torrent ducks (*Merganetta armata*), differences that appear to be unrelated to variation in respiratory capacity but that may be otherwise important for sustaining heart function at high altitudes (Dawson et al. 2016). Cardiac specializations in high-altitude birds may have a transcriptional basis, based on a comparison of cardiac gene expression in late-stage embryos of Tibetan chickens and lowland breeds (Li and Zhao 2009): embryonic hypoxia altered the expression of over 70 transcripts in all chickens, but an additional 12 genes (involved in energy metabolism, signal transduction, transcriptional regulation, cell proliferation, contraction, and protein folding) were differentially expressed in only the highland Tibetan breed. Overall, these findings lend some credence to a previous suggestion that the hypoxaemia tolerance of the heart has a strong influence on the ability to fly at high altitudes (Scheid 1990).

The capacity for O_2 to diffuse from the blood to mitochondria in the flight muscle is also enhanced in high-altitude birds. Andean coot (*Fulica americana peruviana*) populations that reside and were sampled at high altitudes had a higher capillarity and a smaller fibre size in the flight muscle than populations residing at low altitudes (León-Velarde et al. 1993). Because there were no differences in muscle aerobic capacity between coot populations, the increase in O_2 diffusing capacity should serve to improve O_2 transport in hypoxia rather than to match differences in cellular O_2 demands. Similar differences exist between bar-headed geese and lowland waterfowl from a common environment at sea level (Scott et al. 2009a). Mitochondria are also redistributed closer to capillaries in the aerobic fibres of bar-headed geese (Scott et al. 2009a), which reduces intracellular O_2 diffusion distances. These various mechanisms for improving the diffusion capacity for O_2 in the flight muscle should help sustain mitochondrial O_2 supply when hypoxaemia occurs at high altitudes.

In addition to improvements in the capacity to transport O_2 during hypoxia, various features of metabolic O_2 utilization are altered in the flight muscle of high-altitude birds. In bar-headed geese, the flight muscle has an inherently higher aerobic capacity that arises from an increase in the proportional abundance of aerobic fibres (Scott et al. 2009a), without any changes in the abundance and respiratory capacities of mitochondria in individual muscle fibres (Scott et al. 2009a, b). In some other high-altitude taxa, the metabolic capacity of individual muscle fibres was elevated in association with increases in the activity of mitochondrial enzymes (Mathieu-Costello et al. 1998; Dawson et al. 2016). Correspondingly, gene sequences for several mitochondrial enzymes, including NADH dehydrogenase (complex I) (*mt-nd2* and

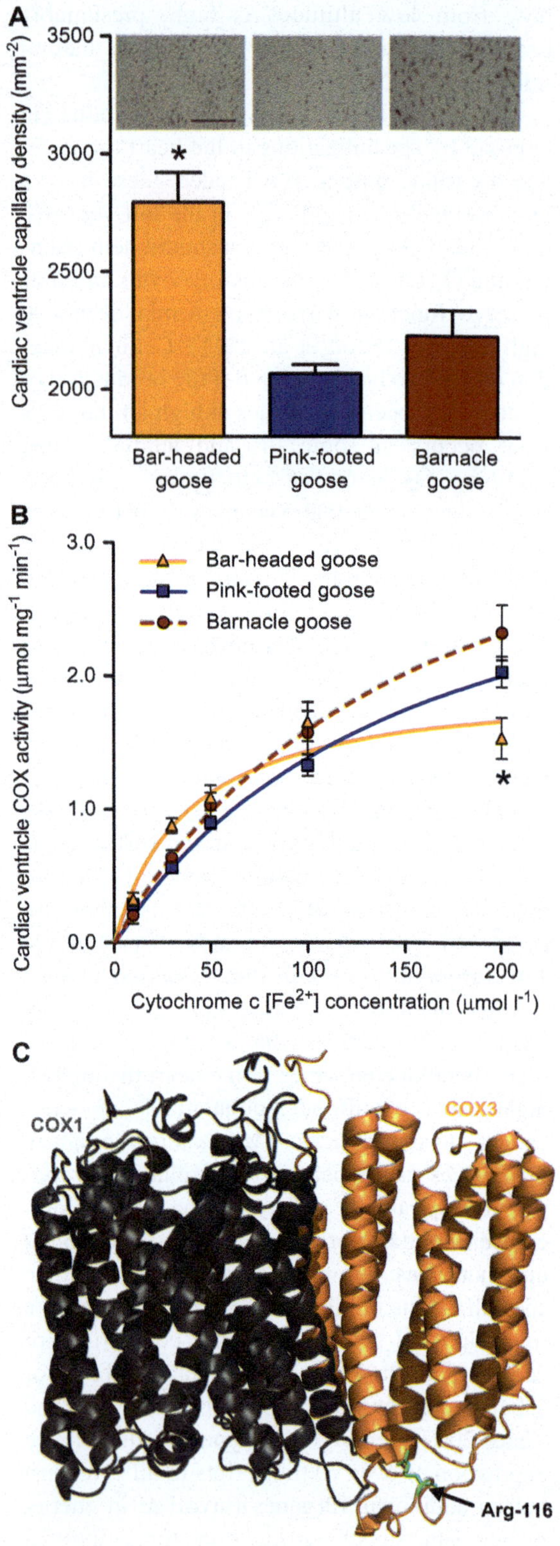

Fig. 4.5 Cardiac adaptations to high altitude in bar-headed geese. (**a**) Capillary density is enhanced in the hearts (*left ventricle*) of bar-headed geese compared to low-altitude geese. Inset are representative images of capillary staining in bar-headed geese (*left*), pink-footed geese (*centre*), and barnacle geese (*right*). *Scale bar* represents 100 μm. (**b**) Cytochrome c oxidase (COX) from the hearts of bar-headed geese has a different maximal activity (lower V_{max}) and substrate kinetics (lower K_m for cytochrome c [Fe^{2+}], cytochrome c in its reduced state) than COX from low-altitude geese. (**c**) COX subunit 3 (COX3) of bar-headed geese contains a single amino-acid mutation at a site that is otherwise conserved across all vertebrates (Trp-116→Arg) and is predicted by structural modelling to alter the interaction between COX3 and COX1. *Asterisk* represents a significant difference from both low-altitude species [modified from Scott (2011) and Scott et al. (2011)]

mt-nd4), cytochrome c oxidase (*cox3*) (Fig. 4.5c), and F_0F_1-ATP synthase (*mt-atp6*), also appear to have been altered in several high-altitude species (Scott et al. 2011; Zhou et al. 2014; Mu et al. 2016). Therefore, increases in aerobic capacity appear to be a convergent strategy for coping with high-altitude environments, which may be important for counterbalancing the inhibitory effects of low O_2 levels on the respiration of individual mitochondria and/or for sustaining thermogenesis in the cold (Hochachka 1985; Lui et al. 2015).

Other aspects of energy turnover also appear to differ in the flight muscle of some high-altitude birds. ATP production is more strongly regulated by creatine kinase in bar-headed geese than in low-altitude waterfowl (Scott et al. 2009b), and the expression of mitochondrial creatine kinase is upregulated by hypoxia in Tibetan chickens (Li and Zhao 2009). A potential consequence of these alterations is that energy supply and demand in the muscle may be better coupled via the creatine kinase shuttle, a system important for moving ATP equivalents around the cell [this system is described in Andrienko et al. (2003)]. An interesting possibility is that bar-headed geese developed a more active shuttle to compensate for the redistribution of mitochondria, which moved these organelles closer to capillaries but further from the contractile elements that constitute the major sites of ATP demand in the flight muscle. In the Tibetan ground tit (*Parus humilis*), genomic analysis suggests that high-altitude adaptation involved positive selection and rapid evolution of genes involved in the hypoxia response, energy metabolism, and skeletal muscle development (Qu et al. 2013, 2015), which could contribute to the changes in O_2 utilization and energy turnover at high altitudes.

4.4 Physiology Amidst the High Peaks

Recent studies on migrating bar-headed geese have provided insight into how the unique physiology of high-fliers is realized during natural flight at high altitudes (Bishop et al. 2015). Most bar-headed geese reach altitudes of 5000–6000 m during the natural migration, where the PO_2 is roughly half of the sea level value (Takekawa et al. 2009; Köppen et al. 2010; Hawkes et al. 2013) (Fig. 4.1). They have occasionally been tracked even higher (two records of birds over 7000 m) (Köppen et al. 2010; Hawkes et al. 2013), and there are anecdotal reports of bar-headed geese flying above the highest peaks in the Himalayas (Swan 1970) where the PO_2 is only one third of that at sea level. This is consistent with the results from theoretical models of O_2 supply and demand across altitudes, which predicted that the maximum altitude at which geese can transport enough O_2 to sustain the metabolic costs of flight may be as high as 8900 m above sea level (Hawkes et al. 2014). The level of hypoxia at these elevations reduces maximal O_2 uptake rates in lowland humans substantially, and even basal metabolism is a challenge to maintain atop the highest peaks (West 2010). Bar-headed geese may therefore be able to maintain the high metabolic rates needed for flapping flight in air that severely limits aerobic metabolism in many lowland animals (Thomas et al. 1995). In fact, it becomes progressively more difficult to generate lift as air density decreases, such that average heart rates increase with elevation, and geese spend a significant amount of time flying at near maximal heart rates (Bishop et al. 2015). This may be especially true during the northward migration from India onto the Tibetan Plateau, because it requires bar-headed geese to ascend for prolonged durations and maintain the longest sustained climbs recorded to date (Hawkes et al. 2011). Climbing flight is a much greater metabolic challenge than level flight, generally requiring higher heart rates and wingbeat frequencies, such that there is a positive relationship between heart rate and rate of ascent (Bishop et al. 2015). It would seem advantageous for bar-headed geese to make use of upslope tailwinds during ascent to help overcome this challenge (Butler and Woakes 1980; Butler 2010), but bar-headed geese often migrate at night and in the early morning when they must fly into the predominant headwinds travelling downslope (Hawkes et al. 2011).

4.5 Conclusions and Perspectives

The ability of birds to fly at high altitudes is critically dependent on the effective transport of O_2 from hypoxic air to all of the tissues of the body. Part of this effectiveness comes from many characteristics that distinguish the O_2 transport pathway of all birds in general from that of other vertebrates. Although not themselves adaptations for high-altitude flight, these characteristics were undoubtedly an important basis upon which high-altitude adaptation could proceed. As it did so, unique specializations appear to have arisen at every step of the O_2 transport pathway of high-fliers to facilitate their impressive exercise performance. However, much of what we know about the physiological basis for high-altitude flight has been gleaned from a small number of compelling high-altitude taxa, and it is not yet certain whether many of the examples above are broadly representative of high-altitude birds or are sufficient to entirely explain high-altitude flight.

One area we know relatively little about is the relative roles of genetic adaptation versus phenotypic plasticity in the ability of birds to fly at high altitudes. Most of the previous work aimed at revealing the unique attributes of high-fliers compared birds in a common environment at sea level. These studies were a useful first step in elucidating inherent and heritable differences, but it is probable that acclimatization to high-altitude hypoxia also shapes the physiology and flight capacity of highland residents (Cheviron et al. 2008; Lague et al. 2016). This could also be true of elevational migrants that spend time staging higher than their native altitudes before they cross high mountain ranges. However, not all hypoxia responses are beneficial, some are in fact maladaptive (Storz et al. 2010), so future work is needed to understand the intricacies of how adaptation and plasticity interact in high-flying birds.

We know less about the uniquely derived specializations for coping with low barometric pressure, cold, and dry air than we do about those for coping with hypoxia. Birds that are adapted to high altitudes have larger wings to help offset the detrimental effects of low air density on lift generation (Feinsinger et al. 1979; Lee et al. 2008; Gutierrez-Pinto et al. 2014). This reduces the power output required to fly at elevation, but it does not completely eliminate the need for highland birds to flap harder or more frequently at elevation than lowland birds at sea level (Altshuler and Dudley 2003; Bishop et al. 2015). Highland birds may also have a higher thermogenic capacity for increasing metabolism to generate heat in the cold, due to either adaptation and temperature-induced plasticity (Lindsay et al. 2009; Zheng et al. 2014; Swanson and Vezina 2015; Zhang et al. 2015a, b). Nevertheless, the importance of temperature adaptations for dealing with cold during flight is unclear, because the active flight muscles already generate a lot of heat (Torre-Bueno 1976; Ward et al. 1999). Water loss during flight is high enough to constrain flight duration at sea level (Engel et al. 2006) and is therefore expected to be a major issue at elevation that could drive the evolution of unique water-saving strategies. Some migratory birds catabolize proteins in order to release protein-bound water during long flights (Gerson and Guglielmo 2011), and it is possible that this mechanism may be used (perhaps even enhanced) during long migrations at high altitudes. Therefore, we still have much to learn about the unique physiology of the high-fliers.

Climate change is projected to have a large effect on avian communities (Gasner et al. 2010). Species distributions are moving to higher elevations as their historical climate envelopes (defined by temperature and humidity) shift upslope, which could have particularly catastrophic effects on the abundance of current highland species whose ranges cannot be pushed further upwards (Jankowski et al. 2012; Flousek et al. 2015; Ocampo-Penuela and Pimm 2015). However, it is possible that altitudinal clines in variables that will not be affected by climate change (i.e. hypoxia and low barometric pressure) could limit the upward movement of lowland populations and potentially give a competitive advantage to birds that are already well adapted to high altitudes. This does not appear to be the case for the effects of barometric pressure on hummingbird distributions, based on a study that combined climate envelope modelling with an analysis of how altitude affects flight biomechanics (Buermann et al. 2011), but the same may not be true for hypoxia. Disentangling the relative influences of these variables, through a combination of integrative mechanistic studies in the laboratory and eco-physiological studies in the field, is key to understanding the potential effects of global change on avian physiology and ecology.

Acknowledgements This work was supported by funds from McMaster University, the Natural Sciences and Engineering Research Council of Canada (NSERC), and the Canada Research Chairs Program.

References

Altshuler DL, Dudley R. Kinematics of hovering hummingbird flight along simulated and natural elevational gradients. J Exp Biol. 2003;206:3139–47.

Altshuler DL, Dudley R. The physiology and biomechanics of avian flight at high altitude. Integr Comp Biol. 2006;46:62–71.

Andrienko T, Kuznetsov AV, Kaambre T, Usson Y, Orosco A, Appaix F, Tiivel T, Sikk P, Vendelin M, Margreiter R, et al. Metabolic consequences of functional complexes of mitochondria, myofibrils and sarcoplasmic reticulum in muscle cells. J Exp Biol. 2003;206:2059–72.

Aon MA, Cortassa S, O'Rourke B. Redox-optimized ROS balance: a unifying hypothesis. Biochim Biophys Acta. 2010;1797:865–77.

Bernstein MH, Duran HL, Pinshow B. Extrapulmonary gas exchange enhances brain oxygen in pigeons. Science. 1984;226:564–6.

Bishop CM, Spivey RJ, Hawkes LA, Batbayar N, Chua B, Frappell PB, Milsom WK, Natsagdorj T, Newman SH, Scott GR, et al. The roller coaster flight strategy of bar-headed geese conserves energy during Himalayan migrations. Science. 2015;347:250–4.

Black CP, Tenney SM. Oxygen transport during progressive hypoxia in high altitude and sea level waterfowl. Respir Physiol. 1980;39:217–39.

Buermann W, Chaves JA, Dudley R, McGuire JA, Smith TB, Altshuler DL. Projected changes in elevational distribution and flight performance of montane Neotropical hummingbirds in response to climate change. Glob Chang Biol. 2011;17:1671–80.

Butler PJ. High fliers: the physiology of bar-headed geese. Comp Biochem Physiol A Mol Integr Physiol. 2010;156:325–9.

Butler PJ, Woakes AJ. Heart rate, respiratory frequency and wing beat frequency of free flying barnacle geese *Branta leucopsis*. J Exp Biol. 1980;85:213–26.

Carey C, Morton ML. Aspects of circulatory physiology of montane and lowland birds. Comp Biochem Physiol A Comp Physiol. 1976;54:61–74.

Chai P, Dudley R. Limits to vertebrate locomotor energetics suggested by hummingbirds hovering in heliox. Nature. 1995;377:722–5.

Cheviron ZA, Whitehead A, Brumfield RT. Transcriptomic variation and plasticity in rufous-collared sparrows (*Zonotrichia capensis*) along an altitudinal gradient. Mol Ecol. 2008;17:4556–69.

Cieri RL, Farmer CG. Unidirectional pulmonary airflow in vertebrates: a review of structure, function, and evolution. J Comp Physiol B. 2016; doi:10.1007/s00360-016-0983-3.

Dawson NJ, Ivy CM, Alza L, Cheek R, York JM, Chua B, Milsom WK, McCracken KG, Scott GR. Mitochondrial physiology in the skeletal and cardiac muscles is altered in torrent ducks, *Merganetta armata*, from high altitudes in the Andes. J Exp Biol. 2016;219:3719–28. doi:10.1242/jeb.142711.

Dodd GAA, Scott GR, Milsom WK. Ventilatory roll off during sustained hypercapnia is gender specific in pekin ducks. Respir Physiol Neurobiol. 2007;156:47–60.

Engel S, Biebach H, Visser GH. Water and heat balance during flight in the rose-colored starling (*Sturnus roseus*). Physiol Biochem Zool. 2006;79:763–74.

Faraci FM. Adaptations to hypoxia in birds: how to fly high. Annu Rev Physiol. 1991;53:59–70.

Feinsinger P, Colwell RK, Terborgh J, Budd S. Elevation and the morphology, flight energetics, and foraging ecology of tropical hummingbirds. Am Nat. 1979;113:481–97.

Flousek J, Telensky T, Hanzelka J, Reif J. Population trends of central European montane birds provide evidence for adverse impacts of climate change on high-altitude species. PLoS One. 2015;10:e0139465.

Galen SC, Natarajan C, Moriyama H, Weber RE, Fago A, Benham PM, Chavez AN, Cheviron ZA, Storz JF, Witt CC. Contribution of a mutational hot spot to hemoglobin adaptation in high-altitude Andean house wrens. Proc Natl Acad Sci USA. 2015;112:13958–63.

Gasner MR, Jankowski JE, Ciecka AL, Kyle KO, Rabenold KN. Projecting the local impacts of climate change on a Central American montane avian community. Biol Conserv. 2010;143:1250–8.

Geiseler SJ, Ludvigsen S, Folkow LP. K_{ATP}-channels play a minor role in the protective hypoxic shutdown of cerebellar activity in eider ducks (*Somateria mollissima*). Neuroscience. 2015;284:751–8.

Gerson AR, Guglielmo CG. Flight at low ambient humidity increases protein catabolism in migratory birds. Science. 2011;333:1434–6.

Gou X, Li N, Lian L, Yan D, Zhang H, Wei Z, Wu C. Hypoxic adaptations of hemoglobin in Tibetan chick embryo: high oxygen-affinity mutation and selective expression. Comp Biochem Physiol B Biochem Mol Biol. 2007;147:147–55.
Grubb BR. Allometric relations of cardiovascular function in birds. Am J Physiol. 1983;245:H567–72.
Gutierrez-Pinto N, McCracken KG, Alza L, Tubaro P, Kopuchian C, Astie A, Cadena CD. The validity of ecogeographical rules is context-dependent: testing for Bergmann's and Allen's rules by latitude and elevation in a widespread Andean duck. Biol J Linn Soc. 2014;111:850–62.
Hawkes LA, Balachandran S, Batbayar N, Butler PJ, Frappell PB, Milsom WK, Tseveenmyadag N, Newman SH, Scott GR, Sathiyaselvam P, et al. The trans-Himalayan flights of bar-headed geese (*Anser indicus*). Proc Natl Acad Sci USA. 2011;108:9516–9.
Hawkes LA, Balachandran S, Batbayar N, Butler PJ, Chua B, Douglas DC, Frappell PB, Hou Y, Milsom WK, Newman SH, et al. The paradox of extreme high-altitude migration in bar-headed geese *Anser indicus*. Proc Biol Sci. 2013;280:20122114.
Hawkes LA, Butler PJ, Frappell PB, Meir JU, Milsom WK, Scott GR, Bishop CM. Maximum running speed of captive bar-headed geese is unaffected by severe hypoxia. PLoS One. 2014;9:e94015.
Hochachka PW. Exercise limitations at high altitude: the metabolic problem and search for its solution. In: Gilles R, editor. Circulation, respiration, and metabolism. Berlin: Springer; 1985. p. 240–9.
Irwin DE, Irwin JH. Siberian migratory divides: the role of seasonal migration in speciation. In: Greenberg R, Marra PP, editors. Birds of two worlds: the ecology and evolution of migration. Baltimore: Johns Hopkins University Press; 2005. p. 27–40.
Jankowski JE, Londoño GA, Robinson SK, Chappell MA. Exploring the role of physiology and biotic interactions in determining elevational ranges of tropical animals. Ecography. 2012;36:1–12.
Jenni L, Jenni-Eiermann S. Fuel supply and metabolic constraints in migrating birds. J Avian Biol. 1998;29:521–8.
Jessen T-H, Weber RE, Fermi G, Tame J, Braunitzer G. Adaptation of bird hemoglobins to high altitudes: demonstration of molecular mechanism by protein engineering. Proc Natl Acad Sci USA. 1991;88:6519–22.
Köppen U, Yakovlev A, Barth R, Kaatz M, Berthold P. Seasonal migrations of four individual bar-headed geese *Anser indicus* from Kyrgyzstan followed by satellite telemetry. J Ornithol. 2010;151:703–12.
Lague SL, Chua B, Farrell AP, Wang Y, Milsom WK. Altitude matters: differences in cardiovascular and respiratory responses to hypoxia in bar-headed geese reared at high and low altitudes. J Exp Biol. 2016; doi:10.1242/jeb.132431.
Lee SY, Scott GR, Milsom WK. Have wing morphology or flight kinematics evolved for extreme high altitude migration in the bar-headed goose? Comp Biochem Physiol C Pharmacol Toxicol Endocrinol. 2008;148:324–31.
León-Velarde F, Sanchez J, Bigard AX, Brunet A, Lesty C, Monge C. High altitude tissue adaptation in Andean coots: capillarity, fiber area, fiber type and enzymatic activities of skeletal muscle. J Comp Physiol B. 1993;163:52–8.
Li M, Zhao C. Study on Tibetan chicken embryonic adaptability to chronic hypoxia by revealing differential gene expression in heart tissue. Sci China C Life Sci. 2009;52:284–95.
Lindsay CV, Downs CT, Brown M. Physiological variation in amethyst sunbirds (*Chalcomitra amethystina*) over an altitudinal gradient in winter. J Exp Biol. 2009;212:483–93.
Ludvigsen S, Folkow LP. Differences in in vitro cerebellar neuronal responses to hypoxia in eider ducks, chicken and rats. J Comp Physiol A Neuroethol Sens Neural Behav Physiol. 2009;195:1021–30.
Lui MA, Mahalingam S, Patel P, Connaty AD, Ivy CM, Cheviron ZA, Storz JF, McClelland GB, Scott GR. - High-altitude ancestry and hypoxia acclimation have distinct effects on exercise capacity and muscle phenotype in deer mice. Am J Physiol Regul Integr Comp Physiol. 2015;308:R779–91.
Maginniss LA, Bernstein MH, Deitch MA, Pinshow B. Effects of chronic hypobaric hypoxia on blood oxygen binding in pigeons. J Exp Zool. 1997;277:293–300.
Maina JN. Development, structure, and function of a novel respiratory organ, the lung-air sac system of birds: to go where no other vertebrate has gone. Biol Rev. 2006;81:545–79.
Maina JN. The design of the avian respiratory system: development, morphology and function. J Ornithol. 2015;156:S41–63.
Mathieu-Costello O. Histology of flight: tissue and muscle gas exchange. In: Sutton JR, Coates G, Remmers JE, editors. Hypoxia: the adaptations. Toronto, ON: BC Decker; 1990. p. 13–9.
Mathieu-Costello O, Agey PJ, Wu L, Szewczak JM, MacMillen RE. Increased fiber capillarization in flight muscle of finch at altitude. Respir Physiol. 1998;111:189–99.
McCracken KG, Barger CP, Bulgarella M, Johnson KP, Sonsthagen SA, Trucco J, Valqui TH, Wilson RE, Winker K, Sorenson MD. Parallel evolution in the major haemoglobin genes of eight species of Andean waterfowl. Mol Ecol. 2009a;18:3992–4005.
McCracken KG, Bulgarella M, Johnson KP, Kuhner MK, Trucco J, Valqui TH, Wilson RE, Peters JL. Gene flow in the face of countervailing selection: adaptation to high-altitude hypoxia in the betaA hemoglobin subunit of yellow-billed pintails in the Andes. Mol Biol Evol. 2009b;26:815–27.
McCracken KG, Barger CP, Sorenson MD. Phylogenetic and structural analysis of the HbA (α^A/β^A) and HbD (α^D/β^A) hemoglobin genes in two high-altitude

waterfowl from the Himalayas and the Andes: bar-headed goose (*Anser indicus*) and Andean goose (*Chloephaga melanoptera*). Mol Phylogenet Evol. 2010;56:649–58.

McGuire JA, Witt CC, Remsen JV, Corl A, Rabosky DL, Altshuler DL, Dudley R. Molecular phylogenetics and the diversification of hummingbirds. Curr Biol. 2014;24:910–6.

McWilliams SR, Guglielmo C, Pierce B, Klaassen M. Flying, fasting, and feeding in birds during migration: a nutritional and physiological ecology perspective. J Avian Biol. 2004;35:377–93.

Meir JU, Milsom WK. High thermal sensitivity of blood enhances oxygen delivery in the high-flying bar-headed goose. J Exp Biol. 2013;216:2172–5.

Mu C-Y, Su Y-H, Wang B, Huang Z-Y, Chen Y, Li Y, Liu R, Xu Q, Chen G-H, Zhao W-M. The complete mitochondrial genome of *Anser indicus* (Aves, Anseriformes, Anatidae). Mitochondrial DNA A DNA Mapp Seq Anal. 2016; doi:10.3109/19401736.2015.1015005.

Natarajan C, Projecto-Garcia J, Moriyama H, Weber RE, Munoz-Fuentes V, Green AJ, Kopuchian C, Tubaro PL, Alza L, Bulgarella M, et al. Convergent evolution of hemoglobin function in high-altitude Andean waterfowl involves limited parallelism at the molecular sequence level. PLoS Genet. 2015;11:e1005681.

Ocampo-Penuela N, Pimm SL. Elevational ranges of montane birds and deforestation in the western Andes of Colombia. PLoS One. 2015;10:e0143311.

Peters GW, Steiner DA, Rigoni JA, Mascilli AD, Schnepp RW, Thomas SP. Cardiorespiratory adjustments of homing pigeons to steady wind tunnel flight. J Exp Biol. 2005;208:3109–20.

Piiper J, Scheid. *Maximum* gas transfer efficacy of models for fish gills, avian lungs and mammalian lungs. Respir Physiol. 1972;14:115–24.

Qu Y, Zhao H, Han N, Zhou G, Song G, Gao B, Tian S, Zhang J, Zhang R, Meng X, et al. Ground tit genome reveals avian adaptation to living at high altitudes in the Tibetan plateau. Nat Commun. 2013;4:2071.

Qu Y, Tian SL, Han NJ, Zhao HW, Gao B, Fu J, Cheng YL, Song G, Ericson PGP, Zhang YE, et al. Genetic responses to seasonal variation in altitudinal stress: whole-genome resequencing of great tit in eastern Himalayas. Sci Rep. 2015;5:14256.

Scheid P. Avian respiratory system and gas exchange. In: Sutton JR, Coates G, Remmers JE, editors. Hypoxia: the adaptations. Toronto, ON: BC Decker; 1990. p. 4–7.

Scott GR. Elevated performance: the unique physiology of birds that fly at high altitudes. J Exp Biol. 2011;214:2455–62.

Scott GR, Milsom WK. Flying high: a theoretical analysis of the factors limiting exercise performance in birds at altitude. Respir Physiol Neurobiol. 2006;154:284–301.

Scott GR, Milsom WK. Control of breathing and adaptation to high altitude in the bar-headed goose. Am J Physiol Regul Integr Comp Physiol. 2007;293: R379–91.

Scott GR, Milsom WK. Control of breathing in birds: implications for high altitude flight. In: Glass ML, Wood SC, editors. Cardio-respiratory control in vertebrates: comparative and evolutionary aspects. Berlin: Springer; 2009. p. 429–48.

Scott GR, Cadena V, Tattersall GJ, Milsom WK. Body temperature depression and peripheral heat loss accompany the metabolic and ventilatory responses to hypoxia in low and high altitude birds. J Exp Biol. 2008;211:1326–35.

Scott GR, Egginton S, Richards JG, Milsom WK. Evolution of muscle phenotype for extreme high altitude flight in the bar-headed goose. Proc R Soc Lond B Biol Sci. 2009a;276:3645–53.

Scott GR, Richards JG, Milsom WK. Control of respiration in flight muscle from the high-altitude bar-headed goose and low-altitude birds. Am J Physiol Regul Integr Comp Physiol. 2009b;297:R1066–74.

Scott GR, Schulte PM, Egginton S, Scott ALM, Richards JG, Milsom WK. Molecular evolution of cytochrome c oxidase underlies high-altitude adaptation in the bar-headed goose. Mol Biol Evol. 2011;28:351–63.

Scott GR, Hawkes LA, Frappell PB, Butler PJ, Bishop CM, Milsom WK. How bar-headed geese fly over the Himalayas. Physiology. 2015;30:107–15.

Stalnov O, Ben-Gida H, Kirchhefer AJ, Guglielmo CG, Kopp GA, Liberzon A, Gurka R. On the estimation of time dependent lift of a European starling (*Sturnus vulgaris*) during flapping flight. PLoS One. 2015;10(9):e0134582.

Storz JF, Scott GR, Cheviron ZA. Phenotypic plasticity and genetic adaptation to high-altitude hypoxia in vertebrates. J Exp Biol. 2010;213:4125–36.

Swan LW. Goose of the Himalayas. Nat Hist. 1970;70:68–75.

Swanson DL, Vezina F. Environmental, ecological and mechanistic drivers of avian seasonal metabolic flexibility in response to cold winters. J Ornithol. 2015;156:S377–88.

Takekawa JY, Heath SR, Douglas DC, Perry WM, Javed S, Newman SH, Suwal RN, Rahmani AR, Choudhury BC, Prosser DJ, et al. Geographic variation in bar-headed geese *Anser indicus*: connectivity of wintering areas and breeding grounds across a broad front. Wildfowl. 2009;59:100–23.

Thomas SP, Follette DB, Thomas GS. Metabolic and ventilatory adjustments and tolerance of the bat *Pteropus poliocephalus* to acute hypoxic stress. Comp Biochem Physiol A Physiol. 1995;112:43–54.

Torre-Bueno JR. Temperature regulation and heat dissipation during flight in birds. J Exp Biol. 1976;65:471–82.

Tucker VA. Respiratory physiology of house sparrows in relation to high-altitude flight. J Exp Biol. 1968;48:55–66.

Ward S, Rayner JM, Moller U, Jackson DM, Nachtigall W, Speakman JR. Heat transfer from starlings *sturnus vulgaris* during flight. J Exp Biol. 1999;202:1589–602.

Ward S, Bishop CM, Woakes AJ, Butler PJ. Heart rate and the rate of oxygen consumption of flying and walking barnacle geese (*Branta leucopsis*) and bar-headed geese (*Anser indicus*). J Exp Biol. 2002;205:3347–56.

Weber JM. The physiology of long-distance migration: extending the limits of endurance metabolism. J Exp Biol. 2009;212:593–7.

Weber RE, Hiebl I, Braunitzer G. High altitude and hemoglobin function in the vultures *Gyps rueppellii* and *Aegypius monachus*. Biol Chem Hoppe Seyler. 1988;369:233–40.

West JB. Comparative physiology of the pulmonary blood-gas barrier: the unique avian solution. Am J Physiol Regul Integr Comp Physiol. 2009;297: R1625–34.

West JB. American medical research expedition to Everest. High Alt Med Biol. 2010;11:103–10.

Wilson MH, Newman S, Imray CH. The cerebral effects of ascent to high altitudes. Lancet Neurol. 2009;8:175–91.

Wilson RE, Peters JL, McCracken KG. Genetic and phenotypic divergence between low- and high-altitude populations of two recently diverged cinnamon teal subspecies. Evolution. 2013;67:170–84.

Zhang J, Hua ZQ, Tame JRH, Lu GY, Zhang RJ, Gu XC. The crystal structure of a high oxygen affinity species of haemoglobin (bar-headed goose haemoglobin in the oxy form). J Mol Biol. 1996;255:484–93.

Zhang YF, Carter T, Eyster K, Swanson DL. Acute cold and exercise training up-regulate similar aspects of fatty acid transport and catabolism in house sparrows (*Passer domesticus*). J Exp Biol. 2015a;218:3885–93.

Zhang YF, Eyster K, Liu JS, Swanson DL. Cross-training in birds: cold and exercise training produce similar changes in maximal metabolic output, muscle masses and myostatin expression in house sparrows (*Passer domesticus*). J Exp Biol. 2015b;218:2190–200.

Zheng WH, Li M, Liu JS, Shao SL, Xu XJ. Seasonal variation of metabolic thermogenesis in Eurasian tree sparrows (*Passer montanus*) over a latitudinal gradient. Physiol Biochem Zool. 2014;87:704–18.

Zhou TC, Shen XJ, Irwin DM, Shen YY, Zhang YP. Mitogenomic analyses propose positive selection in mitochondrial genes for high-altitude adaptation in galliform birds. Mitochondrion. 2014;18:70–5.

Molecular Aspects of Avian Lung Development

5

Rute S. Moura and Jorge Correia-Pinto

Abstract

The pulmonary system develops from a series of complex events that involve coordinated growth and differentiation of distinct cellular compartments. After lung specification of the anterior foregut endoderm, branching morphogenesis occurs generating an intricate arrangement of airways. This process depends on epithelial-mesenchymal interactions tightly controlled by a network of conserved signaling pathways. These signaling events regulate cellular processes and control the temporal-spatial expression of multiple molecular players that are essential for lung formation. Additionally, remodeling of the extracellular matrix establishes the appropriate environment for the delivery of diffusible regulatory factors that modulate these cellular processes. In this chapter, the molecular mechanisms underlying avian lung development are thoroughly revised. Fibroblast growth factor, WNT, sonic hedgehog, transforming growth factor-β, bone morphogenetic protein, vascular endothelial growth factor, and regulatory mechanisms such as microRNAs control cell proliferation, differentiation, and patterning of the embryonic chick lung. With this section, we aim to provide a snapshot of the current knowledge of the molecular aspects of avian lung development.

Keywords

Chick lung • ECM • FGF • WNT • SHH • Epithelial-mesenchymal interactions

Contents

R.S. Moura (✉)
Life and Health Sciences Research Institute (ICVS), School of Medicine, University of Minho, 4710 Braga, Portugal

ICVS/3B's – PT Government Associate Laboratory, 4710-057 Braga/Guimarães, Portugal
e-mail: rutemoura@med.uminho.pt

J. Correia-Pinto
Life and Health Sciences Research Institute (ICVS), School of Medicine, University of Minho, 4710 Braga, Portugal

ICVS/3B's – PT Government Associate Laboratory, 4710-057 Braga/Guimarães, Portugal

Department of Pediatric Surgery, Hospital de Braga, 4710-243 Braga, Portugal

J.N. Maina (ed.), *The Biology of the Avian Respiratory System*, DOI 10.1007/978-3-319-44153-5_5

5.1 General Considerations

The avian respiratory tract is a remarkably efficient system that accurately responds to the high metabolic rate of birds. It is constituted by the parabronchial lung that is involved in gas exchange and air sacs which control air movements. During development, the larynx, trachea, and lungs originate from the gut and are constituted by a thin layer of endoderm surrounded by dense mesoderm. To obtain a fully functional organ, a series of highly regulated molecular interactions must occur between these two cellular compartments to give rise to the parabronchial tree (endoderm-derived) and the muscles, connective tissue, and blood and lymphatic vessels (mesoderm-derived) (Bellairs and Osmond 2014).

In the chicken embryo, *Gallus gallus*, the embryonic lung springs up from the laryngotracheal groove around day 3 of incubation as a small, ridgelike outgrowth. After fusing along the ventral midline, it divides into right and left primordial bud looking like a pair of sacs lying on either side of the esophagus (Maina 2003). In the developing lung bud, the primary bronchus (mesobronchus) grows distally, and secondary bronchi sprout laterally from its dorsal surface into the surrounding mesenchyme, in a craniocaudal sequence. This lateral (or monopodial) branching is similar to the domain branching observed during mammalian lung development (Metzger et al. 2008). The initial stages of chick lung development are named according to the number of secondary buds formed: b1 stage corresponds to lungs with only one secondary bronchus, b2 stage presents two secondary bronchi, b3 stage presents three secondary bronchi per mesobronchus, and so on (Moura et al. 2011). Later, secondary bronchi branch and invade the mesenchyme to originate parabronchi (tertiary bronchi) that will then anastomose and finally connect the secondary bronchi. In the chick lung, the air sacs, a cyst-like structure, are formed ventrally during development as dilations of the mesobronchi and are named according to their position along the craniocaudal axis of the primary bronchus (cervical, intra-clavicular, thoracic, or abdominal) (Bellairs and Osmond 2014). Altogether, these processes depend on epithelial-mesenchymal interactions that rely on the activity of conserved signaling pathways that ultimately regulate cell proliferation, migration, and differentiation and that contribute to proper lung development.

5.2 Lung Specification

The epithelial compartment of the avian respiratory tract has its origin in endoderm progenitor cells of the foregut. This primitive structure possesses regional anterior-posterior (AP) information due to the differential expression of unique transcription factors (in visceral mesoderm and endoderm) that defines discrete domains that will then give rise to different organ primordia. The establishment of respiratory cell fate, in the case of the mammalian lung, is associated with the endodermal expression of *ttf-1* (thyroid transcription factor-1, or *nkx2.1*) that provides AP and dorsoventral (DV) positional information to the gut endoderm (Lazzaro et al. 1991; Minoo et al. 1999). Similarly, *nkx2.1* endodermal expression distinguishes the respiratory progenitors as early as the HH14 stage in the chick embryo (Sakiyama et al. 2003). Additionally, *tbx4* (a member of the T-box transcription factor family) is specifically expressed in the mesoderm surrounding the prospective lung primordium, indicating that it is involved in the initial specification of lung mesoderm (Sakiyama et al. 2003). After the establishment of the lung primordium,

TBX4 can induce endodermal budding by eliciting *fgf10* (fibroblast growth factor 10) expression, thus defining its AP mesodermal expression domain, and triggers *nkx2.1* expression, hence inducing endoderm differentiation; in fact, *tbx4-fgf10* expression overlaps with the posterior border of the *nkx2.1* expression domain at early lung development. Conversely, *nkx2.1* expression before the appearance of the lung bud primordium does not seem to be regulated by the *tbx4-fgf10* system. Furthermore, the *tbx4-fgf10* system is essential for the anatomic separation of the respiratory tract and the esophagus, namely, for the formation of the tracheoesophageal septum (Sakiyama et al. 2003).

nkx2.1 is expressed in the respiratory epithelial cells (trachea and main bronchi) of the early lung bud and all through embryonic lung development (Zeng et al. 1998). The hepatocyte nuclear factor 3β (HNF-3β) that plays a major role in foregut endoderm formation is also expressed in endodermal derivatives at the onset of lung bud formation. It was demonstrated that HNF-3β can activate *nkx2.1* in vitro. The fact that they coexist spatially in early lung organogenesis may indicate a possible role of these transcription factors in the specification of respiratory epithelium along the foregut axis (Zeng et al. 1998).

5.3 Dorsal-Ventral Patterning of the Chick Lung

The avian lung is organized in two distinct morphological and physiological regions, the bronchial tree and the air sacs (cyst-like) that appear dorsally and ventrally, respectively. The regional cyst-branch difference is defined early in development through the coordinated mesenchymal expression of *hoxb* genes (Sakiyama et al. 2000). Moreover, the disparity between the dorsal and ventral side of the chick lung might also be explained by changes in the diffusion rate of morphogens (namely, FGF10) as a result of local extracellular matrix modifications (Miura et al. 2009).

5.3.1 Hox Genes

hox genes, which are clustered on four chromosomes, are a family of transcription factors that specify local differences along the AP axis, therefore guiding the axial regionalization of the vertebrate embryo. There is an association between their location in the chromosome and the timing of their expression during embryogenesis (temporal collinearity) and the position of their expression domain along the AP axis (spatial collinearity). This characteristic temporal-spatial distribution is crucial for patterning the vertebrate body plan since it provides the correct identity to each vertebra (Wellik 2007).

In the chick respiratory tract, *hoxb-5* to *hoxb-9* genes are expressed in restricted domains and regulate the establishment of regional morphological subdivisions. Before the commencement of bronchial branching, the mesenchymal expression domain of *hoxb-5*, *hoxb-6*, *hoxb-7*, and *hoxb-8* exhibits a nested pattern with 5′-located genes on the cluster displaying a more ventral-distal restricted domain. Moreover, in the lung primordium, *hoxb-6* expression domain defines the identity of ventral vs. dorsal pulmonary mesenchyme that, consequently, presents different inductive ability regarding the epithelium (air saclike vs. branching morphogenesis, respectively). At later stages of lung development, the expression domains of these genes seem to correlate with the morphological subdivisions of the bronchial tree and the air sacs. *hoxb* genes might be upstream regulators of the cytodifferentiation of both mesenchyme and epithelium, and might have a role in the determination of the appropriate morphological subdivisions along the proximodistal axis of the lung (Sakiyama et al. 2000).

5.3.2 Diffusion Gradient

The pattern difference between dorsal and ventral chick lung can be explained by the different levels of mesenchymal FGF10, which are a consequence of different morphogen diffusion rates on both sides. FGF10 protein, a chemoattractant of lung epithelium (Bellusci et al. 1997; Park et al. 1998), is present in the dorsal (branched) lung mesenchyme forming a proximal-distal gradient, whereas its expression is weak in the ventral (cystic) mesenchyme. Additionally, the ventral tissue is looser than the dorsal one because it has lower cell density, causing FGF10 to diffuse faster in that region (Miura et al. 2009). On the other hand, the presence of elevated levels of heparan sulfate proteoglycan (HSPG) on the dorsal side, when compared to the ventral region, limits FGF10 diffusion (causing its accumulation). The dorsal-ventral difference in tissue architecture and HSPG concentration contributes to the difference of FGF10 diffusion coefficient between the two regions and is responsible for the developmental dichotomy that leads to branching morphogenesis dorsally and cyst formation ventrally. These findings suggest a possible mechanism as how HOXb transcription factors regulate cyst-branch morphology, probably through the regulation of HSPG expression levels and, eventually, cell proliferation in chick lung mesenchyme cells.

5.4 Branching Morphogenesis

After the establishment of the lung primordium, primary bronchi lengthen caudally and, from the epithelium, secondary bronchi bud laterally into the adjacent mesenchyme in a process known as branching morphogenesis. The new branches appear on the dorsal side of the primary bronchus in an anterior-posterior sequence whereas the mesobronchus continues to extend distally. It has been shown by tissue recombination that the mesenchyme governs the branching pattern of the epithelial primordium of the chick lung (Hilfer et al. 1985), suggesting that epithelial-mesenchymal interactions are crucial for chick lung development. In vitro lung explant cultures have been widely used to determine the role of different growth factors in epithelial branching and mesenchyme morphogenesis. One of the first examples is the work of Goldin and Opperman (1980) who reported the induction of supernumerary tracheal buds by epidermal growth factor (EGF). It has been successively demonstrated that tissue cross talk relies on the activation of conserved signaling pathways that control cellular processes such as extracellular matrix remodeling, growth, and differentiation, in a temporal-spatial manner.

5.4.1 Secondary Bud Initiation

The epithelium of the mesobronchus must go through a synchronized bending process to originate a new secondary bud, which requires the generation of forces conveyed among adjacent cells. It has been demonstrated that monopodial branching is initiated by the contraction of the apical/lumen side of epithelial cells (Kim et al. 2013). Apical constriction is driven by a contractile actomyosin network that triggers the narrowing of cellular apices of the dorsal airway epithelium, which causes tissue deformation and culminates with the appearance of a new bud. In fact, the inhibition of actomyosin contractility reduced the initiation of new buds. Moreover, in vitro FGF signaling inhibition by SU5402 (an FGF receptor antagonist) inhibits lung branching (Moura et al. 2011; Kim et al. 2013) and blocks apical constriction (Kim et al. 2013). Taken together these results suggest that, in the avian lung, apical actomyosin contraction might be FGF signaling dependent. Curiously, despite the high proliferation levels observed along the primary bronchus, blocking cell cycle progression in the mesobronchus did not affect the bud initiation process; nevertheless, proliferation might have a morphogenetic role within and adjacent to the new branches. Notwithstanding, apical constriction and differential cell proliferation are not enough to completely shape the developing lung bud (Kim et al. 2013), pointing to a potential physical role for the mesenchyme (Blanc et al. 2012).

5.4.2 Extracellular Matrix

For branching morphogenesis to occur, the extracellular matrix (ECM) undergoes a series of remodeling processes to suit epithelial growth. The particular composition of ECM throughout lung development directly affects the availability and, consequently, the activity of soluble factors since it influences not only diffusion rates but also the accessibility of their cognate receptors.

The basement membrane, which lies at the interface between epithelial and mesenchymal compartment, has two main components: the basal lamina, directly adjacent to the basal surface of the epithelial cells, and the fibrillar meshwork containing type I collagen. During branching morphogenesis, the basal lamina becomes thinner at the tip of newly formed branches due to reduced glycosaminoglycan (GAG) synthesis. In fact, decreased levels of hyaluronic acid (HA), chondroitin sulfate (CS), keratan sulfate (KS), and collagen are observed at the tip (Gallagher 1986; Abbott et al. 1991). On the other hand, in the interbud region, there is an increase in collagen deposition and GAG accumulation due to stabilization by the collagen (Gallagher 1986). Regarding the mesenchymal compartment, GAGs and glycoproteins (GPs) present a distinct distribution pattern throughout development. At early branching stages, mesenchymal ECM displays visible amounts of HA in the growing regions, near the main bronchus and the branch tip, and this pattern is repeated every time a new bud is formed. As the new branch grows, the surrounding mesenchyme becomes enriched in CS. Later in development, differentiated secondary bronchi accumulate mainly GPs (Becchetti et al. 1988).

At points where a secondary branch will emerge, mesenchymal cells adjoining lung epithelium are flattened, and this feature seems to be correlated with bud initiation. It has been suggested that modification of cell shape might be responsible for the loss of collagen observed in the basement membrane, probably due to the presence of collagenase. This structural weakening facilitates the bulging of new buds. After bud initiation, the new branch must elongate into the surrounding mesenchyme which encompasses cell proliferation. The accumulation of tenascin in the mesenchyme at the incipient bud tip, and in the distal mesobronchus tip, unveils a potential role for this molecule in bud extension (Abbott et al. 1991).

Throughout lung development, GAG/GP levels and distribution (in both mesenchymal ground substance and basement membrane) vary in a spatial-temporal manner to enable epithelial morphogenesis and, later on, the differentiation of thin structures (like the air-blood barrier). There are several molecules capable of changing ECM composition and that hence contribute to the regulation of the branching pattern. For instance, it has been demonstrated, in vitro, that polyamines (PAs) promote sulfate GAG accumulation in the mesenchymal compartment and have a positive effect on lung branching. It has been suggested that PAs are involved in the signal transduction cascade of TGFβ1 (transforming growth factor beta 1) (Evangelisti et al. 1997); TGFβ1 stimulates mesenchyme sulfate GAG deposition but has an opposite effect on lung branching (Stabellini et al. 2001).

On the other hand, ECM remodeling also involves degradation processes through the action of lysosomal enzymes, glycosidases that specifically degrade GAGs. In the chick lung, glycosidase activity is tightly associated with epithelial proliferation, bronchial tubule lengthening, and infiltration of the surrounding mesenchyme. There is a relationship between HA/CS/HS mesenchymal levels and the expression of glycosidases that specifically degrade each of these glycosaminoglycans and alter non-sulfated/sulfated GAG ratio (Stabellini et al. 2002; Calvitti et al. 2004). Enzyme activity, and consequently ECM features, is regulated by growth factors such as TGFβ2 and interleukin-1 (IL-1) that are known to promote tissue stabilization or growth, respectively. In fact, the expression pattern of these growth factors colocalizes with specific glycosaminoglycans that support tissue stabilization and growth (Stabellini et al. 2002, 2007; Calvitti et al. 2004).

To summarize, glycosaminoglycan content differs along the craniocaudal axis of the embryonic chick lung contributing to the regulation of cellular events (such as cell proliferation and differentiation). This distinctive spatial-temporal distribution creates the precise extracellular environment needed for the availability and activation of specific growth factors for the lung morphogenesis process.

5.4.3 FGF Signaling

Fibroblast growth factor (FGF) signaling pathway plays critical roles not only during organogenesis but also in the adult organism. FGF canonical ligands are secreted proteins that act in a paracrine fashion by interacting with neighboring FGF receptors (FGFRs). Canonical FGF ligands bind to HSPGs that function as cofactors for their interaction with FGFRs. FGFR1–FGFR4 are receptor tyrosine kinases that activate intracellular signaling cascades that are, in its turn, regulated by specific proteins as, for instance, Sprouty (SPRY) (reviewed by Ornitz and Itoh 2015).

Fibroblast growth factor signaling pathway, specifically FGF10 and its cognate receptor FGFR2, plays a crucial role in lung development (reviewed by El Agha and Bellusci 2014). In the chick lung, the *tbx4-fgf10* system is involved in the induction of endodermal budding, and it is essential for the formation of the tracheoesophageal septum (please consult Section 5.2). Additionally, *fgf10*, *fgfr1–fgfr4*, and *spry2* are expressed at early stages of chick lung branching (Moura et al. 2011). *fgf10* and *fgfr2* expression pattern is in agreement with the mammalian fetal lung suggesting a similar role for mesenchymal FGF10 as a proliferative factor that stimulates distal epithelial growth through the activation of epithelial FGFR2 (Fig. 5.1a, b). Furthermore, in vitro FGF signaling inhibition by SU5402 (an FGF receptor antagonist) elicited the formation of secondary bronchi with a cystic shape without an increase in their number, and a disruption of the mesenchymal tissue (Moura et al. 2011). Moreover, *fgfr1* is ubiquitously expressed and may be responsible for capturing proliferative factors in both compartments; *fgfr3* has a more proximal expression, whereas *fgfr4* is nearly absent at early stages; *spry2* is expressed in the distal tip supporting its association with FGF10 signaling center (Fig. 5.1c). This study demonstrates the importance of FGFRs in the epithelial-mesenchymal interactions that determine epithelial branching and mesenchyme growth in early chick lung development. It is plausible to believe that the disruption of the mesenchymal scaffold contributes to the cystic phenotype if one considers that a lower cell density substantially contributes to ventral cyst formation since it increases FGF diffusion rate (Miura et al. 2009). Another member of the FGF family, FGF2, has been described to be diffusely expressed in epithelial and mesenchymal lung cells from as early as day 3.5 (Maina et al. 2003). This expression pattern is in agreement to what is described for fetal rat lung (Han et al. 1992), unveiling a potential, yet unknown role for this ligand in the avian lung.

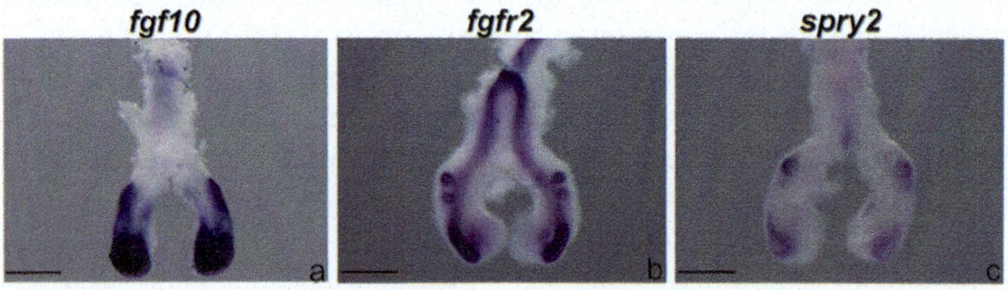

Fig. 5.1 Expression of FGF signaling members in the embryonic chick lung. Whole-mount in situ hybridization of stage b2 lungs probed with *fgf10* (**a**), *fgfr2* (**b**), and *spry2* (**c**). *fgf10* is present in the mesenchyme adjacent to the emerging new buds and in the distal mesenchyme surrounding the main bronchus; *fgfr2* is expressed in the distal epithelium of the main bronchus and secondary bronchi; *spry2* is present in the peri-epithelial mesenchyme bordering the main bronchus and secondary bronchi. *Scale bar*: 500 μm. Adapted from Moura et al. (2011)

5.4.4 SHH Signaling

The hedgehog (HH) signaling pathway is crucial for embryonic development and influences the organogenesis of several organs, specifically the lung. Cell surface transmembrane receptor Patched (PTCH) constitutively represses HH signaling because it blocks the activity of G protein-coupled transmembrane protein Smoothened (SMO). Hedgehog ligands, such as sonic hedgehog (SHH), are secreted proteins that bind PTCH, thus releasing SMO inhibition and allowing the expression of glioblastoma (GLI) zinc finger transcription factors that can either activate or inhibit transcription of target genes. Ligand availability regulates SHH signaling pathway since membrane proteins such as PTCH and HHIP (Hedgehog-interacting protein) bind to SHH and limit its diffusion (reviewed by Briscoe and Thérond 2013).

In the chick embryonic lung, *shh* is present in the trachea at embryonic day (E)6 (Davey et al. 2014) and in the proximal and distal epithelium at E10 (Loscertales et al. 2008). Moreover, in earlier stages (E4.5–E5.5) all the canonical elements of SHH signaling pathway are expressed in the same cellular compartments as their mammalian counterparts, although their proximodistal distribution is slightly changed. *shh* is expressed in the epithelium of the main bronchus and secondary bronchi, but it is absent from the tip of the growing buds (Fig. 5.2a). *smo*, *ptch1* (Fig. 5.2b), *hhip*, and *gli1* (Fig. 5.2c) expression mirror *shh* pattern, as it occurs in the mouse fetal lung (Moura et al. 2016). It is expected that SHH signaling is involved in the epithelial-mesenchymal interactions that regulate chick lung branching. In fact, in lung explants, exogenous SHH protein supplementation induces pulmonary hyperplasia mainly due to the expansion of mesenchyme. In vitro inhibition by cyclopamine that inhibits HH signaling through direct interaction with SMO causes a loss of lung epithelial branching which leads to pulmonary hypoplasia (Loscertales et al. 2008). In *talpid*^{3} chicken mutants, which have a defective SHH signaling, lung morphogenesis is seriously disturbed. *talpid*^{3} lungs do not express *shh* and display a hypoplastic phenotype with severe abnormalities in both epithelial and mesenchymal compartments (Davey et al. 2014). Altogether, these phenotypes disclose a role for SHH signaling in chick lung branching as it occurs in the mammalian lung, although it might not be required for the formation of the tracheoesophageal septum (Litingtung et al. 1998; Pepicelli et al. 1998). In the mammalian lung, FGF and SHH signaling are engaged in a twisted negative feedback interaction that is crucial to secondary branching (Pepicelli et al. 1998; Chuang et al. 2003). In the chick lung, *shh* is not a downstream target of FGF signaling as it occurs in the mammalian lung (Moura et al. 2016). Notwithstanding, the absence of *shh* epithelial expression coincides with *fgf10* expression regions, whereas the presence/activity of *shh* corresponds to pulmonary areas without *fgf10* mesenchymal expression. FGF-Shh signaling interplay in the mammalian lung promotes the same local *fgf10* expression differences that

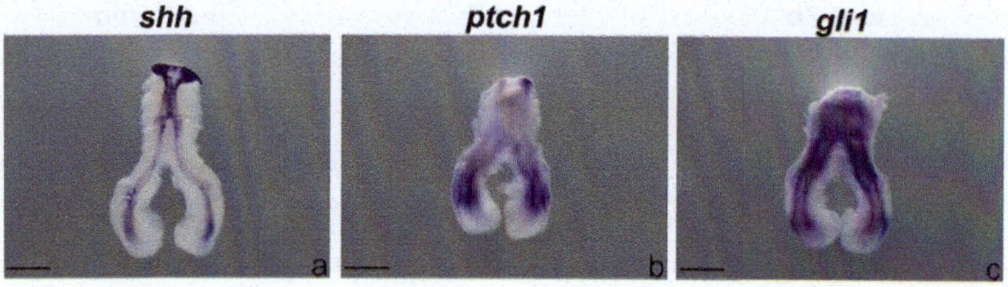

Fig. 5.2 Expression of SHH signaling members in the embryonic chick lung. Whole-mount in situ hybridization of stage b2 lungs probed with *shh* (**a**), *ptch1* (**b**), and *gli1* (**c**). *shh* is expressed in the tracheal region and in the epithelium except the distal tip of the main bronchus and secondary bronchi; *ptch1* and *gli1* are detected in the peri-epithelial mesenchyme of the main bronchial tree, and they are absent from the distal mesenchyme. *Scale bar*: 500 μm. Adapted from Moura et al. (2016)

will contribute to proper branching (Chuang et al. 2003). Despite the fact that *shh* is not FG10-dependent, the regional differences are maintained in the chick lung. The chick lung seems to have some features that are species-specific regarding SHH signaling and that may probably be because avian secondary bronchi undergo epithelial outgrowth without tip splitting.

5.4.5 WNT Signaling

The canonical WNT/β-catenin pathway regulates numerous developmental processes, as for instance cell fate and morphogenesis. WNT-secreted glycoproteins bind to Frizzled (FZD) transmembrane receptors and their co-receptors LRP5/LRP6, low-density lipoprotein receptor-related protein (LRP). The WNT-FZD-LRP complex triggers a series of intracellular reactions that finally leads to the stabilization of cytoplasmic β-catenin and its translocation to the nucleus. Once in the nucleus, it binds to TCF transcriptional complexes and regulates gene expression. This signaling pathway is modulated by secreted Frizzled-related proteins (SFRPs) and Dickkopf (DKK) proteins (reviewed by Baarsma et al. 2013). The noncanonical Wnt signaling is β-catenin independent and activates planar cell polarity (PCP) and WNT/Ca^{2+} pathways, either via FZD receptors or other receptors, including the orphan tyrosine kinase ROR2 (Semenov et al. 2007).

Several WNT ligands (*wnt-1*, *wnt-2b*, *wnt-3a*, *wnt-5a*, *wnt-7b*, *wnt-9a*) are expressed in early stages of chick lung development (Moura et al. 2014). Their expression is, overall, in agreement with the mammalian counterparts and uncovers a similar function for these ligands in the avian pulmonary system. For instance, *wnt-7b* is expressed in the developing airway epithelium (Fig. 5.3a) as it occurs in the mouse model (Shu et al. 2002). Likewise, *lrp5*, *lrp6*, *sfrp1*, *dkk1*, *β-catenin* (Fig. 5.3b), and *axin2* (Fig. 5.3c) are also expressed at early stages of chick lung development. Moreover, phospho/non-phospho forms of LRP6 and β-catenin are present in this embryonic tissue, which proves that WNT signaling is active. In vitro inhibition of canonical WNT signaling leads to an impairment of lung branching as it happens with the mammalian lung. Collectively, these results indicate that canonical WNT signaling is implicated in the molecular mechanisms required for chick lung branching morphogenesis (Moura et al. 2014). Branching morphogenesis is an intricate process that relies on a network of conserved signaling pathways that regulate each other's components to fine-tune the overall process. In this sense, a WNT-FGF cross talk is not unforeseen. In fact, WNT signaling inhibition induces a decrease of *spry2* (the downstream target of FGF signaling) in the chick lung, and it acts upstream FGF signaling (Moura et al. 2014).

Noncanonical WNT signaling seems to be involved in mid-developmental stages of chick lung development. In the developing avian lung, WNT-5a acts noncanonically, through the ROR2 receptor, to regulate pulmonary distal airway and vasculature development (Loscertales et al.

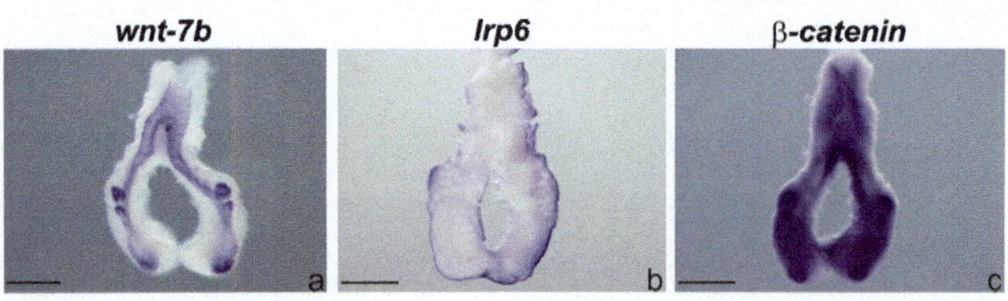

Fig. 5.3 Expression of WNT signaling members in the embryonic chick lung. Whole-mount in situ hybridization of stage b2 lungs probed with *wnt-7b* (**a**), *lrp6* (**b**), and *β-catenin* (**c**). *wnt-7b* is expressed all over the pulmonary epithelium of the main bronchus, markedly in the distal tip and the secondary bronchi; *lrp6* is present in lung mesothelium; *β-catenin* is expressed throughout all lung mesenchyme and in the epithelial tip of secondary buds and main bronchus. *Scale bar*: 500 μm. Adapted from Moura et al. (2014)

2008). Misexpression of *wnt-5a* results in changes of the vascular pattern, lung hypoplasia, and altered expression patterns of *shh*, *fgf10*, *bmp4* (bone morphogenetic protein 4), *fibronectin*, and *vegf* (vascular endothelial growth factor). *wnt-5a* most likely guides pulmonary vascular patterning through the regulation of VEGF pathway via its effect on fibronectin. Branching dysfunction is probably due to an increase in fibronectin levels and a decrease in cell adhesion. The hypoplastic phenotype is rescued by SHH supplementation or inhibition of fibronectin function. These results indicated that *wnt5a* acts upstream of *shh* (in addition to *fgf10* and *bmp4*) and that fibronectin levels are regulated directly by the WNT ligand and indirectly through its regulation of SHH. *wnt-5a* might also have a role in tracheal morphogenesis and, eventually, in the inhibition of branching in this particular region since it is expressed in the tracheal mesenchyme around E5 (Sakiyama et al. 2000; Moura et al. 2014). In fact, *wnt-5a*$^{-/-}$ mouse lungs display a reduced number of cartilage rings that cause the reduction of the trachea size (Li et al. 2002).

5.4.6 TGFβ-BMP Signaling

The transforming growth factor-β (TGFβ) superfamily of secreted cytokines plays critical roles in a diverse set of cellular processes during embryogenesis as well as in mature tissues. Bone morphogenetic proteins (BMPs) constitute the largest subgroup of the TGFβ superfamily. TGFβ and BMP ligands bind to specific membrane receptors that transmit the signal to intracellular SMAD proteins that then trigger the appropriate cellular response by regulating the transcription of target genes. Typically, these cytokines act as inhibitory morphogens (Massagué 2012).

In the embryonic chick lung, TGFβ1 and BMP4 influence branching and growth of the airways in a different manner (Gleghorn et al. 2012). In vitro supplementation with recombinant TGFβ1 or a TGFβ receptor inhibitor causes a decrease/increase in branching, respectively, and in both cases a reduction in lung size. On the other hand, BMP4 supplementation has no effect on branching, but it diminishes organ growth; BMP signaling inhibition significantly reduces branching and causes an increase in overall lung growth. In conclusion, TGFβ1 preferentially prevents lung branching, while BMP4 affects lung growth, and together contribute to defining the dynamics of lung morphogenesis. Despite the alterations in the gross morphology of the lungs, the relative positioning of secondary buds along the primary bronchi remains unaltered which means that branch sites scale according to the size of the organ (rather than being established at fixed positions) (Gleghorn et al. 2012). *tgfβ1* has also been described as a potential target of regulating nuclear factor-kB (NF-kB) (Muraoka et al. 2000). NF-kB increased mesenchymal activity leads to a decrease in epithelial branching, a decrease in cellular proliferation, and an increase in mesenchymal *tgfβ1* expression which most likely exerts an inhibitory action on the adjacent epithelium. This mechanism is in agreement with TGFβ1 inhibitory role described by Gleghorn et al. (2012). Conversely, hyperactivation of mesenchymal NF-kB repressed mesenchymal *fgf10* and *bmp4* expression which contributes to epithelial branching inhibition (Muraoka et al. 2000). This mechanism is a clear example of the epithelial-mesenchymal interactions that take place during branching morphogenesis, in which the transcriptional regulation of growth factors produced in the proximal mesenchyme affects epithelial branching of the neighboring region.

5.4.7 miRNAs

MicroRNAs (miRNAs) have emerged as important regulators of development. These small non-coding RNAs function posttranscriptionally by interacting with the 3′ untranslated region of specific mRNAs in a sequence-specific manner and, therefore, cause mRNA degradation or translational inhibition. Mature miRNAs are formed by a multistep process that involves different enzymes and proteins: DROSHA/DGCR8 nuclear complex, exportin-5 nuclear membrane protein, and cytoplasmic DICER enzyme (reviewed by Ivey and Srivastava 2015). There is also an alternative noncanonical miRNA biosynthesis pathway that can be DROSHA/DGCR8 independent or DICER independent (Abdelfattah et al. 2014).

It has been recently shown that miRNA processing machinery is expressed in the developing chick lung, supporting the previously recognized regulatory role of this mechanism in epithelial and mesenchymal morphogenesis (Moura et al. 2015). The dorsal and distal mesenchymal expression of the biogenesis machinery may well indicate a potential regulatory role in mesenchymal morphogenesis, through the regulation of specific genes. *exportin-5* and *dicer1* are also expressed on the apical side of the distal lung epithelium and emerging secondary bronchi, which points toward the existence of a noncanonical miRNA pathway in the chick embryonic lung (Moura et al. 2015).

Sanford et al. (2016) have revealed that *miR-449a* is expressed in the distal lung epithelium, and not in the mesenchyme, at embryonic day 12. Moreover, they showed that *cdc20b* (a proxy for chick miR-449a) expression was highest at E20 and that transcription factor N-Myc (a predicted target for this miRNA and related with the proliferation of undifferentiated progenitors) was negatively correlated with CDC20B/miR-449a. miR-449a overexpression, by *in ovo* retroviral infection, disrupted lung growth leading to lung hypoplasia. At E13, chick lungs displayed a decreased number of airways. At E15, less severe affected lungs presented reduced proliferation levels and compromised airway epithelial differentiation. This study unveiled a role for miR-449a in the regulation of chick lung differentiation and proliferation most likely through N-Myc (Sanford et al. 2016).

miR-15a seems to be associated with the mechanisms that are responsible for hypoxia-induced changes in lung development, probably through the translational inhibition of the antiapoptotic protein BCL-2 (Hao et al. 2014). *miR-15a* expression has its peak around E19-20 that coincides with the formation of the crosscurrent gas exchange system (CCGS); at the same time, BCL-2 protein levels are diminished. In the embryonic chick lung, the expression of *miR-15a* is induced by hypoxia stress, whereas BCL-2 protein levels decline. Hao et al. (2014) demonstrated that chicken miR-15a posttranscriptionally silences chick *bcl-2*. In this scenario, antiapoptotic mechanisms are inhibited causing an increase in mesenchymal apoptosis. The reduction in the mesenchymal compartment is necessary for the development of a thin blood-gas barrier that determines, together with CCGS, the avian lung air-diffusing capacity. The hypoxia-induced miR-15a-*bcl-2* system mediates CCGS/BGB establishment and is essential for understanding the adaptational development of the chick lung (Hao et al. 2014).

5.5 Surfactant Synthesis

Pulmonary surfactant is constituted by a combination of lipids and proteins that reduce surface tension, therefore, preventing mammalian lung alveoli to collapse at the end of expiration. The avian lung does not exhibit alveoli. Nonetheless, it contains the surfactant that functions to sustain airflow in the air capillaries (Bernhard et al. 2001).

The major class of phospholipid in avian surfactant is phosphatidylcholine (PC), and, quantitatively, the most important subcomponent is disaturated phosphatidylcholine (DSPC) (Fujiwara et al. 1970), namely, dipalmitoylphosphatidylcholine (PC16:0/16:0). Regarding protein content, it is enriched in surfactant protein B (SP-B), whereas SP-C is absent (Bernhard et al. 2001). Recently, SP-A has been identified as an extracellular component of the lung lining fluid of the avian tertiary bronchi (Zhang et al. 2016). The differences in surfactant protein content most likely reflect the particulars of the avian tubular lung.

At the molecular level, pneumocytes can be distinguished by specific protein markers as, for instance, SP-B for cuboidal type II epithelial cells and aquaporin 5 for type I pneumocytes from E17 onward (Bjørnstad et al. 2014), similarly to the mammalian lung. The transcription of the surfactant genes is controlled by nuclear proteins including TTF-1 and HNF-3β, which can bind to regulatory regions of these target genes (Clevidence et al. 1993; Zhang et al. 1997). In late stages of chick lung development (E15), SP-A and SP-B expression pattern overlaps with TTF-1 and HNF-3β suggesting that the transcription regulatory mechanism might be conserved in the avian lung (Zeng et al. 1998).

At the cellular level, flattened type I pneumocytes that facilitate gas exchange line the avian respiratory membrane. Regarding cuboid (type II) pneumocytes, they are visible around E17 and located in the atrial walls, air sacs, and parabronchi but not in the air capillaries (Bernhard et al. 2001; Maina 2003). It has been demonstrated that avian surfactant, produced by type II cells, is capable of efficiently adsorbing and extending into an air/liquid interface like the air capillaries where it exerts its role (Bernhard et al. 2001). Surfactant-producing cells also present storage organelles for surfactant (known as lamellar bodies) that are visible around E16, increase at E18, and decrease post-hatching (Hylka 1989). Glycogen granules are also present in this cell type, and they are depleted between E14 and post-hatching which support the concept that they may contribute with the precursor material to form pulmonary DSPC; nevertheless, the synthesis of DSPC near the end of incubation may rely upon other substrates as lipids of yolk (Hylka 1989).

5.5.1 Hormonal Regulation

Around E18, before pipping, both PC and DSPC contents increase which indicates the beginning of breathing. This increase coincides with a peak of corticoids, thyroid hormone, and prolactin. Hylka and Doneen (1983) have demonstrated that corticosterone inhibits cellular proliferation (leading to a reduction in lung size) and stimulates surfactant phospholipid synthesis (increasing total pulmonary phospholipid and PC content), comparable to mammals. Moreover, the removal of pituitary by hypophysectomy triggers a decrease in lipid content in the embryonic lung, probably due to delayed appearance of lamellar bodies (Hylka and Doneen 1983). Besides, pituitary hormones seem to be responsible for acquisition of maximum content of glycogen by the lung before day 16 (Hylka 1989).

Thyroid hormones are known to control organ growth and development. Nonetheless, thyroid hormone (TH) has no effect on surfactant production, despite the described increase of *thrβ* (TH receptor-β) mRNA at E19 (Forrest et al. 1990). This feature might indicate that TH acts by enhancing the sensitivity to glucocorticoids or that its effects were already maximal at the ages tested meaning that further stimulation with exogenous THs does not result in an additional effect (De Groef et al. 2013). Nonetheless, in a chemically induced model of hypothyroidism, lung maturation was hindered, specifically pneumocyte and vascular differentiation (Bjørnstad et al. 2016). TRβ1 expression levels are elevated in lungs of methimazole (MMI)-treated embryos; on the other hand, Kruppel-like factor 2 (*klf2*) expression levels remain unaltered. KLF2 is a TH-dependent transcription factor involved in type I pneumocyte differentiation program in the mammalian lung (Pei et al. 2011), but that seems to be TH-independent in the avian lung. Notwithstanding, in MMI-induced hypothyroidism, type I (and type II) pneumocyte-specific cell markers are reduced which points to an impairment in pneumocyte differentiation. Moreover, specific miRNAs showed upregulated upon MMI treatment highlighting their role as regulators of developmental lung processes (Bjørnstad et al. 2016).

On the other hand, the embryonic chick lung might function as an extrapituitary production site of pituitary hormones. For instance, luteinizing hormone (LH) has been detected in the trachea and lung of E3–E7 embryos (Shirasawa et al. 1996) and β-TSH (thyrotropin) in the bronchus of E7 developing lung (Murphy and Harvey 2001). Growth hormone (GH) and its receptor (GHR) are also present in the embryonic chick lung as early as E7 (Beyea et al. 2005). Overall these results suggest a potential, so far undetermined, autocrine/paracrine role for these hormones in lung development, before the appearance of the pituitary gland.

5.5.2 Regulation by Oxygen Levels

Surfactant levels are regulated by fetal oxygen tension, and exposure to abnormal levels of oxygen has diverse effects depending on the developmental window in which the insult occurs. In normoxic conditions (21% O_2), disaturated phospholipid (DSPL) content increases between E16 and E19 which reflects the physiologic increase in surfactant production in this developmental window. Mild hypoxic conditions (17% O_2), from E10 forward, speed up surfactant maturation when compared to normoxic conditions. Moreover, corticosterone levels increase, while T3 levels remain unaltered in response to hypoxia, so it is possible that the maturation process is mediated by glucocorticoids (Blacker et al. 2004).

Furthermore, it has been shown that both chronic prenatal hypoxia (15% O_2) and hyperoxia (60% O_2) occurring during late stages of development (from E15 onwards) trigger surfactant synthesis (at E19) which points to an acceleration of lung maturation. In these hypoxic conditions (at E16), expression of *vegf* isoforms 122, 146, and 190 increases transitorily before the stimulation of surfactant production disclosing a potential, yet still concealed, role for VEGF in this process. Conversely, in hyperoxic conditions *vegf* expression levels remained unaltered. On the other hand, chronic long-term hypoxia (from E6 ahead) does not significantly affect surfactant synthesis and VEGF expression. The maturation of the surfactant system displays a high level of plasticity since it can adjust according to fetal oxygenation (Been et al. 2010).

5.6 Hypoxic Adaptation

Embryonic development is highly dependent on oxygen levels. Hypoxia can lead to profound developmental abnormalities, but the severity of these anomalies depends on the duration and onset of the insult (Zhang and Burggren 2012), so hypoxic adaptation is crucial for the survival of avian embryos in such adverse conditions. Chicken populations that usually live at high altitudes have developed mechanisms that help them to cope with lower oxygen concentrations. For instance, the expression in the embryonic lung of endothelial nitric oxide synthase (eNOS), which is responsible for the synthesis of nitric oxide (NO, a potent vasodilator), is higher in a highland breed when compared to a lowland breed (Peng et al. 2012). The authors suggested that, from the genetic point of view, the highland chicken lung may have more efficient respiration capability compared with the lowland chicken lung against high-altitude hypoxia due to NO-induced vasodilation although embryonic lungs do not perform ventilation function until pipping at E19 (Peng et al. 2012).

At the cellular level, hypoxia increases oxidative stress and accumulation of reactive oxygen species (ROS) (Duranteau et al. 1998). The catalytic products of heme oxygenase 1 (HO-1) have a cytoprotective role under hypoxia since they can scavenge ROS. The *ho-1* pulmonary expression has its highest level at E19 probably due to the degree of tissue hypoxia reached in the chicken embryo at late stages of embryonic development (Gou et al. 2014). Moreover, when subjected to hypoxic conditions (13% O_2), *ho-1* expression shows upregulated in lung tissues especially at E19 of highland chickens. HO-1 has an indirect role in erythropoiesis since it supplies iron, from heme recycling, needed for the formation of new red blood cells. Taking this into consideration, the augmented levels of HO-1 as a response to hypoxia contribute to relieving hypoxic damage since they contribute to enhancing blood oxygen transport capacity (Gou et al. 2014).

Several studies have reported that embryonic lung mass is virtually unaffected by hypoxic incubation which reflects the absolute need of a proper respiratory surface area to survive (Chan and Burggren 2005; Lewallen and Burggren 2015). Moreover, lung morphology is not significantly altered under hypoxic conditions (15% O_2) since pulmonary morphometric parameters as well as blood vessel density are very similar to normoxic conditions. Nonetheless, at the

Table 5.1 Chick lung signaling molecules: expression patterns and known/putative functions

Symbol	Name	Expression pattern	Function	References
TTF-1 (or NKX2.1)	Thyroid transcription factor-1	Endoderm Epithelial cells	Distinguishes respiratory progenitors (HH14) Epithelial marker Regulates transcription of surfactant genes	Sakiyama et al. (2003), Zhang et al. (1997) and Zeng et al. (1998)
TBX4	T-box transcription factor family member	Mesoderm	Specification of lung mesoderm Induction of endodermal budding (*fgf10*-dependent)	Sakiyama et al. (2003)
HNF-3β	Hepatocyte nuclear factor 3β	Endoderm	Specification of respiratory epithelium Regulates transcription of surfactant genes	Zhang et al. (1997) and Zeng et al. (1998)
HOXb-5 to HOXb-9	Homeobox transcription factors	Mesenchyme	Dorsal-ventral patterning	Sakiyama et al. (2000)
Actomyosin	Filamentous actin & phosphorylated myosin light chain	Apical surface of emerging bud epithelium	Initiation of apical bud constriction (FGF-dependent)	Kim et al. (2013)
GAGs GPs	Glycosaminoglycans Glycoproteins	Extracellular matrix	Differential spatial-temporal levels contribute to epithelial and mesenchymal morphogenesis	Gallagher (1986), Becchetti et al. (1988), Abbott et al. (1991), Evangelisti et al. (1997), Stabellini et al. (2001, 2002, 2007) and Calvitti et al. (2004)
FGF2	Fibroblast growth factor 2	Epithelium and mesenchyme		Maina et al. (2003)
FGF10	Fibroblast growth factor 10	Mesenchyme	Lung bud outgrowth	Moura et al. (2011)
FGFR1	Fibroblast growth factor receptor 1	Epithelium and mesenchyme		
FGFR2	Fibroblast growth factor receptor 2	Epithelium	FGF10 receptor	
FGFR3	Fibroblast growth factor receptor 3	Proximal peri-epithelial mesenchyme		
SPRY2	Sprouty 2	Distal peri-epithelial mesenchyme	FGF downstream target	
HSPG	Heparan sulfate proteoglycan	Mesenchyme	Dorsal-ventral patterning	Miura et al. (2009)
SHH	Sonic hedgehog	Trachea and epithelium (except distal regions)	Branching morphogenesis	Loscertales et al. (2008), Davey et al. (2014) and Moura et al. (2016)
PTCH1	Patched 1	Mesenchyme (except distal regions)		Moura et al. (2016)
SMO	Smoothened	Mesenchyme		

(continued)

Table 5.1 (continued)

Symbol	Name	Expression pattern	Function	References
HHIP	Hedgehog-interacting protein	Mesenchyme (except distal regions)		
GLI1	Glioblastoma zinc finger transcription factor 1	Mesenchyme (except distal regions)		
WNT-5a (noncanonical WNT pathway)	Wingless-related MMTV integration site-5a	Trachea (E5) Epithelium (E11 onwards)	Tracheal morphogenesis Vascular patterning (via VEGF-fibronectin) Pulmonary distal airway morphogenesis	Sakiyama et al. (2000), Loscertales et al. (2008) and Moura et al. (2014)
WNT-7b (canonical WNT pathway)	Wingless-related MMTV integration site-7b	Epithelium	Lung branching	Moura et al. (2014)
LRP6	Low-density lipoprotein receptor-related protein 6	Mesothelium		
β-catenin	β-catenin	Mesenchyme and epithelial tips		
TGFβ1	Transforming growth factor-β	Mesenchyme	Lung branching and lung growth Target of nuclear factor-kB	Muraoka et al. (2000) and Gleghorn et al. (2012)
BMP4	Bone morphogenetic protein 4	Mesenchyme	Lung growth	
NF-kB	Nuclear factor-kB	Mesenchyme	Epithelial branching (*fgf10* and *bmp4* dependent)	Muraoka et al. (2000)
miR-449a	MicroRNA-449a	Distal lung epithelium	Lung branching: regulation of differentiation and proliferation	Sanford et al. (2016)
miR-15a	MicroRNA-15a		Hypoxia-induced adaptation	Hao et al. (2014)

molecular level, the expression of hypoxia-inducible factor (HIF-1) is increased and, consequently, *vegf* transcription is upregulated. In hyperoxic conditions (30% O_2), the *vegf* expression is also upregulated at E18 (Lewallen and Burggren 2015). VEGF is a key factor in promoting pulmonary angiogenesis and vasculogenesis and contributes to enhancing perfusion and tissue oxygenation (reviewed by Woik and Kroll 2015). However, in the case of the avian lung, its presence does not correlate with an increase in blood vessel formation, and in this context, its role is poorly understood.

5.7 Concluding Remarks

The molecular mechanisms underlying avian lung development are being increasingly more explored. Nonetheless, they are still less studied than mammalian ones. Table 5.1 compiles the expression pattern of chick lung's signaling molecules, transcription factors, and other molecular factors mentioned in this chapter as well as their putative or known role. The chick model is a suitable and appealing animal model for research since it circumvents major ethical issues; is affordable, with shorter gestation times; and is more easily maintained/manipulated than mammalian models. Despite the major structural morphological differences, when compared to mouse/rat adult lung, early avian pulmonary development presents several molecular similarities. On the other hand, molecular differences may account for the specifics of the avian tubular lung. Nevertheless, further studies are still needed to dissect and to unveil mechanisms responsible for the divergence between mammals and birds.

Acknowledgments Funding sources: FEDER funds, through the Competitiveness Factors Operational Program (COMPETE), and by National funds, through the Foundation for Science and Technology (FCT), under the scope of the project POCI-01-0145-FEDER-007038; project NORTE-01-0145-FEDER-000013, supported by the Northern Portugal Regional Operational Program (NORTE 2020), under the Portugal 2020 Partnership Agreement, through the European Regional Development Fund (FEDER).

References

Abbott LA, Lester SM, Erickson CA. Changes in mesenchymal cell-shape, matrix collagen and tenascin accompany bud formation in the early chick lung. Anat Embryol. 1991;183(3):299–311. doi:10.1007/BF00192217.

Abdelfattah AM, Park C, Choi MY. Update on non-canonical microRNAs. Biomol Concepts. 2014;5(4):275–87. doi:10.1515/bmc-2014-0012.

Baarsma HA, Königshoff M, Gosens R. The WNT signaling pathway from ligand secretion to gene transcription: molecular mechanisms and pharmacological targets. Pharmacol Ther. 2013;138(1):66–83. doi:10.1016/j.pharmthera.2013.01.002.

Becchetti E, Evangelisti R, Stabellini G, Pagliarini A, del Borrello E, Calastrini C, Carinci P. Developmental heterogeneity of mesenchymal glycosaminoglycans (GAG) distribution in chick embryo lung anlagen. Am J Anat. 1988;181(1):33–42. doi:10.1002/aja.1001810105.

Been JV, Zoer B, Kloosterboer N, Kessels CG, Zimmermann LJ, van Iwaarden JF, Villamor E. Pulmonary vascular endothelial growth factor expression and disaturated phospholipid content in a chicken model of hypoxia-induced fetal growth restriction. Neonatology. 2010;97(3):183–9. doi:10.1159/000252970.

Bellairs R, Osmond M. Atlas of Chick Development. 3rd ed. Oxford: Academic Press, Elsevier; 2014.

Bellusci S, Grindley J, Emoto H, Itoh N, Hogan BL. Fibroblast growth factor 10 (FGF10) and branching morphogenesis in the embryonic mouse lung. Development. 1997;124(23):4867–78.

Bernhard W, Gebert A, Vieten G, Rau GA, Hohlfeld JM, Postle AD, Freihorst J. Pulmonary surfactant in birds: coping with surface tension in a tubular lung. Am J Physiol Regul Integr Comp Physiol. 2001;281(1):R327–37.

Beyea JA, Olson DM, Vandergriend RA, Harvey S. Expression of growth hormone and its receptor in the lungs of embryonic chicks. Cell Tissue Res. 2005;322(3):379–92. doi:10.1007/s00441-005-0040-0.

Bjørnstad S, Paulsen RE, Erichsen A, Glover JC, Roald B. Type I and II pneumocyte differentiation in the developing fetal chicken lung: conservation of pivotal proteins from birds to human in the struggle for life at birth. Neonatology. 2014;105(2):112–20. doi:10.1159/000355346.

Bjørnstad S, Samara A, Erichsen A, Paulsen RE, Glover JC, Roald B. Hampered lung maturation in methimazole-induced hypothyroidism in fetal chicken: morphological and molecular correlates to human fetal development. Neonatology. 2016;110(2):83–92. doi:10.1159/000444656.

Blacker HA, Orgeig S, Daniels CB. Hypoxic control of the development of the surfactant system in the chicken: evidence for physiological heterokairy. Am J Physiol Regul Integr Comp Physiol. 2004;287(2):R403–10. doi:10.1152/ajpregu.00399.2003.

Blanc P, Coste K, Pouchin P, Azaïs JM, Blanchon L, Gallot D, Sapin V. A role for mesenchyme dynamics in mouse lung branching morphogenesis. PLoS One. 2012;7(7):e41643. doi:10.1371/journal.pone.0041643.

Briscoe J, Thérond PP. The mechanisms of hedgehog signalling and its roles in development and disease. Nat Rev Mol Cell Biol. 2013;14(7):416–29. doi:10.1038/nrm3598.

Calvitti M, Baroni T, Calastrini C, Lilli C, Caramelli E, Becchetti E, Carinci P, Vizzotto L, Stabellini G. Bronchial branching correlates with specific glycosidase activity, extracellular glycosaminoglycan accumulation, TGF beta(2), and IL-1 localization during chick embryo lung development. J Histochem Cytochem. 2004;52(3):325–34. doi:10.1177/002215540405200303.

Chan T, Burggren W. Hypoxic incubation creates differential morphological effects during specific developmental critical windows in the embryo of the chicken (*Gallus gallus*). Respir Physiol Neurobiol. 2005;145(2–3):251–63. doi:10.1016/j.resp.2004.09.005.

Chuang PT, Kawcak TN, McMahon AP. Feedback control of mammalian hedgehog signaling by the hedgehog-binding protein, Hip1, modulates Fgf signaling during branching morphogenesis of the lung. Genes Dev. 2003;17:342–7. doi:10.1101/gad.1026303.

Clevidence DE, Overdier DG, Tao W, Qian X, Pani L, Lai E, Costa RH. Identification of nine tissue-specific transcription factors of the hepatocyte nuclear factor 3/forkhead DNA-binding-domain family. Proc Natl Acad Sci U S A. 1993;90(9):3948–52. doi:10.1073/pnas.90.9.3948.

Davey MG, McTeir L, Barrie AM, Freem LJ, Stephen LA. Loss of cilia causes embryonic lung hypoplasia, liver fibrosis, and cholestasis in the talpid3 ciliopathy mutant. Organogenesis. 2014;10(2):177–85. doi:10.4161/org.28819.

De Groef B, Grommen SV, Darras VM. Hatching the cleidoic egg: the role of thyroid hormones. Front Endocrinol (Lausanne). 2013;4:63. doi:10.3389/fendo.2013.00063.

Duranteau J, Chandel NS, Kulisz A, Shao Z, Schumacker PT. Intracellular signaling by reactive oxygen species during hypoxia in cardiomyocytes. J Biol Chem. 1998;273(19):11619–24. doi:10.1074/jbc.273.19.11619.

El Agha E, Bellusci S. Walking along the fibroblast growth factor 10 route: a key pathway to understand the control and regulation of epithelial and mesenchymal cell-lineage formation during lung development and repair after injury. Scientifica. 2014;2014:538379. doi:10.1155/2014/538379.

Evangelisti R, Valeno V, Bodo M, Bosi G, Stabellini G, Carinci P. Involvement of polyamines in the action of transforming growth factor beta and interleukin-1 on cultured chick embryo fibroblasts. Cell Biochem Funct. 1997;15(1):47–51. doi:10.1002/(SICI)1099-0844(199703)15:1<47::AID-CBF719>3.0.CO;2-F.

Forrest D, Sjöberg M, Vennström B. Contrasting developmental and tissue-specific expression of alpha and beta thyroid hormone receptor genes. EMBO J. 1990;9(5):1519–28.

Fujiwara T, Adams FH, Nozaki M, Dermer GB. Pulmonary surfactant phospholipids from Turkey lung: comparison with rabbit lung. Am J Phys. 1970;218(1):218–25.

Gallagher BC. Basal laminar thinning in branching morphogenesis of the chick lung as demonstrated by lectin probes. J Embryol Exp Morphol. 1986;94:173–88.

Gleghorn JP, Kwak J, Pavlovich AL, Nelson CM. Inhibitory morphogens and monopodial branching of the embryonic chicken lung. Dev Dyn. 2012;241(5):852–62. doi:10.1002/dvdy.23771.

Goldin GV, Opperman LA. Induction of supernumerary tracheal buds and the stimulation of DNA synthesis in the embryonic chick lung and trachea by epidermal growth factor. J Embryol Exp Morphol. 1980;60:235–43.

Gou W, Peng J, Wu Q, Zhang Q, Zhang H, Wu C. Expression pattern of heme oxygenase 1 gene and hypoxic adaptation in chicken embryos. Comp Biochem Physiol B Biochem Mol Biol. 2014;174:23–8. doi:10.1016/j.cbpb.2014.05.005.

Han RN, Liu J, Tanswell AK, Post M. Expression of basic fibroblast growth factor and receptor: immunolocalization studies in developing rat fetal lung. Pediatr Res. 1992;31(5):435–40. doi:10.1203/00006450-199205000-00004.

Hao R, Hu X, Wu C, Li N. Hypoxia-induced miR-15a promotes mesenchymal ablation and adaptation to hypoxia during lung development in chicken. PLoS One. 2014;9(6):e98868. doi:10.1371/journal.pone.0098868.

Hilfer SR, Rayner RM, Brown JW. Mesenchymal control of branching pattern in the fetal mouse lung. Tissue Cell. 1985;17(4):523–38. doi:10.1016/0040-8166(85)90029-1.

Hylka VW. Ultrastructural and biochemical evidence of glycogen in the developing lung of the chick embryo: possible contribution to surfactant. Comp Biochem Physiol A Comp Physiol. 1989;93(4):677–83. doi:10.1016/0300-9629(89)90483-0.

Hylka VW, Doneen BA. Ontogeny of embryonic chicken lung: effects of pituitary gland, corticosterone, and other hormones upon pulmonary growth and synthesis of surfactant phospholipids. Gen Comp Endocrinol. 1983;52(1):108–20. doi:10.1016/0016-6480(83)90163-6.

Ivey KN, Srivastava D. microRNAs as developmental regulators. Cold Spring Harb Perspect Biol. 2015;7(7):a008144. doi:10.1101/cshperspect.a008144.

Kim HY, Varner VD, Nelson CM. Apical constriction initiates new bud formation during monopodial branching of the embryonic chicken lung. Development. 2013;140(15):3146–55. doi:10.1242/dev.093682.

Lazzaro D, Price M, de Felice M, Di Lauro R. The transcription factor TTF-1 is expressed at the onset of thyroid and lung morphogenesis and in restricted regions of the foetal brain. Development. 1991;113:1093–104.

Lewallen MA, Burggren WW. Chronic hypoxia and hyperoxia modifies morphology and VEGF concentration of the lungs of the developing chicken (*Gallus gallus* variant *domesticus*). Respir Physiol Neurobiol. 2015;219:85–94. doi:10.1016/j.resp.2015.08.004.

Li C, Xiao J, Hormi K, Borok Z, Minoo P. Wnt5a participates in distal lung morphogenesis. Dev Biol. 2002;248(1):68–81. doi:10.1006/dbio.2002.0729.

Litingtung Y, Lei L, Westphal H, Chiang C. Sonic hedgehog is essential to foregut development. Nat Genet. 1998;20(1):58–61. doi:10.1038/1717.

Loscertales M, Mikels AJ, Hu JK, Donahoe PK, Roberts DJ. Chick pulmonary Wnt5a directs airway and vascular tubulogenesis. Development. 2008;135(7):1365–76. doi:10.1242/dev.010504.

Maina JN. A systematic study of the development of the airway (bronchial) system of the avian lung from days 3 to 26 of embryogenesis: a transmission electron microscopic study on the domestic fowl, *Gallus gallus* variant *domesticus*. Tissue Cell. 2003;35(5):375–91. doi:10.1016/S0040-8166(03)00058-2.

Maina JN, Madan AK, Alison B. Expression of fibroblast growth factor-2 (FGF-2) in early stages (days 3–11) of the development of the avian lung, *Gallus gallus* variant *domesticus*: an immunocytochemical study. J Anat. 2003;203(5):505–12. doi:10.1046/j.1469-7580.2003.00236.x.

Massagué J. TGFβ signalling in context. Nat Rev Mol Cell Biol. 2012;13(10):616–30. doi:10.1038/nrm3434.

Metzger RJ, Klein OD, Martin GR, Krasnow MA. The branching programme of mouse lung development. Nature. 2008;453(7196):745–50. doi:10.1038/nature07005.

Minoo P, Su G, Drum H, Bringas P, Kimura S. Defects in tracheoesophageal and lung morphogenesis in Nkx2.1 (−/−) mouse embryos. Dev Biol. 1999;209(1):60–71. doi:10.1006/dbio.1999.9234.

Miura T, Hartmann D, Kinboshi M, Komada M, Ishibashi M, Shiota K. The cyst-branch difference in developing chick lung results from a different morphogen diffusion coefficient. Mech Dev. 2009;126(3–4):160–72. doi:10.1016/j.mod.2008.11.006.

Moura RS, Carvalho-Correia E, da Mota P, Correia-Pinto J. Canonical Wnt signaling activity in early stages of chick lung development. PLoS One. 2014;9(12):e112388. doi:10.1371/journal.pone.0112388.

Moura RS, Coutinho-Borges JP, Pacheco AP, Damota PO, Correia-Pinto J. FGF signaling pathway in the developing chick lung: expression and inhibition studies. PLoS One. 2011;6(3):e17660. doi:10.1371/journal.pone.0017660.

Moura RS, Silva-Gonçalves C, Vaz-Cunha P, Correia-Pinto J. Expression analysis of Shh signaling members in early stages of chick lung development. Histochem Cell Biol. 2016;146(4):457–66. doi:10.1007/s00418-016-1448-1.

Moura RS, Vaz-Cunha P, Silva-Gonçalves C, Correia-Pinto J. Characterization of miRNA processing machinery in the embryonic chick lung. Cell Tissue Res. 2015;362(3):569–75. doi:10.1007/s00441-015-2240-6.

Muraoka RS, Bushdid PB, Brantley DM, Yull FE, Kerr LD. Mesenchymal expression of nuclear factor-kappaB inhibits epithelial growth and branching in the embryonic chick lung. Dev Biol. 2000;225(2):322–38. doi:10.1006/dbio.2000.9824.

Murphy AE, Harvey S. Extrapituitary beta TSH and GH in early chick embryos. Mol Cell Endocrinol. 2001;185(1–2):161–71. doi:10.1016/S0303-7207(01)00615-3.

Ornitz DM, Itoh N. The fibroblast growth factor signaling pathway. Wiley Interdiscip Rev Dev Biol. 2015;4(3):215–66. doi:10.1002/wdev.176.

Park WY, Miranda B, Lebeche D, Hashimoto G, Cardoso WV. FGF-10 is a chemotactic factor for distal epithelial buds during lung development. Dev Biol. 1998;201(2):125–34. doi:10.1006/dbio.1998.8994.

Pei L, Leblanc M, Barish G, Atkins A, Nofsinger R, Whyte J, Gold D, He M, Kawamura K, Li HR, Downes M, Yu RT, Powell HC, Lingrel JB, Evans RM. Thyroid hormone receptor repression is linked to type I pneumocyte-associated respiratory distress syndrome. Nat Med. 2011;17(11):1466–72. doi:10.1038/nm.2450.

Peng JF, Ling Y, Gou WY, Zhang H, Wu CX. Identification of chicken eNOS gene and differential expression in highland versus lowland chicken breeds. Poult Sci. 2012;91(9):2275–81. doi:10.3382/ps.2012-02197.

Pepicelli CV, Lewis PM, McMahon AP. Sonic hedgehog regulates branching morphogenesis in the mammalian lung. Curr Biol. 1998;8(19):1083–6. doi:10.1016/S0960-9822(98)70446-4.

Sakiyama J, Yamagishi A, Kuroiwa A. Tbx4-Fgf10 system controls lung bud formation during chicken embryonic development. Development. 2003;130(7):1225–34. doi:10.1242/dev.00345.

Sakiyama J, Yokouchi Y, Kuroiwa A. Coordinated expression of hoxb genes and signaling molecules during development of the chick respiratory tract. Dev Biol. 2000;227(1):12–27. doi:10.1006/dbio.2000.9880.

Sanford EL, Choy KW, Donahoe PK, Tracy AA, Hila R, Loscertales M, Longoni M. MiR-449a affects epithelial proliferation during the pseudoglandular and canalicular phases of avian and mammal lung development. PLoS One. 2016;11(2):e0149425. doi:10.1371/journal.pone.0149425.

Semenov MV, Habas R, Macdonald BT, He X. SnapShot: noncanonical Wnt signaling pathways. Cell. 2007;131(7):1378. doi:10.1016/j.cell.2007.12.011.

Shirasawa N, Shiino M, Shimizu Y, Nogami H, Ishii S. Immunoreactive luteinizing hormone (ir-LH) cells in the lung and stomach of chick embryos. Cell Tissue Res. 1996;283(1):19–27. doi:10.1007/s004410050508.

Shu W, Jiang YQ, Lu MM, Morrisey EE. Wnt7b regulates mesenchymal proliferation and vascular development in the lung. Development. 2002;129:4831–42.

Stabellini G, Calvitti M, Baroni T, Marinucci L, Calastrini C, Carinci P, Becchetti E. Glycosidases during chick embryo lung development and their colocalization with proteoglycans and growth factors. Eur J Histochem. 2002;46(1):41–52. doi:10.4081/1653.

Stabellini G, Calvitti M, Becchetti E, Carinci P, Calastrini C, Lilli C, Solmi R, Vizzotto L, Baroni T. Lung regions differently modulate bronchial branching development and extracellular matrix plays a role in regulating the development of chick embryo whole lung. Eur J Histochem. 2007;51(1):33–41. doi:10.4081/1009.

Stabellini G, Locci P, Calvitti M, Evangelisti R, Marinucci L, Bodo M, Caruso A, Canaider S, Carinci P. Epithelial-mesenchymal interactions and lung branching morphogenesis. Role of polyamines and transforming growth factor beta1. Eur J Histochem. 2001;45(2):151–62. doi:10.4081/1625.

Wellik DM. Hox patterning of the vertebrate axial skeleton. Dev Dyn. 2007;236(9):2454–63. doi:10.1002/dvdy.21286.

Woik N, Kroll J. Regulation of lung development and regeneration by the vascular system. Cell Mol Life Sci. 2015;72(14):2709–18. doi:10.1007/s00018-015-1907-1.

Zeng X, Yutzey KE, Whitsett JA. Thyroid transcription factor-1, hepatocyte nuclear factor-3β and surfactant protein A and B in the developing chick lung. J Anat. 1998;193(3):399–408. doi:10.1046/j.1469-7580.1998.19330399.x.

Zhang H, Burggren WW. Hypoxic level and duration differentially affect embryonic organ system development of the chicken (*Gallus gallus*). Poult Sci. 2012;91(12):3191–201. doi:10.3382/ps.2012-02449.

Zhang W, Cuperus T, van Dijk A, Skjødt K, Hansen S, Haagsman HP, Veldhuizen EJ. Developmental regulation of chicken surfactant protein A and its localization in lung. Dev Comp Immunol. 2016;61:80–7. doi:10.1016/j.dci.2016.03.010.

Zhang L, Whitsett JA, Stripp BR. Regulation of Clara cell secretory protein gene transcription by thyroid transcription factor-1. Biochim Biophys Acta. 1997;1350(3):359–67. doi:10.1016/S0167-4781(96)00180-7.

Development of the Airways and the Vasculature in the Lungs of Birds

6

Andrew N. Makanya

Abstract

Generally, the vertebrate lung has its origin from the endoderm in the region of the primitive foregut, where the epithelium gives rise to the airway system and the gas-exchanging units and the mesenchyme forms the connective tissue, muscles, and vessels. The lung starts as a primordium, which splits into a left and right bud each of which forms the respective lung. In birds, the lung buds form the primary bronchi from which the secondary bronchi (SB) arise. The parabronchi (PB) sprout from the SB and occupy specific locations within the lung. Atria start as excavations with attenuating cells and give rise to infundibula and finally to air capillaries. The mechanisms underlying the formation of the remarkably thin blood–gas barrier (BGB) closely resemble those of exocrine secretion, but occur in a programmed, time-limited manner. In general, they result in cutting or decapitation of the cell until the required thickness is attained. In the first step, the high columnar epithelium undergoes dramatic size reduction and loses morphological polarization by two main processes: secarecytosis (cell decapitation by cutting) and peremerecytosis (cell decapitation by squeezing, spontaneous constriction, or pinching off). Secarecytosis has at least two facets: transcellular double-membrane formation followed by separation between such membranes or cell cutting by vesiculation. Both processes lead to formation of a thin BGB. Blood vessel formation in the avian lung occurs concomitantly with formation of the airway system. There is close interaction between the budding endoderm and the surrounding mesenchyme, where their crosstalk leads to development and differentiation of the components of the functional lung. Blood vessel formation starts with vasculogenesis where blood islands are formed. The islands then form blood vessels that expand further through sprouting, and once a network is established, it is augmented and remodeled through intussusceptive angiogenesis.

Keywords

Avian lung • Airway development • Vasculogenesis • Angiogenesis

Contents

A.N. Makanya (✉)
Department of Veterinary Anatomy and Physiology, University of Nairobi, Riverside Drive,
P. O. Box 30197-00100, Nairobi, Kenya
e-mail: makanya63@yahoo.com

J.N. Maina (ed.), *The Biology of the Avian Respiratory System*, DOI 10.1007/978-3-319-44153-5_6

Abbreviations

AC	Air capillary
BC	Blood capillary
bFGF	Basic fibroblast growth factor
BGB	Blood–gas barrier
BMP-4	Bone morphogenetic protein-4
FGF	Fibroblast growth factor
GATA	GATA transcription factor
HNF-3	Hepatocyte nuclear factor 3
IA	Intussusceptive angiogenesis
LD	Laterodorsal secondary bronchus
MV	Medioventral secondary bronchi
NPB	Neopulmonic parabronchi
PB	Parabronchus
PDGF	Platelet-derived growth factor
PO	Posterior secondary bronchi
PPB	Paleopulmonic parabronchi
SA	Sprouting angiogenesis
SB	Secondary bronchus
Shh	Sonic hedgehog
SMA	Smooth muscle actin
TFG-β	Transforming growth factor β
TGF-β1	Transforming growth factor β1
TTF-1	Thyroid transcription factor 1
VEGF	Vascular endothelial growth factor
WNT5a	Wingless-type 5

6.1 Introduction

The avian lung differs in fundamental aspects from the gas exchangers of all the other vertebrates in that it is largely noncompliant and depends on the bellows-like action of the air sacs for its ventilation (Bellairs and Osmond 1998; King and McLelland 1984). In addition, it presents the thinnest and most uniform blood–gas barrier of any vertebrate described (Maina and West 2005). The avian lung has a volume, which is about three quarters that of mammals of comparable size, whereas the gas exchange surface area is 15% greater (Maina et al. 1989). This means that during lung development, the bird has to pack more surface area in a smaller noncompliant lung. This phenomenon calls for ingenious mechanisms of cell attenuation. Notably, the air capillaries in birds grow by radiating into the mesenchymal tissue (Maina 2004a), quite unlike the alveoli in mammalian lungs, which are formed by successive and alternating phases of air space expansion and septation (Duncker 1990; Makanya et al. 2001; Schittny and Burri 2003).

6.2 Early Lung Development

While the basic events of lung development in vertebrates are preserved, the avian lung shows remarkable differences in accomplishment of its final structure. The lung, together with the trachea, larynx, extrapulmonary bronchi, and air sacs, is derived from the endoderm of the primitive foregut. The first sign is establishment of the laryngotracheal groove, which becomes visible in the midline of the floor of the pharynx during embryonic day 3–4 (E3–E4) in the domestic chicken, posterior to the fourth pharyngeal pouches (Bellairs and Osmond 1998). Unless otherwise stated, the current discussion is largely based on the domestic chicken, where extensive studies have been conducted.

6.2.1 Development of the Airway System

The air conduit system of the avian lung, which extends from the intrapulmonary primary bronchi to the air capillaries, originates from the ventral aspect of the primitive foregut. From the lung bud stage, the developing tube grows into the surrounding mesenchyme and divides into left and right buds, which give rise to the respective lungs. On the mesobronchus, secondary bronchi start from specific areas and grow into the pulmonary mesenchyme (Fig. 6.1), occupying their predetermined locations (Sakiyama et al. 2003; Makanya and Djonov 2008). The parabronchi sprout in a similar manner to occupy their specific locations (Fig. 6.2).

6.2.2 The Secondary Bronchi and the Parabronchi

An update of the various categories of secondary bronchi in the chicken lung, their names, topography, and numbers has been documented (Makanya and Djonov 2008). Similar data are available for the duck lung (Makanya et al. 2014). Early at E8 in the chick lung, the secondary bronchi as well as the parabronchi have started to sprout. These are separated by abundant mesenchymal tissue, and by E11 the number of air conduits increases at the expense of the mesenchyme. At E13 the laterodorsal secondary bronchi (LDs), the posterior secondary bronchi (PO), and the medioventral secondary bronchi (MV) are evident. The MVs already have parabronchial branches (PB) at this time showing that they are the first secondary bronchi to emerge. Parabronchial branches from MV grow caudally to anastomose with those of the first two LDs. The rest of the LDs, that is numbers 3–10, grow dorsolaterally and form parabronchial branches that grow toward the dorsal border where they anastomose with similar branches from MVs (Makanya and Djonov 2008). Anastomoses of the paleopulmonic parabronchi occur by approaching conduits sending slender

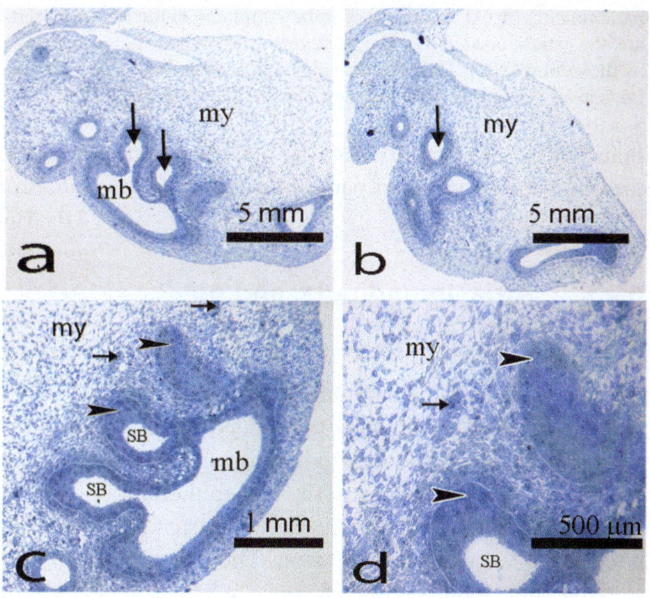

Fig. 6.1 Semithin sections showing the developing chicken lung at E8. (**a**, **b**) At this stage the lung mesenchyme (my) has a few secondary bronchi (*arrows*) sprouting from the mesobronchus (mb). (**c**, **d**) At a higher magnification, the secondary bronchi are seen to grow as epithelial cords (*arrowheads*) into the mesenchyme (my), while some have a lumen (SB). A few developing blood vessels (*arrows*) are also present

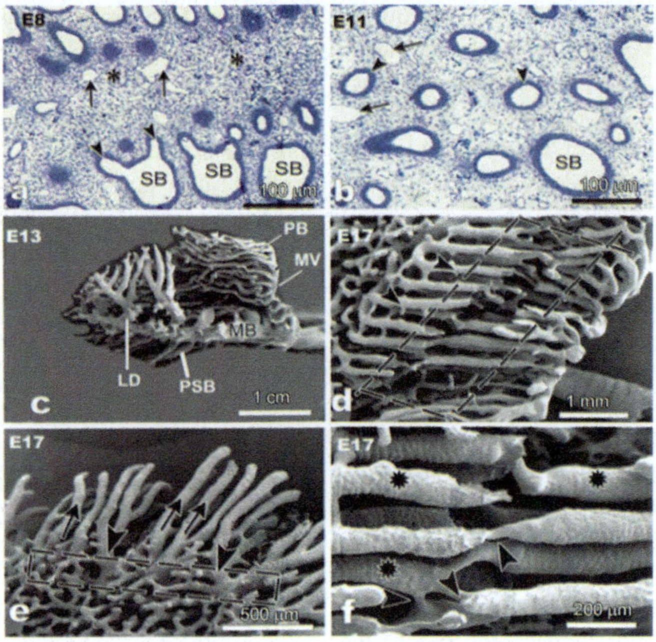

Fig. 6.2 Semithin sections (**a**, **b**) and SEM micrographs (**c–f**) illustrating the sprouting secondary bronchi and parabronchi in the chicken embryo lung between E8 and E17. (**a**, **b**) Early at E8 the SB with sprouting parabronchi (*arrowheads*) are already present (**a**). These are separated by abundant mesenchymal tissue (*asterisks*) with a few empty spaces (*arrows*), which probably contain tissue fluid during lung development. By E11 (**b**), the number of air conduits (*arrowheads*) increases at the expense of the mesenchyme. A few empty spaces (*arrows*) are also evident. (**c–f**) Intratracheal mercox casts showing developmental patterns of the secondary bronchi and parabronchi. (**c**) At E13 the secondary bronchi evident are the laterodorsal (LD), the posterior (PSB), and the medioventral (MV). The MVs already have parabronchial branches (PB) at this time showing that they are the first secondary bronchi to emerge. (**d**) Parabronchial branches from MVs grow caudally to anastomose with those of the first two LDs. The hatched rectangle shows the zone of anastomoses. (**e**) The rest of the LDs grow dorsolaterally and form parabronchial branches (*arrows*) that grow toward the dorsal border of the lung where they anastomose with their cognates from MVs. Note the boundary between the formative parabronchi of paleopulmonic region (*arrowheads*) and zone of parabronchial anastomoses between the two regions (*hatched rectangle*). (**f**) Anastomoses of the paleopulmonic parabronchi occurs by approaching conduits (*asterisks*) sending slender branches (*arrowheads*) to one or more of the targeted cognates (Modified from Makanya and Djonov (2008), with permission)

branches to one or more of similar branches from the approaching group of parabronchi (Fig. 6.2).

6.2.3 Atria, Infundibula, and Air Capillaries

At E10, the parabronchial epithelium consists of tall columnar cells with smooth rounded luminal apices. Occasional club-like protrusions of variable sizes appear, and by E12, the epithelium becomes rugged with club-like apical protrusions at different stages of development. There is progressive development of the atria as shallow depressions form on the luminal surface of the parabronchi and respiratory (i.e., exchanging) SB. The incipient atria on the mucosal surface are separated by ridges of the mucosa, and by E14 the depressions are converted to circular excavations surrounded by irregularly shaped cells (Fig. 6.3). The atrial excavations at E15 are much deeper and the surrounding cells more regular in height and broadened, presenting a general flat surface. The atria by E16 are broadened tremendously and show secondary depressions, which are the incipient infundibula. The floor of the nascent infundibula is characterized by sparse tall cells or portions

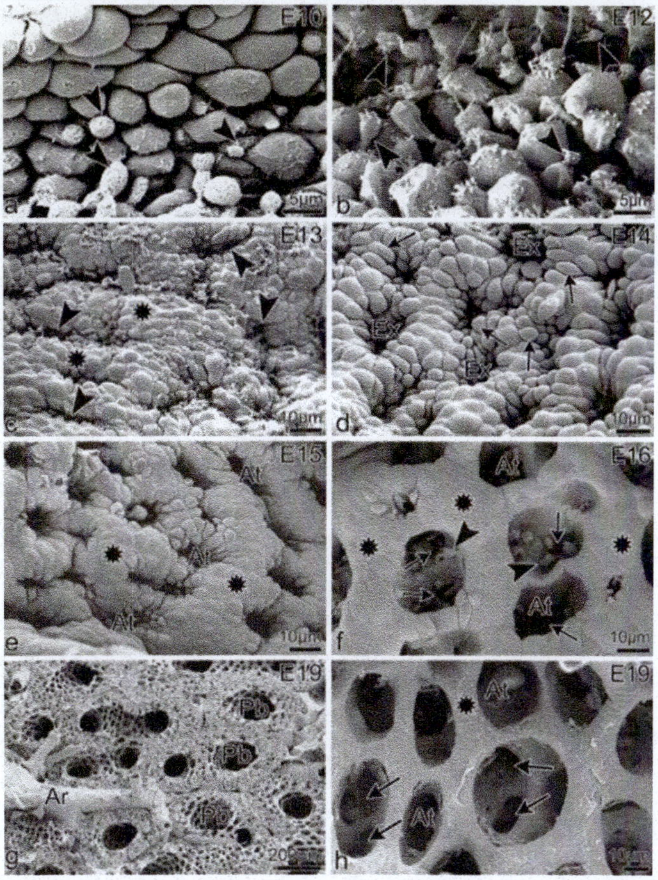

Fig. 6.3 Scanning electron micrographs of the chicken lung illustrating the various changes characterizing the developing parabronchi between E10 and E19. (**a, b**) At E10, the parabronchial epithelium consists of tall columnar cells with smooth rounded luminal apices. Occasional club-like protrusions of variable sizes are apparent (*arrowheads* in **a**). By E12, the epithelium is rather rugged, and club-like apical protrusions at different stages of development are abundant (*arrowheads* in **b**). (**c, d**) At E13 and E14, the progressive development of the atria is apparent. These inaugurate as shallow depressions on the mucosal surface (*arrowheads* in **c**) separated by ridges of the mucosa (*asterisks* in **c**) and give the mucosa a folded appearance. By E14 (**d**), the depressions are converted to circular excavations (Ex) surrounded by irregularly shaped cells (*arrows*). (**e, f**) The atrial excavations (At) at E15 are much deeper, and the surrounding cells more regular in height and broadened, presenting a general flat surface (*asterisks* in **e**). By E16 (**f**), the atria are broadened and have secondary depressions, the incipient infundibula (*arrows* in **f**). The floor of the nascent infundibula is characterized by sparse tall cells or portions thereof, projecting conspicuously above the general surface of the atrial floor (*arrowheads* in **f**). The mucosa between the atria is now remarkably smooth (*asterisks*). (**g, h**) The pulmonary tissue at E19 generally resembles the structure of adult birds with almost evenly spaced parabronchi (Pb) lined by numerous openings to the atria (At). The interatrial septa (*asterisk* in **h**) are largely thinned out and separate the entrances to the contiguous atria (*arrows* in **h**). Notice also an artery (Ar in **g**), giving rise to interparabronchial vessels (Modified from Makanya et al. (2006), with permission from the publishers)

thereof, projecting conspicuously above the general surface of the atrial floor. Such structures are remnants of the attenuating epithelium. The mucosa between the atria is now remarkably smooth, and the pulmonary tissue at E19 generally resembles the structure in adult birds with almost evenly spaced parabronchi lined by numerous openings to the atria. The interatrial septa are largely thinned out and separate the entrances to the contiguous atria.

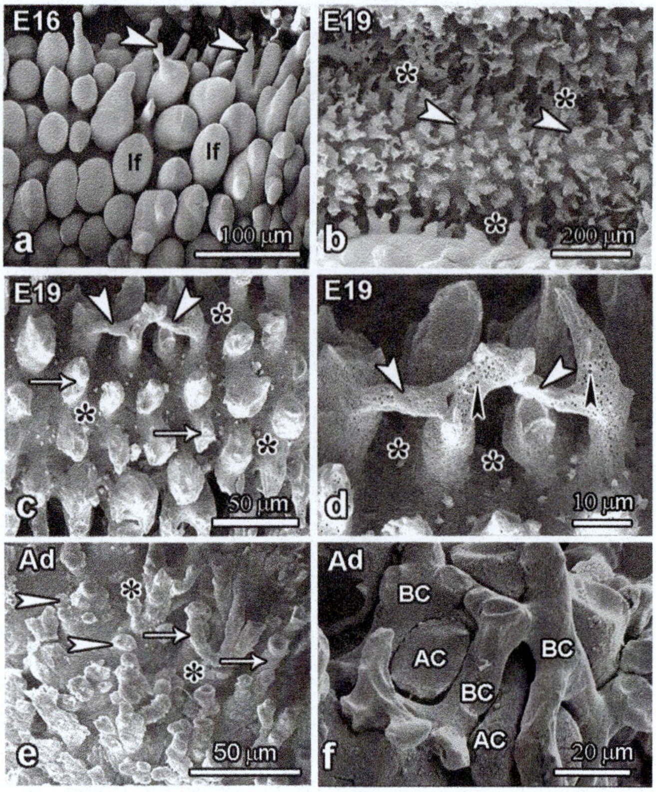

Fig. 6.4 Scanning electron micrographs of intratracheal mercox casts of the lungs of late stage embryos (E16–E19; **a–d**) and adult birds (**e, f**) showing the steps in formation of air capillaries and their branches as well as their definitive structures. (**a**) At E16 the infundibulae (If) appear as smooth mounds on the parabronchial casts with only a few sprouting air capillaries (*arrowheads*). (**b–d**) The ACs by E19 start to form branches, and some of such lateral branches extend to anastomose with neighboring ACs within the same parabronchus (*white arrowheads* in **c** and **d**). *Asterisks* in (**b**) represent the interparabronchial septa, while those ones in (**c**) and (**d**) represent the space occupied by the exchange blood capillaries. The *white arrows* in (**c**) indicate unbranched ACs, while the *black arrowheads* in (**d**) indicate tiny holes left in the ACs by microplicae. (**e, f**) In the adult (Ad) chicken, ACs are largely discrete branching structures that taper toward the periphery (*open arrows* in **e**), while some form globular shapes (*open arrowheads* in **e**). The *asterisks* in (**e**) denote spaces occupied by blood capillaries. A double mercox cast of the airway and vascular conduits shows that the blood capillaries (BC in **f**) tend to form rings around the air capillaries (AC). This essentially forms a crosscurrent system between the ACs and the BCs (Modified from Makanya et al. (2006), with permission from the publishers)

Steps in formation of atria, infundibula, and air capillaries can be followed from intratracheal mercox casts (Fig. 6.4) and critical-point-dried tissues (Fig. 6.5). These techniques allow a 3D visualization of the progressive changes. Atria now appear as domes on the casts of parabronchi, and at E16 the infundibula appear as smooth mounds on the parabronchial casts, but occasionally have sprouting air capillaries (ACs, Fig. 6.4). By E19 air capillaries have started to form branches, and some of such lateral branches extend to anastomose with neighboring ACs within the same parabronchus (Fig. 6.5). Air capillaries from contiguous parabronchi do not anastomose as they are separated by the interparabronchial septa. At later stages tiny holes appear on the casts of air capillaries (Fig. 6.5), and these are indicators of the presence of microplicae (Fig. 6.6) that result from cell attenuation (see below). Air capillaries are largely discrete branching structures that taper toward the periphery of the parabronchus, while some form globular shapes (Fig. 6.5). These ACs interact with the blood capillaries in such a way that a largely

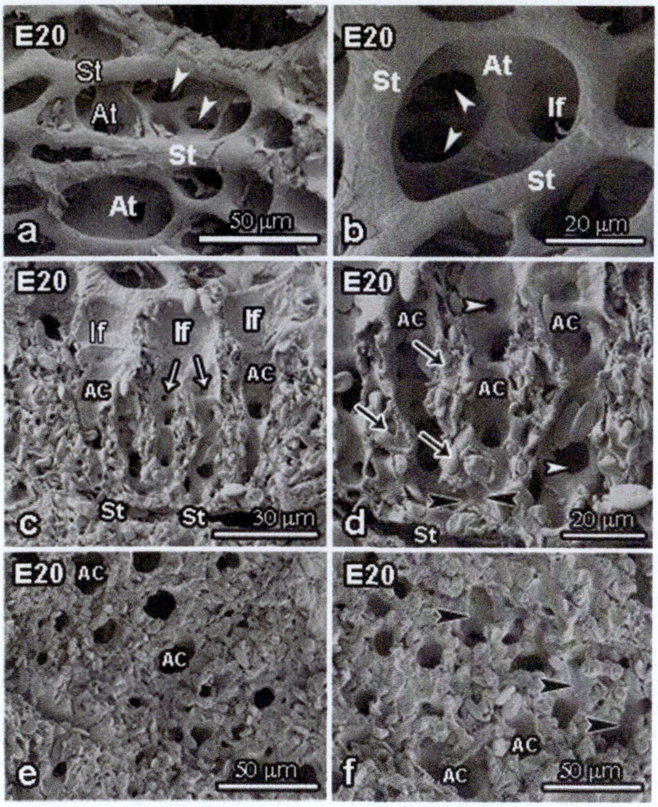

Fig. 6.5 SEM micrographs of E20 critical-point-dried lung tissue illustrating the disposition of the air conduits in the parabronchial gas exchange area. (**a**, **b**) A bird's-eye view of the atria (At) with the intervening interatrial septa (St). Atria open into infundibula (If), which in turn give rise to air capillaries (*white arrowheads*). (**c**, **d**) Sections through the parabronchial mantle showing the distribution of the air conduits. The infundibulae (If) project from the atria and extend for a short distance toward the periphery of the parabronchial mantle, then give rise to air capillaries (*white arrows* in **c**). The individual air capillaries (AC) form smaller branches (*white arrowheads* in **d**) that radiate into the parabronchial tissue. Toward the interparabronchial septum (St), the ACs anastomose (*black arrowheads* in **d**). Notice the erythrocytes (*black arrows* in **d**) lodged in BCs that occur in the septa separating adjacent ACs. (**e**, **f**) Profiles of the peripheral parts of the parabronchial mantle showing transected air capillaries (AC). Notice the anastomoses (*black arrowheads* in **f**) connecting adjacent ACs

crosscurrent system is established (Makanya and Djonov 2009) where the two structures cross each other perpendicularly (Fig. 6.4).

Development of the secondary bronchi and the parabronchi results in defined categories of these structures that occupy specific topographic locations in the lung (Figs. 6.7 and 6.8). Generally, the initial parts of the secondary bronchi are non-respiratory, that is, they have no atrial openings. The first MV has primary and secondary branches which are supported by cartilage and are non-respiratory. In the chick lung (*Gallus gallus variant domesticus*), the MVs are four in number, and together with their parabronchi, they occupy the medioventral cranial quarter of the lung. The dorsolateral cranial quarter is occupied by LD1 and its parabronchi, while LD2 with its parabronchi takes the ventrolateral cranial quarter (Figs. 6.7 and 6.8). The rest of the LDs form the paleopulmonic parabronchi (Fig. 6.8) that occupy the bulk of the lateral aspect of the lung (Makanya and Djonov 2008).

The simplified schematic drawings of the adult chicken lung (Fig. 6.8) illustrate the patterns of the secondary bronchi and the parabronchi. The categories are the medioventral

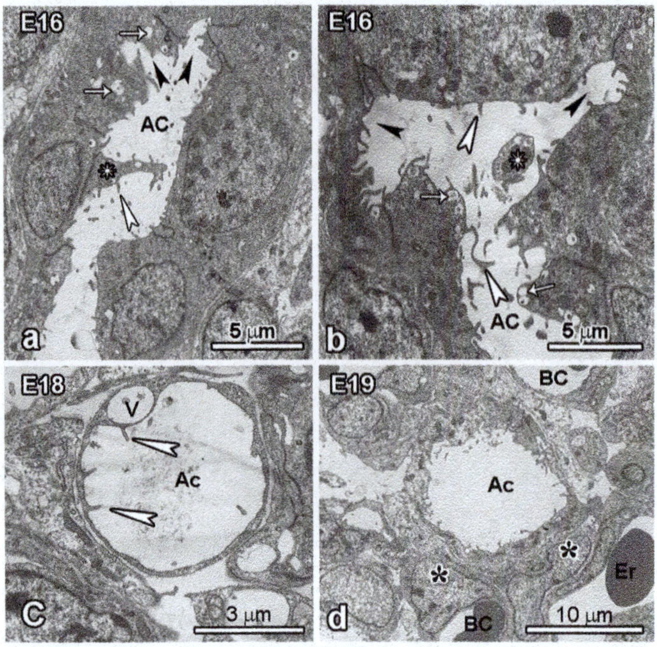

Fig. 6.6 TEM micrographs showing the developing air capillaries at E16–E19. (**a**, **b**) At E16, the air capillaries contain both microplicae (*white arrowheads*) and intraluminal projections (*stars*) some of which appear to be floating bodies (*star* in **b**) in the lumen. Such structures result from cell attenuation processes collectively named secarecytosis, which in part proceed through vesicle formation (*arrows* in **a** and **b**) and their subsequent rapture resulting in formation of microplicae (*white arrowheads*). Furthermore, these attenuation mechanisms appear to be important in formation and canalization of the air capillary branches (see *black arrowheads*). (**c**, **d**) Cross sections of air capillaries at E18 (**c**) and E19 (**d**). Attenuation of the epithelium proceeds through many mechanisms that include vesiculation. Notice a vesicle (V) in the wall of the air capillary and microplicae (*white arrowheads*) that are the result of vesiculation. At this stage the ACs are approximated by blood capillaries (BC) to form a blood–gas barrier, which at this stage is still thick (*asterisks* in **d**). During these late stages of development, BCs are distinguished from ACs due to the presence of erythrocytes (Er in **d**) and the absence of microplicae

secondary bronchi (MV, 4), the lateroventral secondary bronchi (LV, 2), and the saccobronchi (numerous). The smooth nature of the intrapulmonary primary bronchus, MV1, the saccobronchi, and the initial parts of the secondary bronchi indicate that they have no atrial openings (Figs. 6.7 and 6.8). The two categories of parabronchi are neopulmonic (NPB) and paleopulmonic (PPB).

6.3 Formation of the Blood–Gas Barrier

During the early developmental stages, the chicken lung is characterized by loose mesenchyme containing occasional isolated blood vessels and numerous developing parabronchi. The parabronchi undergo dramatic changes affecting both their 3D structure as well as their epithelial morphology. An overview of the changing epithelial morphology in the developing chicken lung between E8 and E18 is presented in Fig. 6.9, while Fig. 6.10 shows the situation in the ostrich at E24. At E8, the epithelium of the parabronchial tubes is cuboidal to low columnar (Fig. 6.9), becoming high columnar by E10, and at the same time, cells acquire pointed apices. From E11, cells with tapering luminal apices and debris of rounded cell blebs are evident. Furthermore, the epithelium is transformed into a pseudostratified one, apparently due to multiplication and enlargement of epithelial cells such that some cells are squeezed out of

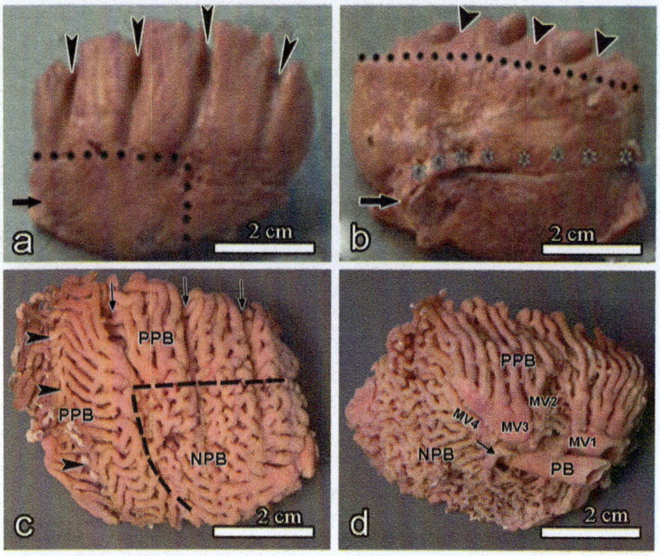

Fig. 6.7 (**a–d**) Macrographs of fixed lung specimens (**a**, **b**), and Technovit casts (**c**, **d**) explicating the lung structure in the adult chicken. (**a**) On the lateral surface, the rib impressions (*arrowheads*) extend down to the level of the long axis of the lung. Note the extent of the neopulmonic region (*dotted line*). The *arrow* indicates the position of the ostium of the abdominal air sac at the caudal part of the lung. (**b**) The medial surface of the lung looks rather smooth, the ventral border (*dotted line*) is rounded, and the rib impressions at the dorsal border are deep (*arrowheads*). The *arrow* indicates the cranial opening of the intrapulmonary primary bronchus. The trajectory of the primary bronchus is indicated by the *asterisks*. (**c**) On the lateral aspect, the cranial and dorsal groups of the paleopulmonic parabronchi (PPB) are evident. The *black arrowheads* indicate the planum anastomosica, where the PPB from LD1 and LD2 meet those from MV1. The *hatched line* delineates the region of neopulmonic parabronchi (NPB). *Black arrows* denote costal sulci, while the white ones indicate the parabronchial branches from MV1. (**d**) Medioventral aspect of the lung showing the primary bronchus (PB), the medioventral secondary bronchi (MV1, MV2, MV3, MV4), the lateroventral secondary bronchus (*arrow*), the paleopulmonic parabronchi (PPB), and the neopulmonic parabronchi (NPB)

phase. By E12, the nipping process is accelerated so that these apical parts of the cells are severed and extruded into the parabronchial lumen (Fig. 6.10). Some of the subepithelial cells change their orientation and come to lie parallel to the basement membrane, forming a continuous layer at early stages, which later become interrupted by narrow gaps through which migrating epithelial cells of the incipient atria emerge (Fig. 6.10). The parabronchi by E16 are no longer simple-surfaced cylinders, because the internal aspects are now interrupted by the formative atria. The atria arise probably by mesenchyme invasion and attenuation of the epithelium without additional proliferation. After formation of the atria, at E18 (Fig. 6.11) the mesenchymal invasion progresses by formation of multiple infundibula covered by very thin epithelium. At the same time point, the mesenchyme is massively reduced so that thin interparabronchial septa containing the major supplying and draining parabronchial vessels are formed. The ostrich lung at E24 closely resembles the chicken lung at E12 (Fig. 6.11) and shows similar attenuation processes.

The events, processes, and mechanisms that transform the developing lung epithelium are best captured in details at the transmission electron microscopy level. The first signs become apparent at late E8 with the cells becoming morphologically polarized, i.e., they are tapered toward the lumen and electron-dense bands separate the tapered apical part from the rest of the cell (Fig. 6.12). By E10, the epithelium becomes tall columnar, tapering is accelerated, and the cells enlarge and appear to squeeze their

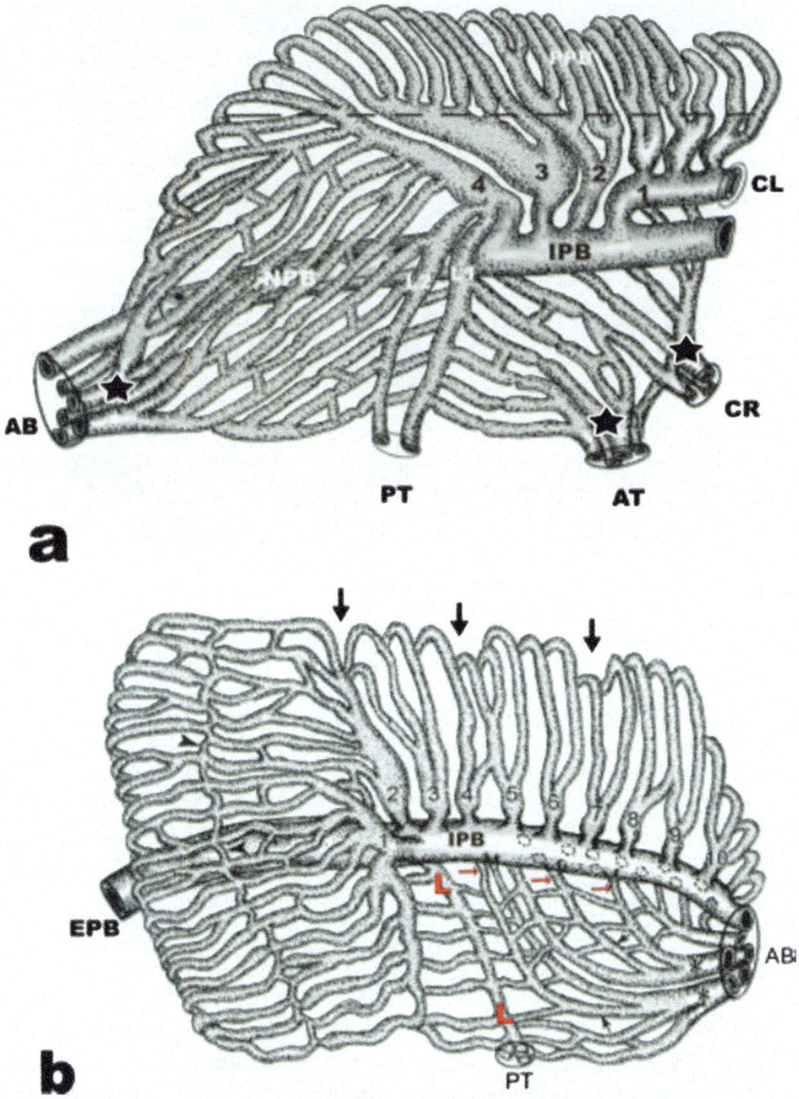

Fig. 6.8 Simplified schematic drawings of the adult chicken lung illustrating the patterns of the secondary bronchi and the parabronchi. (**a**) Medioventral aspect of the lung showing the intrapulmonary primary bronchus (IPB), the medioventral secondary bronchi (1, 2, 3, 4), the LV (L1 and L2), and the sacobronchi (*stars*). Notice the smooth nature of the intrapulmonary primary bronchus, MV1, the sacobronchi, and the initial parts of the secondary bronchi showing that they have no atrial openings. The ostia to the various air sacs are interclavicular (CL), cervical (CR), anterior thoracic (AT), posterior thoracic (PT), and abdominal (AB). The various categories of parabronchi are neopulmonic (NPB) and paleopulmonic (PPB). The *dashed line* shows the medial border. (**b**) The lateral aspect of the lung shows the pattern of the various groups of secondary bronchi and the parabronchi emanating from them. The LDs are labeled 1–10. Notice that the first two LDs furnish the parabronchi that inosculate with those from MV1. The plane of anastomoses is marked by smaller parabronchial branches (*arrowheads*). Notice the position of the LV (L) and the various PO (*thin horizontal arrows*) that continue to form the sacobronchi (*asterisks*) to the abdominal air sac. AB and PT are the ostia of the abdominal and posterior thoracic air sacs, respectively, EPB is extrapulmonary primary bronchus, and IPB is intrapulmonary primary bronchus. The *thick arrows* denote costal sulci and *dashed circles* denote stumps of PO. The neopulmonic parabronchi (*small arrows*) are continuous with the sacobronchi and the PO

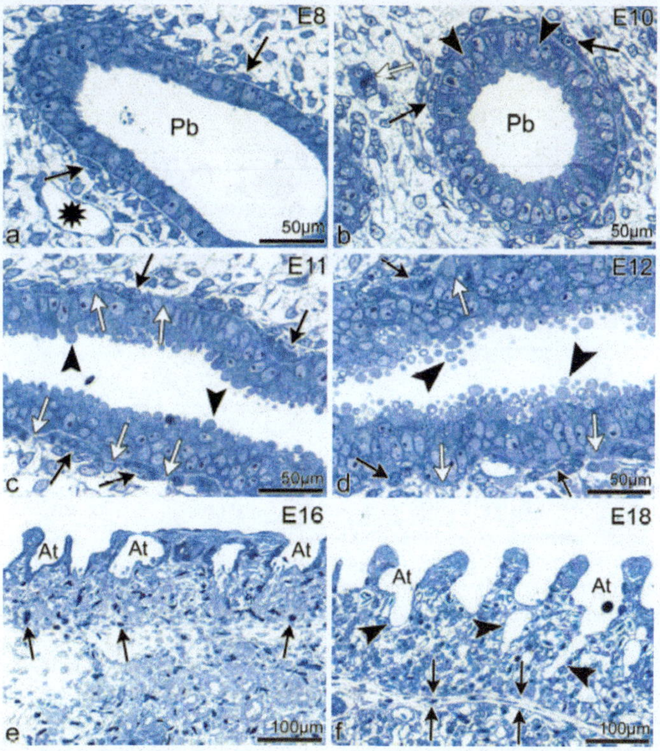

Fig. 6.9 Micrographs of semithin sections showing the structural changes in the parabronchial epithelium of the developing chick embryo lung between embryonic days 8 (E8) and E18. (**a, b**) At E8, the parabronchi (Pb) are lined with a cuboidal epithelium, which became tapered by E10, and transformed to a high columnar one. The first signs of pseudostratification are depicted by apical relocation of some nuclei (*black arrowheads* in **b**). The surrounding mesenchyme contained occasional developing feeding and draining vessels (*asterisk* in **a**) and few blood capillaries (*white arrowhead* in **b**). Subepithelial cells became aligned along the basement membrane (*black arrows* in **a** and **b**). (**c, d**) By E11 (**c**), the parabronchial epithelium is pseudostratified, and the apical parts of the cells appeared club-like (*arrowheads* in **c**). By E12, these apical parts are severed such that they appeared to fall off into the parabronchial lumen (*arrowheads* in **d**). Notably, the arrangement of the subepithelial cells associated with the parabronchial basement membrane (BM) changed their continuous appearance (*dark arrows*), so that, from E11, regions of the BM devoid of such cells are detectable (*white arrows*). (**e, f**) The atria (At) had already formed by E16, the epithelium had thinned out, and the number of vascular profiles (*arrows* in **e**) increased in the mesenchyme. At early E18, the infundibula (*arrowheads* in **f**) started to invade the mesenchyme. The interparabronchial septum is well formed (*arrows* in **f**)

immediate neighbors. The electron-dense bands constitute a double membrane, and, at such double membranes, the apical portion of the cell is separated from the basal part. Apical intercellular junctions open up so that the apical portions (aposomes) of adjacent cells are clearly delineated (Fig. 6.12). The cell separation becomes more prominent at E11. Additionally, tapering apical parts of some cells are squeezed out by the better-endowed neighboring cells. Concomitant with these processes, apical intercellular spaces become broadened, separating the apical processes and in so doing push the apical cell junctions basally (Fig. 6.12). The processes attendant to cell attenuation in the avian lung involve cell cutting (secarecytosis) and cell pinching (peremerecytosis) as described below. The processes of peremerecytosis and secarecytosis have also been captured in the ostrich lung (Fig. 6.10). During these two processes, apical tapering of cells and basal relocation of lateral cell junctions occur as

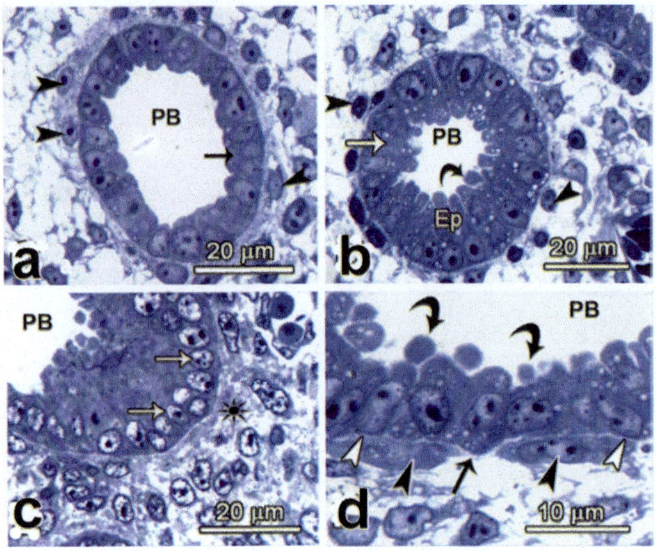

Fig. 6.10 Semithin micrographs (**a–d**) showing the structural changes in the ostrich lung parenchyma at E24. (**a**, **b**) A close-up of individual parabronchial tubes (PB) showing a cuboidal epithelium (*black arrow* in **a**) and a thickened columnar epithelium (Ep, *white arrow* in **b**). Note that in both cases, the nuclei remain in the basal region; the apical part of the cell becomes elongated thus reducing the parabronchial lumen (see PB in **a** and **b**). Subsequently the parts become constricted and severed (see *curved arrow* in **b**). Concomitant with these events, peritubular cells approach the basal aspects of the PB epithelium and align themselves in groups (*black arrowheads*). (**c**, **d**) Micrographs showing a migrating parabronchial tube (**c**) with basal nuclei (*white arrows* in **c**). Notice the absence of cells in the region ahead of the migrating tube (*asterisk* in **c**). Groups of peritubular cells (*black arrowheads* in **d**) appear to restrict migration of tubular cells while at the same time leaving gaps for cell cords to migrate (*black arrow* in **d**). Such cells adhere to a prominent basal membrane (*white arrowheads*). Migration of the epithelial cells results in formation of atria and parabronchial branches and occurs concomitantly with attenuation of the apical parts of the cell (curved *arrows* in **d**)

demonstrated in Fig. 6.11 by transmission electron microscopy and by Alexa–phalloidin staining for filamentous actin (f-actin) in the chicken lung (Fig. 6.12). The processes leading to formation of atria and formation of a thin BGB entail progressive events of cell attenuation such that at E16, protrusions of various shapes and sizes are encountered in the atria and infundibula (Fig. 6.13).

As soon as the infundibula are formed, they give rise to the air capillaries by formation and subsequent fusion of intraepithelial vesicles, as may be inferred from the presence of many short microfolds. Often, cells located at the infundibular entrance at the bottom of the atria show intraluminal projections of different sizes. The atrial epithelium at E16 changes from cuboidal to low cuboidal. Some of the cells have irregularly shaped apical processes jutting out into the lumen. Subsequently, electron-dense bands develop at the bases of the projections, followed by spaces with the ultimate separation of the projection and formation of cell attenuation bodies. The projections appear as club-like processes projecting into the atrial space. Many such projections are severed by spontaneous strangulation, which involves progressive constriction of the basal part of such a portion of a cell. Some of the projections are severed by vesiculation and subsequent fusion as can be deduced from the many short microfolds on the remaining epithelium and on the discharged aposomes (Fig. 6.14).

From E15, a new mechanism augments the process of cell attenuation and this entails formation of many small vesicles in the epithelium

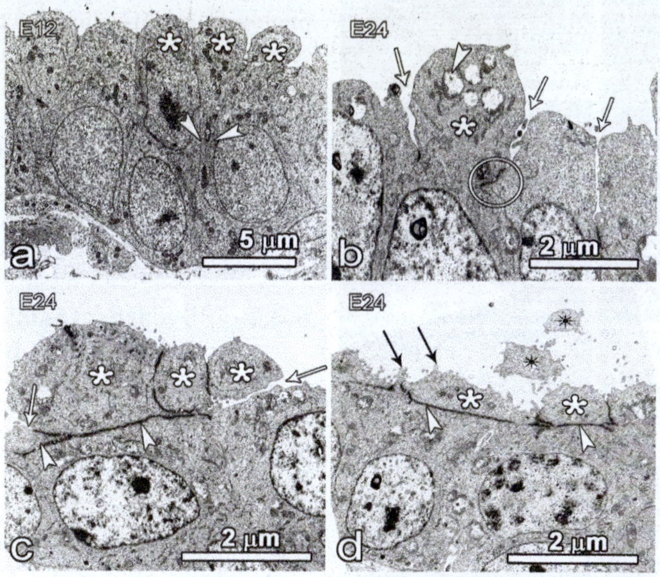

Fig. 6.11 Transmission electron microscopy (TEM) micrographs demonstrating the structural changes in the attenuating epithelia of the chicken lung at E12 and ostrich lung at E24 to E35. (**a**, **b**) The attenuating cells exhibit apical elongation forming aposomes (**a**), which protrude above the general epithelial surface (*asterisks*). Better endowed cells appear to squeeze weaker ones (*white arrowheads* in **a**). There is displacement of tight junctions (*encircled*) basally enlarging the intercellular spaces (*white arrowheads* in **b**). Some lamellar bodies are visible at this stage (*white arrowhead* in **b**). The aposomes are constricted at their basal part, showing the point where they finally get separated from the rest of the cell in the process referred to as peremerecytosis (*white arrowheads* in **a**). (**c**, **d**) Nonsecretory loss of cell portions by cutting has been referred to as secarecytosis. A double membrane is formed at the apical part of the cell, and this separates the aposomes to be lost from the rest of the cell. The double membrane forms the basis for separation (*arrowheads* in **c** and **d**). Notice the points where separation begins (*white arrows* in **c**). The presence of microfolds on the attached aposomes (*black arrows* in **d**) as well as the detached ones (*asterisks*) also indicates that vesicular degeneration is an important process in the dissolution of the discharged aposomes. *Stars* denote the aposomes to be severed

(Fig. 6.14), and in subsequent steps, the vesicles enlarge and their membranes fuse with the apical plasma membrane so that the vesicles rupture releasing their contents into the lumen (Fig. 6.15). Rupturing of the vesicles forms microfolds of varying shapes and sizes on the apical plasma membrane. In an alternative step, contiguous vesicles coalesce and fuse with each other and with the lateral plasma membranes and in so doing cut off large parts of the apical portions of the cells with the result that more aposomal bodies are released into the lumen (Figs. 6.14 and 6.15). The epithelia of air capillaries at E18 still contain many lamellar bodies. Such lamellar bodies are summarily extruded by bulging onto the apical plasma membrane with subsequent rupture of the membrane. Alternative methods of eliminating lamellar bodies entail formation of vesicles underneath the lamellar bodies with subsequent rupture of the vesicles so that the lamellar bodies are discharged together with part of the cytoplasm. The by-products of epithelial attenuation, which include lamellar bodies and cellular debris, are cleared by pulmonary macrophages. Notably at the time of hatching (E21), type II pneumocytes are restricted to the region of atrial openings (not shown). Further attenuation of the now low cuboidal epithelium proceeds through vesiculation as described above, an event that results in many microfolds (Fig. 6.15). The microfolds are subsequently severed by the same process of

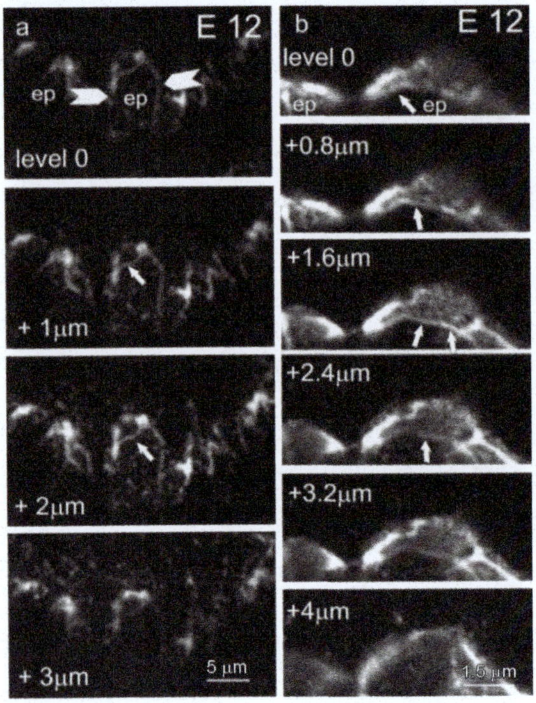

Fig. 6.12 Micrographs of serial sections of two parabronchi from embryonic day 12 (E12) chick embryos labeled with Alexa–phalloidin outlining actin filaments (**a**, **b**). The two series show tapered epithelial cells. Actin filaments delineate the apical and basolateral borders (*arrowheads* in **a**) of individual epithelial cells (ep) and mark the aposomal protrusion (*arrows*). At higher magnification (**b**), the aposomal body and the incomplete separation from its epithelial cell by a narrow actin demarcation line are visible (*arrows* in **b**)

strangulation at the base such that at E20, the epithelium is approximately 1 μm thin and is virtually devoid of microfolds (Fig. 6.15).

6.4 Mechanisms in Epithelial Attenuation During Blood–Gas Barrier Formation

The avian lung uses mechanisms closely related to those of exocrine secretion, albeit in a programmed, time-limited manner to accomplish a thin BGB. In general, they result in cutting or decapitation of the cell until the required thickness is attained. In the first step, the high columnar epithelium of the incipient parabronchi undergoes dramatic size reduction and loses morphological polarization by two main processes: secarecytosis (cell decapitation by cutting) and peremerecytosis (cell decapitation by squeezing, spontaneous constriction, or pinching off). Alternating and successive steps of secarecytosis and peremerecytosis reduce the epithelial height to an efficient functional thickness.

6.4.1 Secarecytosis: Cell Attenuation by Cutting

The processes that entail cell cutting can be grouped under one name, *secarecytosis*, which is a process of cell cutting and subsequent liberation of the severed apical cellular part (Makanya et al. 2006). It has at least two facets: cell double-membrane formation followed by separation between such membranes and cell cutting by vesiculation height to an efficient functional thickness.

6.4.2 Secarecytosis by Double-Membrane Formation

Secarecytosis entails shifting of lateral cell junctions basally so that the apical intercellular spaces open, clearly delineating the apical projection. Subsequently, a transcellular double membrane forms to separate the apical from the basal part. A gap develops between the membranes progressively separating the apical and basal portions of the cell with the result that the apical part of the cell is released into the lumen (Fig. 6.13). This process generally leads to formation of intraluminal blebs of varied shapes and sizes. The blebs that are discharged closely resembled those of aposecretion, safe for the presence of organelles. Although this process occurs predominantly during the earlier stages of cell attenuation, it continues at a lower level during the entire period of embryonic development. Formation of tubules above the dividing membrane parallel to the membrane probably leads

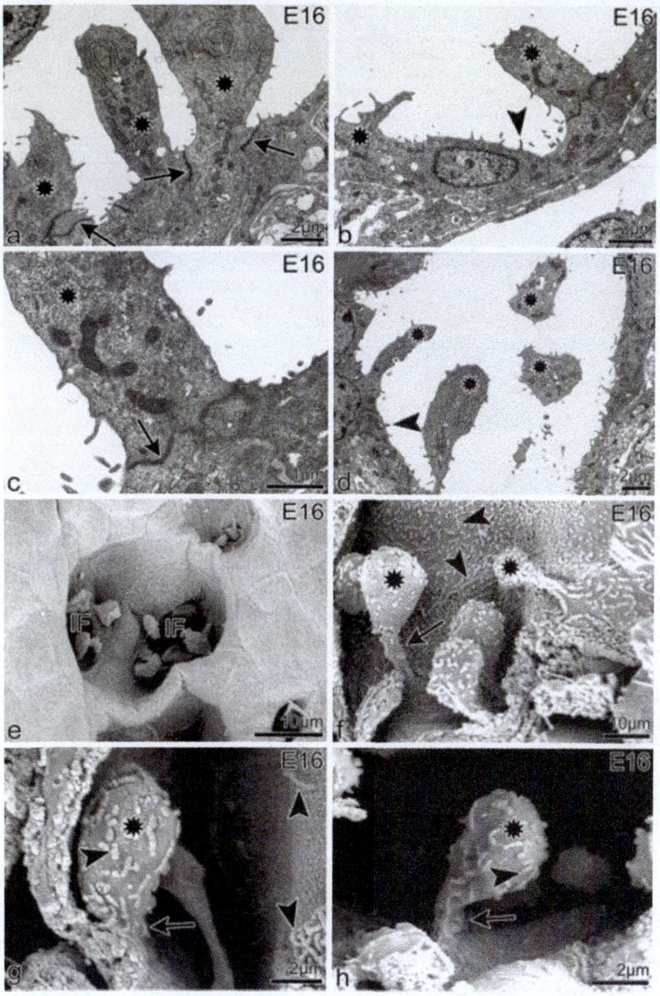

Fig. 6.13 Transmission (**a–d**) and scanning (**e–h**) electron micrographs illustrating the various mechanisms involved in attenuation of the gas exchange epithelia during development of the infundibula and air capillaries at E16. (**a–d**) The epithelium lining the forming atria and infundibula is of a low cuboidal type. Some of the cells send long apical processes into the lumen (*asterisks*). Electron-dense bands (double membranes) then develop at the bases of the projections (*arrows* in **a**, **c**). The two leaves of the double membrane subsequently separate resulting in formation of cell attenuation bodies. Microfolds (*arrowheads* in **b**, **d**) are conspicuous at this stage. (**e–h**) Scanning electron micrographs showing a zoom-in on the club-like intraluminal cell processes encountered during formation of the infundibula (If). These processes are mainly at the base of the atria, whereas the rest of the epithelium is generally flattened. The processes have a rounded apical part (*asterisks* in **f–h**) and a remarkably narrowed basal part (*arrows*). Numerous microfolds (*arrowheads*) are typical on the apical projections and the rest of the cell surface

to separation of the aposome from the basal part of the cell. A similar phenomenon has been described in apocrine glands, where tubules form parallel to and above the membranes and the subsequent detachment of the protrusion occurs (Schaumburg-Lever and Lever 1975; Hashimoto 1978; Zeller and Richter 1990; Gesase and Satoh 2003). Positive staining for f-actin at the interphase between the protruding aposome and the basal part of the cell strongly supports this hypothesis for the developing lung epithelium (Fig. 6.14).

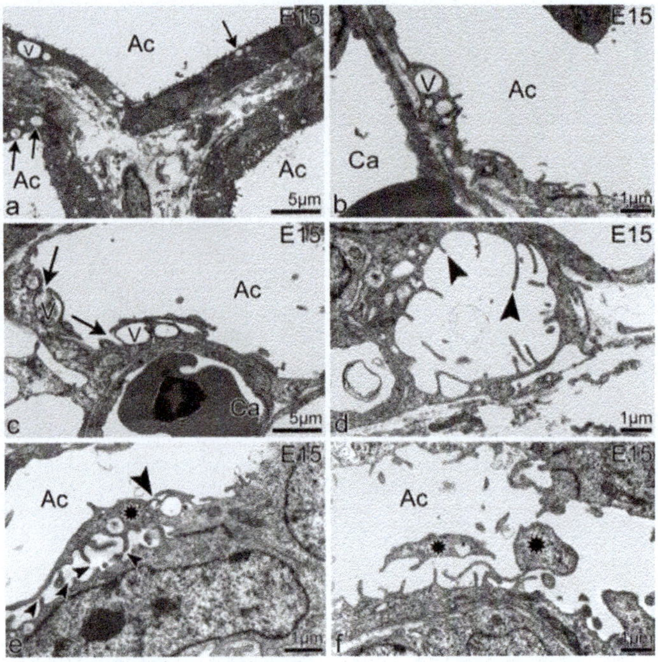

Fig. 6.14 TEM micrographs illustrating the additional mechanisms of epithelial attenuation that occur as from E15. (**a–d**) From E15, the process of cell attenuation entails formation of many vesicles (V and *small arrows* in **a** respectively) in the epithelium. In subsequent steps, the vesicles enlarge (V in **b**) and fuse with the apicolateral plasma membrane (V and *arrows* in **c**) and, in so doing, leave microfolds of varying shapes and sizes (*arrowheads* in **d**). Ac denotes air capillaries, whereas Ca stands for blood capillaries. Note the varied heights and distribution of the apical cell projections (see *arrowheads* in **d**), quite unlike microvilli found in epithelia. (**e, f**) In an alternative step, contiguous vesicles coalesce and fuse with their neighboring cognates (*arrowheads* in **e**) and, in so doing, cut off large parts of the apical portions of the cells (*asterisks* in **e, f**). Although the events described above cannot be unequivocally demonstrated on still pictures, the appearance of vesicles (**a–c**), their subsequent disappearance and formation of microfolds (**d**), and intraluminal blebs (**f**) plausibly underpin the processes of secarecytosis by vesiculation

Secarecytosis by coalescing vesiculation occurs via formation of vesicles in rows below the cell portion to be severed. Such vesicles finally fuse with their neighboring cognates and then with the apicolateral plasma membrane and, in so doing, sever the aposomal projection from the rest of the cell (Fig. 6.15). The latter process occurs much later and appears to target mainly attenuation of the low cuboidal epithelium in the formative atria as well as in the migrating air capillaries. The microfolds result from the fusion of contiguous vesicular membranes (Fig. 6.15), at the interphase between the aposome and the basal part of the cell, hence discharging the aposome. Similar demarcating vesicles have been reported during aposecretion in exocrine glands (Smith and Hearn 1979; Gesase et al. 1996; Gesase and Satoh 2003), with the notion that the demarcating vesicles were morphologically different from secretory vesicles (Smith and Hearn 1979).

A similar process has been described by Satoh et al. (1992) and Gesase et al. (1995), which is dubbed "massive exocytosis" due to the absence of apical protrusions that are characteristic of classic aposecretion. This process was later qualified to nonprotrusion aposecretion by Gesase and Satoh (2003), because portions of the cytoplasm

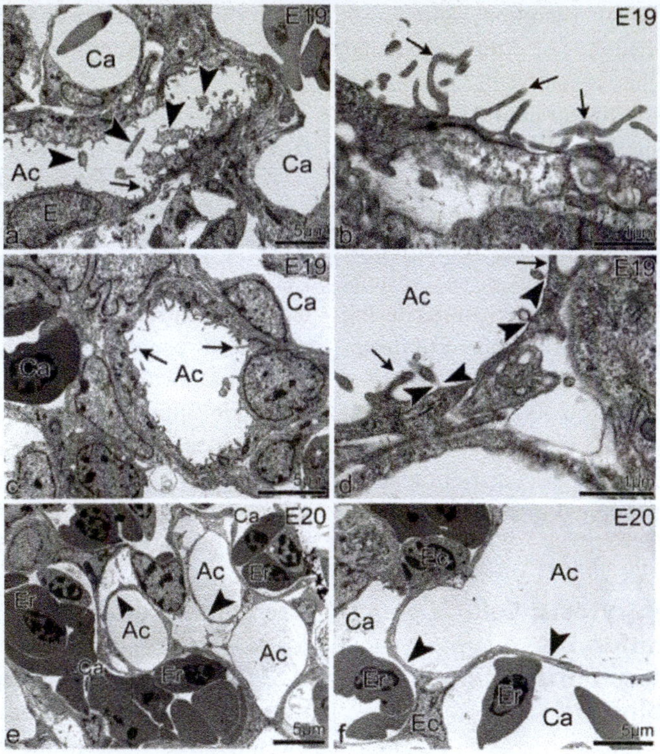

Fig. 6.15 TEM micrographs showing the terminal mechanisms involved in cell attenuation. (**a**, **b**) Continued vesiculation, fusion, and rupture of vesicles results in numerous cell attenuation bodies of varied sizes and shapes (*arrowheads* in **a**) within the air capillaries (Ac) as well as microfolds of irregular shapes and sizes (*arrows*). The blood capillaries (Ca) approach the thinning epithelium (E) to form a BGB. (**c**, **d**) The cell attenuation bodies become fewer in air capillaries (Ac) by E19, and the microfolds (*arrows*) are reduced in number and are much shorter. The microfolds are severed by the process of spontaneous strangulation (*arrowheads* in **d**). (**e**, **f**) By E20, there are virtually no microfolds and the mesenchyme is reduced, and numerous air capillaries (Ac) are closely apposed to blood capillaries (Ca) forming a thin BGB (*arrowheads* **e**). All the microfolds have disappeared at this stage so that the only distinguishing feature between the air capillaries and the blood capillaries is the presence of red blood cells (Er) in the latter. EC denotes endothelial cell

trapped between vesicles were released into the glandular lumen, leaving cuplike concavities.

6.4.3 Secarecytosis by Rupturing Vesiculation

In this case, vesicles are formed at the apical part of the cell; they enlarge, putatively by accumulation of fluid from the cell, and, therefore, tend to protrude toward the lumen above the general cell surface. Subsequently, the vesicular membrane fuses with the apical plasma membrane, releasing its contents into the lumen and forming shallow concavities separated by microfolds. Fusion of such vesicles with the apical plasma membrane results in rupture and formation of tiny microfolds but with concomitant reduction in cell height. The microfolds formed as a result of vesiculation are then severed by progressive thinning and constriction at the base with ultimate formation of a smooth, squamous epithelium (Fig. 6.15). Successive phases of rapturing vesiculation and severance of microfolds plausibly discharge substantial portions of the fluid and solid portions of the cell so that, at the time of hatching, the respiratory epithelium is thin, smooth, and squamous.

During physiological secretion, exocytosis occurs through fusion of the vesicular membranes with the apical cell membrane forming small pores (Kliewer et al. 1985). However, during rupturing vesiculation, the fusion of the vesicular membrane with the apical plasma membrane results in rupture of the vesicle, discharge of the entire contents, and formation of microfolds without regeneration of the vesicle. Notably, numerous microfolds are formed during secarecytosis, but microfolds are noted to disappear during both apocrine and non-apocrine secretion (Gesase and Satoh 2003). Decapitation of microfolds during avian lung development occurs as a final step, through constriction at the base and fusion of the apical membranes.

6.4.4 Peremerecytosis: Cell Decapitation by Squeezing, Constriction, or Pinching Off

This process is morphologically distinguished by apical club-like cell protrusions. In spontaneous peremerecytosis, the cells become tapered, intercellular spaces between the apical parts of the cells widen, and the cells become constricted at the supranuclear region until the apical part is squeezed out. In some cases the better-endowed cognate neighbors squeeze a sandwiched cell out until the apical portion is ejected (Fig. 6.11). In either case, progressive thinning of the stalk of the protrusion results in severing of the aposome. Previously, aposecretion by pinching off has been described in glands (Gesase and Satoh 2003). In classic aposecretion there is bulging of the apical cytoplasm, absence of subcellular structures, and presence of membrane-bound cell fragments, the so-called aposomes (Deyrup-Olsen and Luchtel 1998). Peremerecytosis during development is different in that the discharged blebs contain major organelles; sometimes an entire cell is squeezed out. The mechanisms that cause thinning of the aposomal stalk are unknown (Gesase and Satoh 2003), but actin filaments (Fig. 6.12) are probably involved (Metzler et al. 1992; Aumuller et al. 1999; Stoeckelhuber et al. 2000; Makanya et al. 2006).

6.5 Role of α-SMA-Positive Cells During Parabronchial Development

Early during development (E8 in the chick lung), α-SMA-positive cells become associated with the basal aspects of parabronchial epithelial cells and surround the latter (Fig. 6.16). Such cells are interrupted by gaps through which incipient atria sprout. As soon as the atria are patterned, the α-SMA staining intensity declines dramatically, probably due to the reduction of α-SMA-positive cells by apoptosis and/or their transformation to fibroblasts. The driving mechanisms involved in the pattering of α-SMA-positive cells are unclear. Putatively, α-SMA-positive cells participate in the patterning of the atria but finally are restricted to supporting the interatrial septa in the prehatch embryo but may also participate in peremerecytosis. During milk secretion, for example, myoepithelial cells squeeze the secretory epithelium and, in so doing, facilitate the release of milk into the secretory acinus (Furuya et al. 2004; Macuhova et al. 2004). The association of α-SMA-positive cells with the parabronchial epithelium during the time of secarecytosis and peremerecytosis probably is important in facilitating these aposecretion-like cell attenuation processes.

6.6 Putative Mechanisms in Physiological Secretion, Secarecytosis, and Peremerecytosis

Mechanisms in aposecretion have been reviewed by Gesase and Satoh (2003) and Farkaš (2015), with the notion that the biochemical and physiological pathways regulating aposecretion as well as the plasma membrane dynamics are poorly understood. This is complicated by the fact that, in most occasions, aposecretion is accompanied by exocytosis.

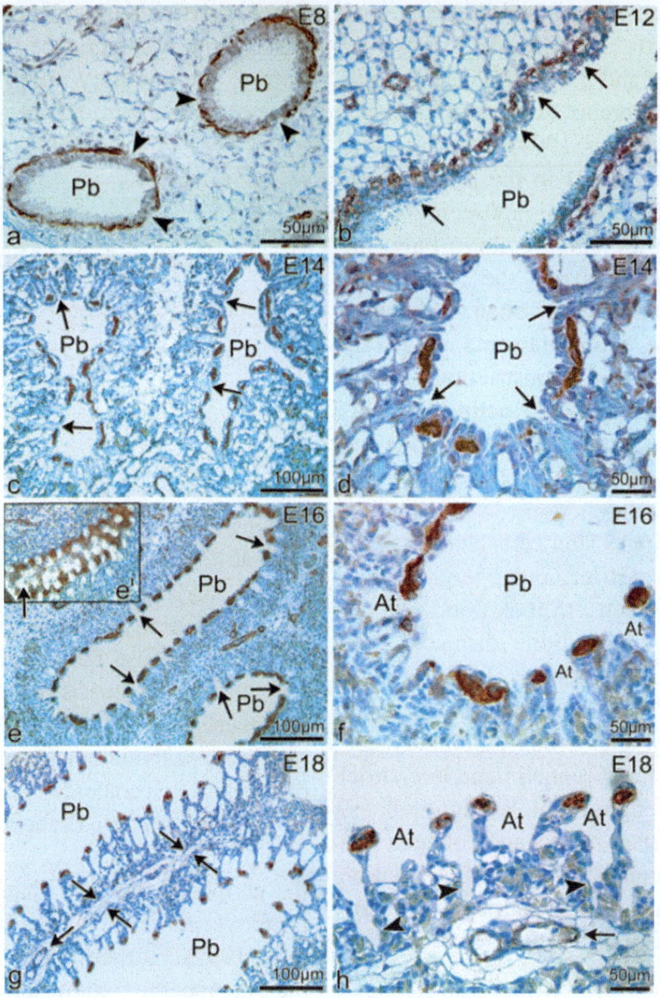

Fig. 6.16 Immunohistochemical staining (*dark brown coloration*) for α-SMA at different developmental stages. (**a**, **b**) At E8, the epithelium is surrounded by a layer of α-SMA-positive mesenchymal cells interrupted by intervening gaps (*arrowheads* in **a**). By E12 (**b**), the gaps between the α-SMA-positive cells are smaller but occur more frequently and are more regularly spaced. The first strands of migrating epithelial cells (*arrows* in **b**), which represent the sprouting atria, invade the mesenchyme through gaps delineated by the α-SMA-positive cells. (**c**, **d**) By E14, columns of parabronchial epithelial cells have penetrated deeper into the surrounding mesenchyme (*arrows*). Note that (**d**) is an inset of (**c**) at higher magnification showing the migrating atrial cells (*arrows*). (**e**, **f**) The atria (At) at E16 form through the gaps between the α-SMA-positive cells (*arrows* in **e**), and the latter cells mark the incipient interatrial septa. The α-SMA-positive cells form a network or honeycomb-like pattern surrounding the parabronchus as depicted in e′ (inset). At a higher magnification (**f**), the formative atria (At) opening into the parabronchial lumen (Pb) are clearly delineated by the dark staining immunopositive cells. (**g**, **h**) By E18, the interparabronchial septum is thin (*arrows* in **g**) but well endowed with blood capillaries. The atria (At in **h**) are well formed, continuous with the infundibula (*arrowheads* in **h**), and separated by prominent interatrial septa whose tips are reinforced with α-SMA-positive cells (*arrows* in **h**). Pb denotes parabronchial lumina. Note that (**h**) is an inset of (**g**) at a higher magnification

Physiological release of surfactant occurs through elliptical cell surface pores averaging 0.2 × 0.4 microns in size on the alveolar luminal side of type II cells. In the attenuating epithelium of the chicken lung, entire lamellar bodies were released either through large apical pores or together with part of the cytoplasm by abscising vesiculation (Makanya et al. 2006).

The plasma membrane dynamics during apocrine secretion have not been extensively investigated (Gesase and Satoh 2003). Participation of cytoskeletal proteins, such as myosin and gelsolin (Aumuller et al. 1999) or even actin (Stoeckelhuber et al. 2003) in the pinching off of the apical protrusion during aposecretion has been implicated. In the chick embryo lung, the presence of actin filaments in the constricting aposome has been demonstrated (Makanya et al. 2006). These actin filaments putatively participate in aposomal constriction during peremerecytosis. Furthermore, actin filaments are known to be associated with the cell adhesion belt (Volberg et al. 1986) and are indicators for distal relocation of cell junctions. During embryonic development, ingressing cells change shape and their apices constrict, putatively through actomyosin contraction (Shook and Keller 2003). Such a constriction displaces the organelles basally in readiness for migration (Shook and Keller 2003). Basement membranes form the support for attenuating pulmonary epithelium. They contain fibronectin and interstitial collagens, which seem to promote migration and proliferation. Basement membrane collagen and laminin stimulate attachment and differentiation and are involved in signaling pathways that promote alveolar epithelial cell differentiation and VEGF expression (reviewed in Makanya 2016), but their direct role in in peremerecytosis and/or secarecytosis is unknown.

6.7 Apoptosis and Formation of the Thin Blood–Gas Barrier

Cell death is not directly involved in the attenuation of the chicken lung epithelium. Apoptosis, however, is very important in the formation of the thin BGB. Since mesenchymal cells undergo apoptosis, giving way to the expanding and migrating epithelial tubes those immediately ahead of the migrating epithelial tubes become the initial targets of apoptosis (Makanya et al. 2006). During atria formation, mesenchymal cell apoptosis is responsible for diminution of the interstitium between the epithelial tubes and, hence, participates in the approximation and subsequent apposition of the attenuated epithelium to the capillary endothelium and the ultimate establishment of a very thin BGB.

6.8 Vasculature Development in the Avian Lung

In the avian embryo, the pulmonary and bronchial vasculature arises from the splanchnic plexus that forms around the foregut before the lung buds develop (DeRuiter et al. 1993). In the chicken embryo, the pulmonary arteries are delineated at approximately E5 when the proximal part of the truncus arteriosus becomes incorporated into the right atrium, and the distal part is divided into the aorta and pulmonary artery by the aorticopulmonary septum (Bellairs and Osmond 1998). Generally, pulmonary vasculature arises by both vasculogenesis (deMello et al. 1997) and angiogenesis (Hansen-Smith 2000). Vasculogenesis starts with blood islands (deMello et al. 1997) that coalesce to form sinusoidal spaces of irregular profiles. Branches sprouting from the central pulmonary vascular trunks connect with the peripheral vascular channels establishing the primitive pulmonary vascular network (deMello et al. 1997). Furthermore, it has been shown that the peripheral pulmonary venous network develops earlier than the pulmonary arterial network (deMello and Reid 2000). In the adult avian lung, the pulmonary arteries run alongside the airways, and the pulmonary veins show a similar branching pattern to the arteries, although separated from them. During early fetal development, the airways act as a template for pulmonary blood vessel development in that the vessels form by vasculogenesis around the branching airways (Hansen-Smith 2000).

Both vasculogenesis and sprouting angiogenesis (SA) are important players in the establishment of the pulmonary vasculature. Indeed, embryonic mesodermal tissue contains angioblasts (Noden 1990), which are capable of migration and initiation of vasculogenesis in the homing sites (Ambler et al. 2001; De La 2003).

6.9 Vasculogenesis

In the avian lung, vasculogenesis has been described by Maina (2004a) and Anderson-Berry et al. (2005). Development of the vasculature in the avian lung results in an elaborate pattern of supplying and draining vessels that culminates in the dense cylindrically arranged capillary networks associated with the parabronchi, the ultimate sites of gas exchange in the avian lung (West et al. 1977; Makanya and Djonov 2009). The chronological events, processes, and mechanisms as well as the molecules that lead to this pattern are beginning to emerge.

6.10 Angiogenesis

Generally, sprouting (Fig. 6.17) precedes intussusceptive (Fig. 6.18) angiogenesis (IA), but the latter process participates in the ultimate multiplication and remodeling of vascular branches (Makanya and Djonov 2009). IA, a novel mechanism of vascular growth, expansion, and remodeling, has received impetus in the recent past, with the indication that it proceeds faster and at lower energetic costs than SA and preserves physiological status since extensive cell proliferation and basal membrane degradation are not required (Burri et al. 2004; Makanya et al. 2009). The quintessence of IA is transvascular pillar formation (Fig. 6.18), and the mechanism comprises three cognates expressed in sequential, albeit overlapping, chronological phases (Djonov and Makanya 2005; Makanya et al. 2009). Intussusceptive microvascular growth is the first process, which greatly expands the vascular network; intussusceptive arborization delineates the classic vascular tree pattern; and intussusceptive branching remodeling modifies the vascular pattern in regard to the physiological requirements (Makanya et al. 2009).

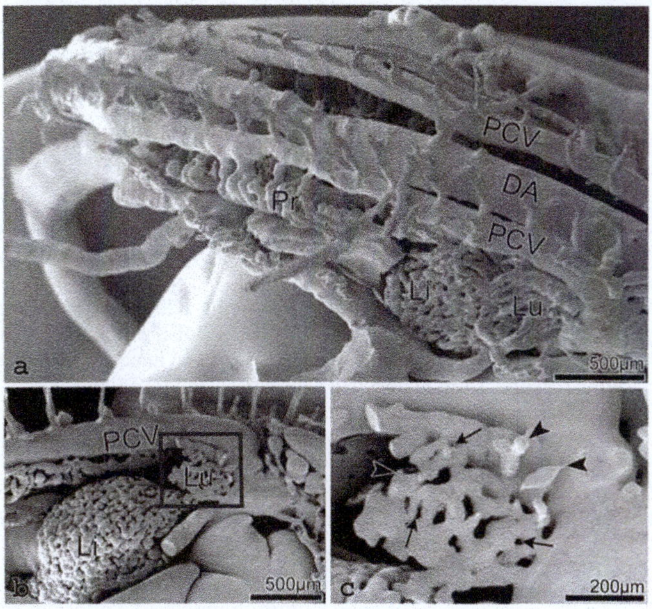

Fig. 6.17 Mercox corrosion casts from E5 demonstrating the developing avian pulmonary vasculature. (**a**) At E5, the lung vasculature comprises only a few sprouting blood vessels that form a rather simple pulmonary plexus (Lu), which is located cranial to and is continuous with the hepatic one (Li). The vascular plexus of the pronephros (Pr) is located in the caudal part of the embryo ventral to the major vessels: *PCV* posterior cardinal vein, *DA* dorsal aorta. (**b**) The lung vasculature later at E5 appears more complex. The vessels have a generally thicker girth and irregular shapes. The vasculature of the liver is also better developed. (**c**) The pulmonary plexus expands by sprouting (*arrowheads*) and intussusceptive angiogenesis. The latter process is represented by tiny holes in the more central parts of the plexus (*arrows*). (**c**) is an inset of (**b**) at higher magnification

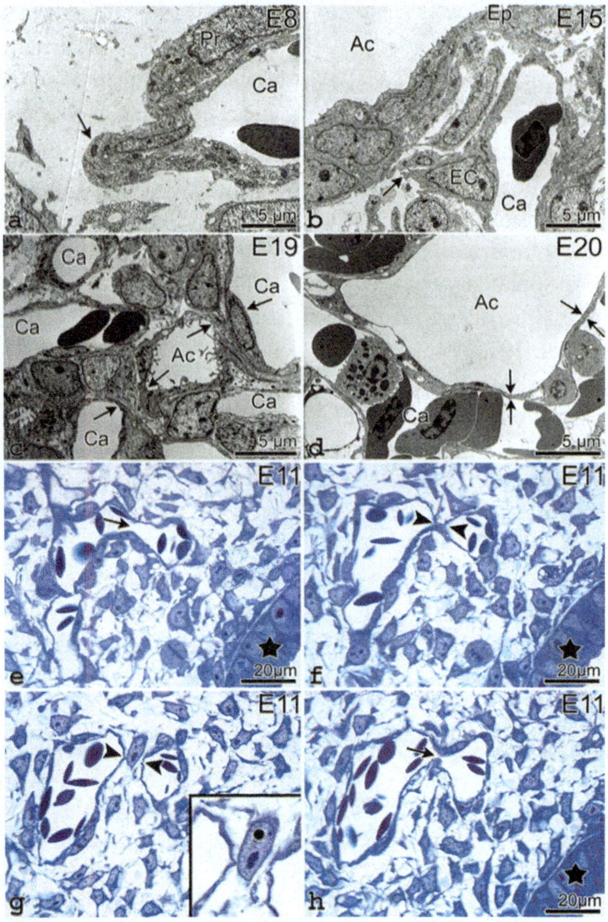

Fig. 6.18 TEM micrographs demonstrating sprouts (**a**, **b**) and formation of the BGB (**c**, **d**) and light micrographs (**e–h**) illustrating the morphology of a pillar and the archetype of intussusceptive angiogenesis. (**a**) An early blood capillary sprout at E8 with a tip cell (*arrow*) and slit-like lumen. Notice the absence of basal membrane and rarefied extracellular matrix surrounding the tip of the sprout. (**b**) A sprout (*arrow*) originating from a polarized endothelial cell (EC) of a blood capillary (Ca) approaching the epithelium (Ep) of an air capillary (Ac) at E15. (**c**) At E19, the blood capillaries approach the developing air capillaries to establish the BGB (*double arrows*), which at this time is relatively thick. (**d**) Dramatic attenuation of the epithelium and reduction of the interstitial tissue result in a remarkably thin BGB (*double arrows*) by E20. (**e–h**) Serial semithin sections showing the extents of a transluminal pillar in a lung capillary at E11. The open lumen of the vessel in (**e**) (*arrow*) is closed by a transluminal pillar in (**f**) and (**g**) (*double arrowheads*) and becomes open again at the end of the pillar (*arrow* in **h**). The parabronchial epithelium (*stars*) is used as a landmark to show that this is the same vessel. A pericyte (inset in **g**) is typically associated with the transluminal tissue pillar. The step section is 3 µm

6.11 Sprouting Angiogenesis

The primordial vasculature is established through both vasculogenesis and sprouting angiogenesis (Makanya et al. 2007; Loscertales et al. 2008), while subsequent vasculature expansion and remodeling proceed mainly through angiogenesis (Anderson-Berry et al. 2005; Maina 2004b; Makanya and Djonov 2009). Sprouting angiogenesis lays down the basic vascular plexus (Fig. 6.19), while intussusceptive angiogenesis remodels the ultimate pulmonary angioarchitecture (Makanya et al. 2007). At the end of the process, a rather unique pattern of vessels surrounding the parabronchi and

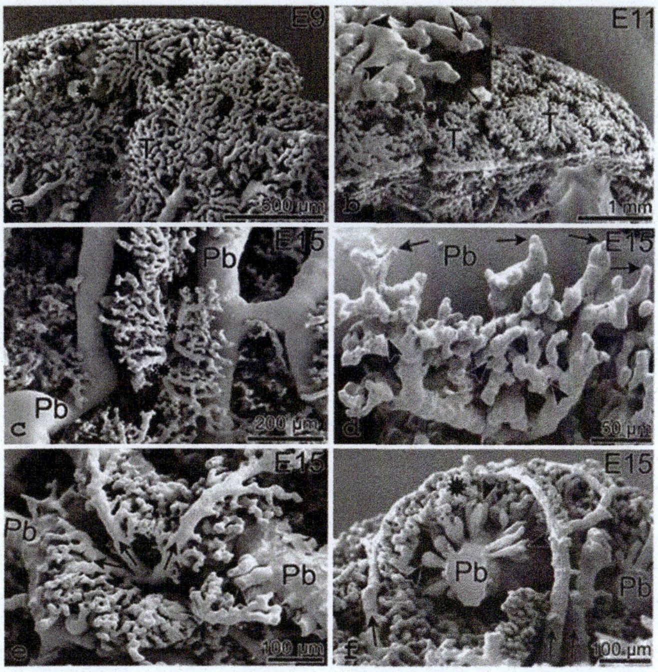

Fig. 6.19 Mercox corrosion casts from the avian lung vasculature (**a**, **b**) and double-injection intratracheal and intravascular mercox casts (**c**–**f**) illustrating the developing pulmonary microvasculature and its relationship to the parabronchial system. (**a**, **b**) At E9 and E11, the pulmonary vasculature is organized into treelike capillary tufts (T) comprising major supplying vessels [formative interparabronchial and parabronchial arteries (*) connected to capillary plexuses]. Vacant spaces between the tufts represent the areas occupied by the parabronchial system. The capillaries at the periphery of the tufts (**b**) expand predominantly by sprouting (*arrows*), whereas the central part grows by intussusception (*arrowheads*). Inset: part of (**b**) at higher magnification. (**c**, **d**) Double-injection mercox casts showing the spatial relation between the developing lung vasculature and the parabronchi (Pb) at E15. The parabronchial arteries (*) give rise to the arterioles that form the dense network of capillaries that surround the Pb. The capillary network expands predominantly by sprouting (*arrows*) in peripheral parts of the tufts, followed by intussusceptive pillar formation (*arrowheads* in **d**). (**d**) is an inset of (**c**). (**d**, **f**) Several arterioles (*arrows*) emerge from a common parabronchial artery and course toward opposite sides to supply adjacent Pb and the atria with incipient infundibulae and air capillaries (*arrowheads*). The parabronchial arteries (*arrows*) supply the plexus from the outside, while the veins receive branches from the inside

a dexterously crafted blood capillary network that interlaces with the equally complex air capillary labyrinth (Maina 1988; Makanya and Djonov 2008) is accomplished.

Air conduits are already well established by the time the smaller parabronchial vessels inaugurate (Fig. 6.19), and these presumably form a guide to the developing blood vessels. Subsequently, the vasculature is formed to entirely engulf the parabronchus (Figs. 6.20, 6.21 and 6.22). Interestingly, rami of the pulmonary artery and vein approach the parabronchi in step and at the same time form a capillary network initially by sprouting (Fig. 6.21). The rami form the interparabronchial arteries, which form the various subsequent generations of blood vessels (Figs. 6.22, 6.23, and 6.24) seen in the adult lung (Table 6.1).

6.12 Intussusceptive Angiogenesis

In the chicken lung, establishment of the vascular network relies largely on capillary migration and transdifferentiation into arterioles and arteries as well as anastomoses, perhaps guided by ephrins

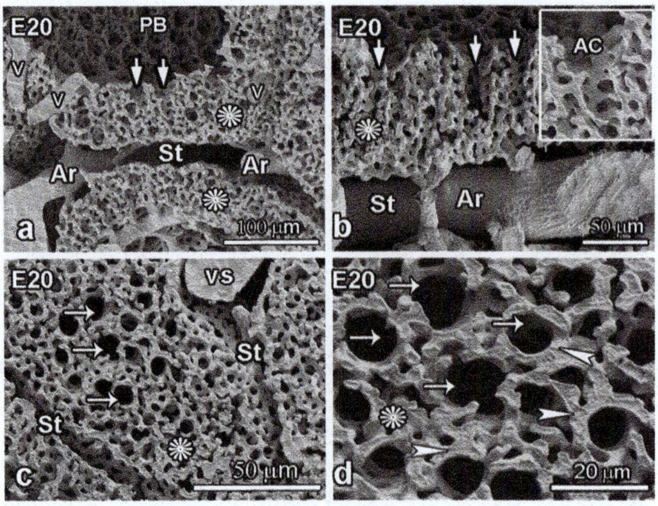

Fig. 6.20 SEM micrographs of intravascular lung casts showing the 3D disposition of blood capillaries around the lumina of air capillaries at E20. (**a**, **b**) Casts sectioned parallel to the long axes of the air capillaries. The ACs (*arrows*) spread from the parabronchial lumen (PB) and pierce the parabronchial capillary mantle toward the septum (St). The interparabronchial septum has the large arteries (Ar). The parabronchial mantle (*asterisks*) is traversed by intraparabronchial veins (V). Inset in (**b**) shows the netlike blood capillary meshwork surrounding an individual air capillary (AC). (**c**) Casts sectioned at the periphery of the parabronchial mantle to show the interparabronchial septum (St) with a large interparabronchial vessel (Vs) and air capillaries (*arrows*) that extend to the periphery of the blood capillary meshwork (*asterisk*). (**d**) A close-up of the transected BCs at the periphery of the parabronchial mantle shows that BCs tend to form complete rings (*arrowheads*) that surround the ACs, accomplishing a crosscurrent system. In between the ACs, the BCs form a dense network (*asterisk*)

and ephrin receptors, which are known to integrate blood vessel and tissue morphogenesis (Adams 2002). Eph receptors are a group of receptors that are activated in response to binding with Eph receptor-interacting proteins (Ephrins). Eph signaling is requisite for SA and blood vessel remodeling during vascular development (Cheng et al. 2002). In addition to flow dynamics, genetic programs control arterial–venous cell fate and blood vessel identity (Adams 2003).

The precise mechanisms and the chronology of events that entail the establishment of the discrete arterial and venous systems and the relationship between the two during development are unclear. Several models for the pulmonary vascular tree establishment have been proposed, but no unequivocal evidence has been adduced for either. A combination of central angiogenesis where large vessels develop centrally and migrate to the lung and fuse with vessels inaugurated through vasculogenesis (distal vasculogenesis) was suggested by deMello et al. (1997). In contrast, Hall and coworkers (2000) contended with just vasculogenesis as the preponderant mechanism, while Parera et al. (2005) argued that only distal angiogenesis was the plausible mechanism for establishing the pulmonary vasculature.

At E5, a simple pulmonary plexus exists the developing lung. This plexus (Fig. 6.17) is made up of a few blood vessels situated close to the more endowed hepatic plexus, an analogy, which underpins the common origin of the two organs from the primitive gut. The lung is topographically cranial to the liver, and its vasculature is also continuous with that of the posterior cardinal vein and also the dorsal aorta. A few hours later, the pulmonary plexus becomes more complex (Fig. 6.17), the vessels having a generally thicker girth and irregular shapes. At this stage, capillary sprouts are visible, particularly in the periphery, as well as sparse holes in the most central regions, representing transluminal tissue pillars, the quintessence of IA.

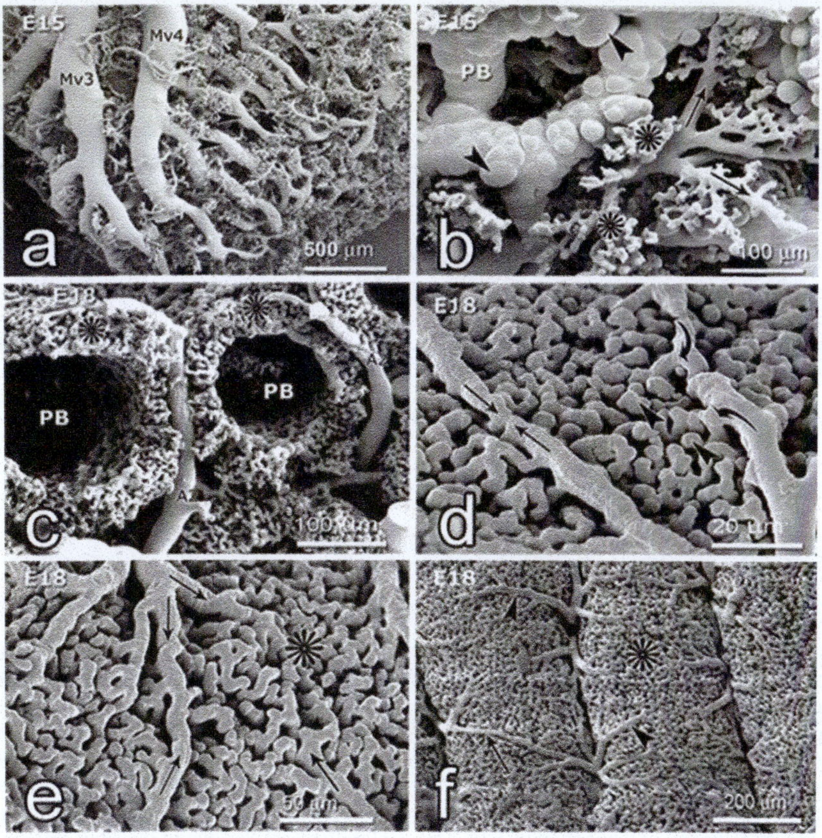

Fig. 6.21 Scanning electron micrographs of double (i.e., intravascular together with intratracheal); (**a**, **b**) and intravascular (**c–f**) casts, showing the establishment of the PB capillary meshwork at various developmental stages. (**a**) Both the secondary bronchi (medioventrals; MV3 and MV4) and the PB (*arrowheads*) are well established, but only a few capillary tufts (*asterisks*) are present. These capillary tufts send branches to surround the formative PB (*arrowheads*) and the secondary bronchi. (**b**) A close-up showing the interaction between the developing vasculature and the formative atria. The atria (*arrowheads*) sprout from the PB ahead of the spreading capillaries (*arrows*), so that they guide the patterning. Note the capillary tufts (*asterisks*) sending branches that engulf the PB. (**c**) At E18, the primitive PB capillary mantle has already formed the typical cylindrical, albeit less geometrically refined, patterns (*asterisks*) surrounding the PB lumen. The interparabronchial vessels, such as the PB artery (A2), are clearly discernible. (**d**) On external aspect, the arterial system is represented by migrating arterioles that come to fuse with their opposite cognates (*opposing arrows*). Notice the capillary sprouts that establish the PB mantle (*arrowheads*). As soon as the arterioles abut, they start giving rise to capillary branches (*curved arrows*) that expand the capillary mantle toward the interior of the capillary mantle. (**e**) The processes of vascular remodeling happen very fast, so that the anastomosed arterioles (*arrows*) are broken down to form the dense capillary meshwork (*asterisk*) characteristic of the PB wall. (**f**) Vascular remodeling results in short PB arterioles (*arrowheads*) that feed the incipient capillary meshwork from the external aspect of the PB (*asterisk*). Notice a constricting remaining anastomosed arteriole (*arrow*)

The extent and intensity of centrifugal sprouting becomes augmented so that at E15 capillary sprouts from contiguous tufts approximate each other and anastomose and by doing so enclose the developing parabronchi. From the aforementioned capillary tufts, larger supplying vessels (arterioles) become delineated (Fig. 6.19) and develop external to the general capillary network. These arterioles approximate each other at the level of the middle long axes of the

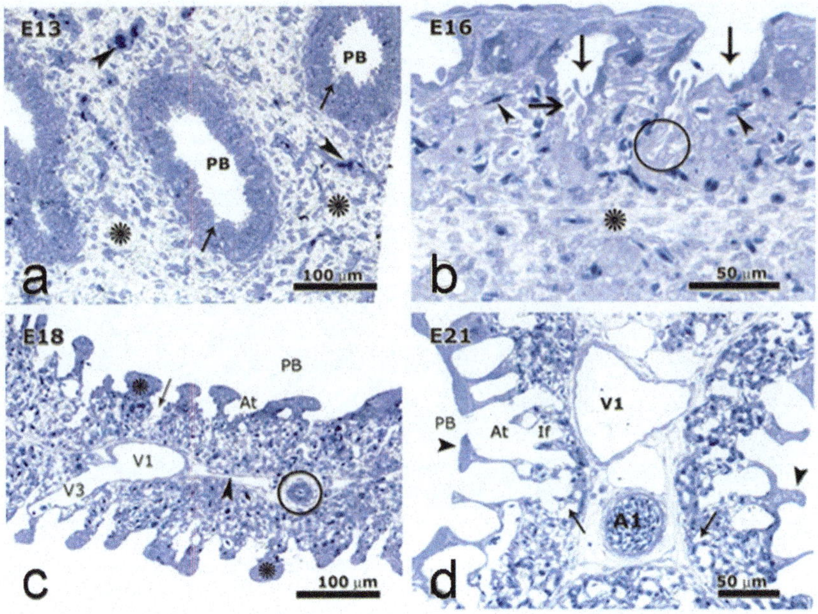

Fig. 6.22 Semithin sections showing the establishment of the PB blood vessels. (**a**) By E13, the PB is the characteristic structure in the lung parenchyma. The tissue of the interparabronchial septum (*asterisks*) at this stage is abundant, and a few vascular profiles (*arrowheads*), as well as formative atria (*arrows*), are encountered. (**b**) The formative atria (*vertical arrows*) start to grow toward the mesenchymal tissue (*asterisks*) of the septum by E16. The interparabronchial septum is clearly delineated, and vascular profiles (*arrowheads*) of the incipient PB vessels are evident. The formative infundibula are indicated with a *horizontal arrow*, while the *circle* indicates a developing air capillary. (**c**) By E18, the PB parenchymal tissue has replaced most of the mesenchyme, so that only a thin interparabronchial septum is evident (*arrowhead*). The interparabronchial vessels include the interparabronchial vein (V1), the intraparabronchial vein (V3), and interparabronchial artery (*encircled*). The intraparabronchial vein traverses the PB wall to join the atrial capillary network. Note the interatrial septa (*asterisks*) projecting into the PB lumina. An atrium and an incipient infundibulum (*arrow*) are also shown. (**d**) The PB parenchymal tissue at E21 closely resembles that of the adult chicken. Gas exchange parenchyma has replaced the mesenchyme, so that only the interparabronchial vessels stand out prominently. These include the interparabronchial vein (V1) and the intraparabronchial artery (A1). Note the interatrial septa (*arrowheads*) projecting into the PB lumina. The atria (At) give rise to infundibula (If), which form air capillaries (*arrow*)

parabronchus and anastomose. At the same time, the capillary plexus is augmented by massive SA with only sparse pillar holes being apparent.

At the ventro-posterior part of the lung, the secondary bronchi give rise to the branching neopulmonic parabronchi, and the treelike capillary tufts grow between the parabronchi. The formative atria are evident on the surface of the parabronchi and appear as rounded mounds of cast protruding from the parabronchus. The development of the blood vessels is deftly harmonized with that of the gas exchange system so that as the blood capillary network forms, the atria as well as the infundibula and air capillaries inaugurate (Fig. 6.19).

The prolonged sprouting phase of the pulmonary vasculature development presents a host of sprout morphophenotypes. As indicated in Fig. 6.18, the capillary sprouts show morphological polarization of the endothelial cells with adluminal protrusions (so-called tip cells). The cell membrane at the tip of the sprout appears amorphous, and the area surrounding the sprouts (sprout mantle) is rather rarefied. Furthermore, such cells have abundant cytoplasm and organelles, and the basement membranes are often absent. The sprouts grow around the developing air conduits and at same time approximate the air capillaries. Microvessel apposition to the

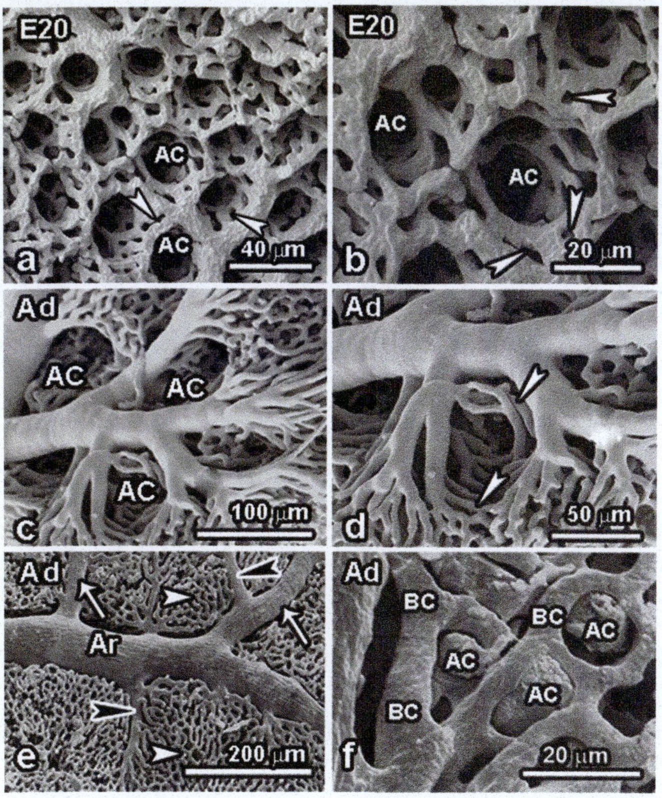

Fig. 6.23 SEM micrographs of intravascular lung casts comparing the spatial arrangement of the blood capillaries around the lumina of air capillaries in the prehatch chick embryo (E20; **a**, **b**) and the adult (**c**, **d**). An intravascular cast and a double cast in the adult are presented in (**e**) and (**f**), respectively. (**a**, **b**) Viewed from the parabronchial lumen, the blood capillaries at E20 are composed of netlike baskets surrounding the air capillaries (AC). These capillaries are of irregular girth, and the casts are characterized by numerous holes (*open arrowheads*), which represent tissue pillars characteristic of intussusceptive angiogenesis. (**c**, **d**) Luminal view of vascular casts in the adult bird reveals a totally different picture with the BCs forming smooth tubular structures (*white arrowheads* in **d**) running perpendicular to the long axes of the ACs and surrounding the ACs in such a manner that a crosscurrent system is established. Notably these capillaries are virtually devoid of tissue pillars, as indicated by the absence of holes on the vascular casts. (**e**, **f**) Overview of the external aspect of the parabronchial vascular casts. Note the close-knit capillary network (*white arrowheads* in **e**) that covers the gas exchange tissue. The large arteries (Ar) run in the interparabronchial septum and give rise to parabronchial arteries (*white arrows*) and parabronchial arterioles (*black arrowheads*). Notice the exchange blood capillaries surrounding the tips of ACs (BC in **f**). Such BCs form arcs around the ACs, highlighting the tendency to maintain a crosscurrent system even at the peripheral parts of the parabronchial mantle

air capillaries leads to progressive attenuation of the blood–gas barrier so that by the time of hatching, a thin squamous epithelium is formed (Fig. 6.15).

During formation of the parabronchial vascular plexus at E17 (Fig. 6.21), arterioles emanating from the parabronchial arteries migrate across the external aspect of the parabronchial capillary plexus and elongate to anastomose with those from the opposite side at the level of the long axis of the parabronchus. Additional lateral arteriolar branches form and give rise to the parabronchial capillary plexus. The primary parabronchial arterioles constrict in the plane of the long axis and are then broken into capillaries, probably by the process of pillar formation. The parabronchial plexus is drained by large interparabronchial veins, which are

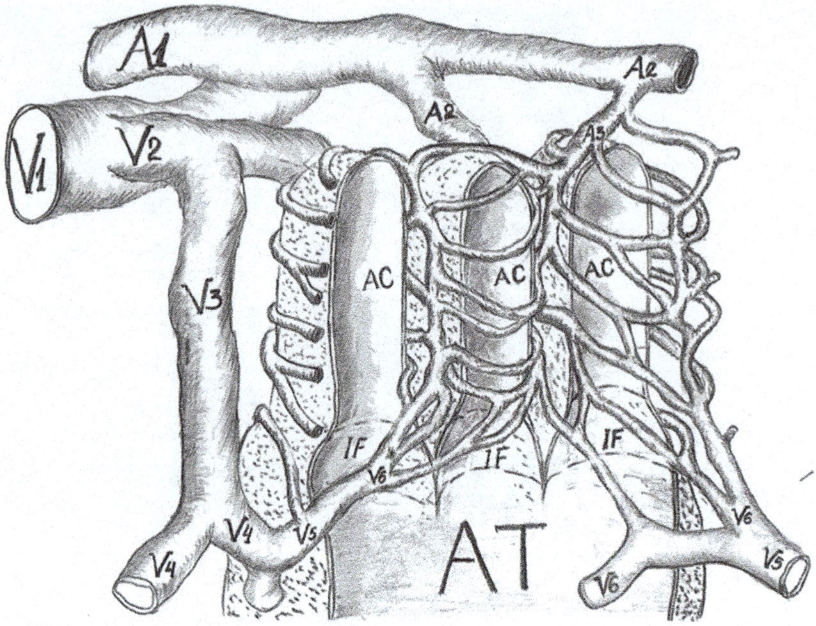

Fig. 6.24 A schema showing the interaction of the BCs with the ACs in the adult chicken. In the interparabronchial septum are the large supplying arteries (A1, A2, and A3) and draining veins (V1 and V2). The smallest draining vessels (V4, V5, and V6) are located on the interior aspect of the PB. They converge to form V3, which traverses the gas exchange mantle to join the large draining vessels. The names of the vessels are as provided above and detailed in Table 6.1. *AT* atrium, *IF* infundibulum

Table 6.1 Vessel generation, name, blood flow direction, and blood oxygenation status in the various categories of blood vessels encountered at the parabronchial unit

Vessel generation	Vessel name	Flow direction	Related air conduits	Blood oxygenation status
A1	Interparabronchial artery	Cocurrent	Parabronchi	Deoxygenated
A2	Parabronchial artery	Crosscurrent	Parabronchi	Deoxygenated
A3	Parabronchial arteriole	Crosscurrent	Parabronchi	Deoxygenated
	Parabronchial capillaries	Crosscurrent	Air capillaries	Oxygenation occurs
V6	Infundibular venule	Crosscurrent, countercurrent, cocurrent	Infundibula, parabronchi	Oxygenated
V5	Septal venule	Crosscurrent, countercurrent, cocurrent	Atria, parabronchi	Oxygenated
V4	Atrial vein	Crosscurrent, countercurrent, cocurrent	Atria, parabronchi	Oxygenated
V3	Intraparabronchial vein	Crosscurrent	Parabronchi	Oxygenated
V2	Parabronchial vein	Crosscurrent	Parabronchi	Oxygenated
V1	Interparabronchial vein	Countercurrent	Parabronchi	Oxygenated

Key: *A* artery, *V* vein

connected to the internal aspect of the parabronchial capillary plexus by short venules. The well-developed arterial vascular tree supplies the outer aspect of the parabronchial plexus via a multitude of parabronchial arterioles (Fig. 6.21).

Toward hatching, the ACs spread from the parabronchial lumen and pierce the parabronchial capillary mantle toward the interparabronchial septum. The latter septum has the large vessels (Figs. 6.20 and 6.21). The parabronchial mantle is also traversed by intraparabronchial veins. Casts sectioned at the periphery of the parabronchial mantle show the interparabronchial septum with large interparabronchial vessels and air capillaries that extend to the periphery of the blood capillary meshwork. Blood capillaries tend to form complete rings that surround the ACs, accomplishing a crosscurrent system (Fig. 6.20).

The parabronchial arterioles split hierarchically into smaller branches that connect to the capillary plexus. Venular branches from the interparabronchial veins pierce through the parabronchial capillary plexus and then successively ramify on the internal surface of the parabronchial capillary plexus (Fig. 6.22). Segregation of the atrial microvasculature occurs through pillar formation. The capillary plexus is organized into an irregular meshwork surrounding the atria. The meshwork expands by insertion and enlargement of a multitude of tissue pillars. Many of the pillars reshape and fuse progressively delineating straight solitary segments around the atria (and infundibula) from the chaotic capillary meshwork. Vascular connections between the atria are severed by transluminal folds and retraction, which results in segregation of individual atrio-infundibular vascular units. Delineation of the gas-exchanging blood capillaries also is driven by pillar formation. The capillary meshwork between the arteriole and the venule is initially irregular with randomly distributed pillars. Subsequently, the pillars rearrange in rows resembling a "necklace" pattern between the arterioles and venules. Pillar elongation and fusion in this axis delineate solitary capillary segments surrounding the atria and air capillaries (Fig. 6.23). Ultimately, a deftly crafted vascular pattern seen in the adult is accomplished where the parabronchial mantle is occupied by the ACs and BCs in a complex pattern where a predominantly crosscurrent system of gas exchange is established (Figs. 6.22, 6.23, and 6.24). The large blood vessels are contained in the parabronchial septum, with only the intraparabronchial vein traversing the parabronchial mantle (Figs. 6.22 and 6.23). The latter vein collects oxygenated blood from the internal aspect of the parabronchus and conveys it directly to the parabronchial vein which further drains into the interparabronchial vein (Figs. 6.22 and 6.24). The names of the various categories of blood vessels, their direction of blood flow, and status of blood oxygenation are presented in Table 6.1.

In conclusion, it is noteworthy that the vasculature of the avian lung inaugurates through the processes of vasculogenesis and angiogenesis. The formation of the basic pattern is accomplished by sprouting, whereas expansion of the vascular patterns, refinement of the finer vascular entities, and formation of the final vascular phenotype typical for the avian lung are by intussusception. These processes are probably supported by interplay between several angiogenic factors with both VEGF and bFGF playing a major role.

6.13 Molecular Control of Avian Lung Development

A detailed account of molecular control of the avian lung development is provided in Chap. 5 of this book and can be visited elsewhere (Maina 2012; Makanya and Djonov 2015; Makanya et al. 2012, 2013; Warburton et al. 2005, 2010). During lung embryogenesis, there are two major phases of the lung tissue developing, the gas transport and exchange system on one hand and the vascular system on the other. These two systems have their own sets of controlling molecules, some of which may overlap, especially when it comes to apposition of the blood capillaries to the epithelium during BGB formation. Generally, lung development is driven by two forces: intrinsic factors that include a host of regulatory molecules and extrinsic forces, the main one being extracellular lung fluid (Demayo et al. 2002). A complex set of morphoregulatory molecules constitutes the intrinsic factors, which can be grouped into three classes: transcription factors (e.g., Nkx2.1 also known as thyroid transcription factor 1 (TTF-1), GATA, and HNF-3);

signaling molecules such as FGF, BMP-4, PDGF, Shh, and TGF-β; as well as extracellular matrix proteins and their receptors (Demayo et al. 2002; Shannon and Hyatt 2004; Warburton et al. 2005). Laminins found in the basement membrane are particularly important in the signaling pathways that result in formation of squamous pneumocytes and pulmonary capillaries, the two major components of the BGB (Makanya 2016). In mammals extrinsic/mechanical forces have been shown to be important for fetal alveolar epithelial cell differentiation. Such forces emanate from fetal lung movements that propel fluid through incipient air conduits (Schittny et al. 2000). On the other hand, the molecules that are known to be most important in angiogenesis have been implicated in the avian lung. In the chicken lung, VEGF is expressed early during sprouting, while bFGF and PDGF-B are elevated during intussusceptive angiogenesis (Makanya et al. 2007; Tomanek et al. 2001) supporting the VEGF/bFGF "two-step" hypothesis, which shows that vascular tube formation is initially induced by VEGF and is totally abolished by neutralizing antibodies to bFGF. PDGF-B is essential for pericyte recruitment (Anderson-Berry et al. 2005; Argraves et al. 2002) and therefore may be important in formation and maturation of the tissue pillars during IA (Makanya et al. 2007, 2009).

References

Adams RH. Vascular patterning by Eph receptor tyrosine kinases and ephrins. Semin Cell Dev Biol. 2002;13:55–60.

Adams RH. Molecular control of arterial-venous blood vessel identity. J Anat. 2003;202:105–12.

Ambler CA, Nowicki JL, Burke AC, Bautch VL. Assembly of trunk and limb blood vessels involves extensive migration and vasculogenesis of somite-derived angioblasts. Dev Biol. 2001;234:352–64.

Anderson-Berry A, O'Brien EA, Bleyl SB, Lawson A, Gundersen N, Ryssman D, Sweeley J, Dahl MJ, Drake CJ, Schoenwolf GC, Albertine KH. Vasculogenesis drives pulmonary vascular growth in the developing chick embryo. Dev Dyn. 2005;233:145–53.

Argraves WS, Larue AC, Fleming PA, Drake CJ. VEGF signaling is required for the assembly but not the maintenance of embryonic blood vessels. Dev Dyn. 2002;225:298–304.

Aumuller G, Wilhelm B, Seitz J. Apocrine secretion—fact or artifact? Ann Anat. 1999;181:437–46.

Bellairs R, Osmond M. The atlas of chick development. New York: Academic; 1998.

Burri PH, Hlushchuk R, Djonov V. Intussusceptive angiogenesis: its emergence, its characteristics, and its significance. Dev Dyn. 2004;231:474–88.

Cheng N, Brantley DM, Chen J. The ephrins and Eph receptors in angiogenesis. Cytokine Growth Factor Rev. 2002;13:75–85.

De La RR. Photonic crystals: microassembly in 3D. Nat Mater. 2003;2:74–6.

Demayo F, Minoo P, Plopper CG, Schuger L, Shannon J, Torday JS. Mesenchymal–epithelial interactions in lung development and repair: are modeling and remodeling the same process? Am J Physiol Lung Cell Mol Physiol. 2002;283:L510–7.

DeRuiter MC, Gittenberger-de Groot AC, Poelmann RE, VanIperen L, Mentink MM. Development of the pharyngeal arch system related to the pulmonary and bronchial vessels in the avian embryo. With a concept on systemic-pulmonary collateral artery formation. Circulation. 1993;87:1306–19.

Deyrup-Olsen I, Luchtel DL. Secretion of mucous granules and other membrane-bound structures: a look beyond exocytosis. Int Rev Cytol. 1998;183:95–141.

Djonov V, Makanya AN. New insights into intussusceptive angiogenesis. EXS. 2005;94:17–33.

Duncker HR. Respirationstrakt. In: Hinrichsen KV, editor. Human-embryologie. Berlin: Springer; 1990. p. 571–606.

Farkaš R. Apocrine secretion: new insights into an old phenomenon. Biochim Biophys Acta. 2015;1850:1740–50.

Furuya K, Akita K, Sokabe M. Extracellular ATP mediated mechano-signaling in mammary glands. Nippon Yakurigaku Zasshi. 2004;123:397–402.

Gesase AP, Satoh Y. Apocrine secretory mechanism: recent findings and unresolved problems. Histol Histopathol. 2003;18:597–608.

Gesase AP, Satoh Y, Ono K. G-protein activation enhances $Ca^{(2+)}$-dependent lipid secretion of the rat Harderian gland. Anat Embryol (Berl). 1995;192:319–28.

Gesase AP, Satoh Y, Ono K. Secretagogue-induced apocrine secretion in the Harderian gland of the rat. Cell Tissue Res. 1996;285:501–7.

Hall SM, Hislop AA, Pierce CM, Haworth SG. Prenatal origins of human intrapulmonary arteries: formation and smooth muscle maturation. Am J Respir Cell Mol Biol. 2000;23:194–203.

Hansen-Smith FM. Capillary network patterning during angiogenesis. Clin Exp Pharmacol Physiol. 2000;27:830–5.

Hashimoto K. The apocrine gland. In: Jarret A, editor. The physiology and pathophysiology of the skin. London: Academic; 1978. p. 1575–96.

King AS, McLelland J. Birds: their structure and function. London: Baillière Tindall; 1984.

Kliewer M, Fram EK, Brody AR, Young SL. Secretion of surfactant by rat alveolar type II cells: morphometric analysis and three-dimensional reconstruction. Exp Lung Res. 1985;9:351–61.

Loscertales M, Mikels AJ, Hu JK, Donahoe PK, Roberts DJ. Chick pulmonary Wnt5a directs airway and vascular tubulogenesis. Development. 2008;135:1365–76.

Macuhova J, Tancin V, Bruckmaier RM. Effects of oxytocin administration on oxytocin release and milk ejection. J Dairy Sci. 2004;87:1236–44.

Maina JN. Scanning electron microscope study of the spatial organization of the air and blood conducting components of the avian lung (*Gallus gallus variant domesticus*). Anat Rec. 1988;222:145–53.

Maina JN. Systematic analysis of hematopoietic, vasculogenetic, and angiogenetic phases in the developing embryonic avian lung, *Gallus gallus variant domesticus*. Tissue Cell. 2004a;36:307–22.

Maina JN. Morphogenesis of the laminated, tripartite cytoarchitectural design of the blood-gas barrier of the avian lung: a systematic electron microscopic study on the domestic fowl, *Gallus gallus variant domesticus*. Tissue Cell. 2004b;36:129–39.

Maina JN. Comparative molecular developmental aspects of the mammalian- and the avian lungs, and the insectan tracheal system by branching morphogenesis: recent advances and future directions. Front Zool. 2012;9:16.

Maina JN, West JB. Thin and strong! The bioengineering dilemma in the structural and functional design of the blood–gas barrier. Physiol Rev. 2005;85:811–44.

Maina JN, King AS, Settle G. An allometric study of pulmonary morphometric parameters in birds, with mammalian comparisons. Philos Trans R Soc Lond Ser B Biol Sci. 1989;326:1–57.

Makanya AN. Membrane-mediated development of the vertebrate blood-gas-barrier. Birth Defects Res C Embryo Today. 2016;108:85–97.

Makanya AN, Djonov V. Development and spatial organization of the air conduits in the lung of the domestic fowl, *Gallus gallus variant domesticus*. Microsc Res Tech. 2008;71:689–702.

Makanya AN, Djonov V. Parabronchial angioarchitecture in developing and adult chickens. J Appl Physiol. 2009;106:1959–69.

Makanya AN, Djonov V. Prenatal and postnatal development of the vertebrate blood-gas barrier. In: Makanya AN, editor. The vertebrate blood-gas barrier in health and disease. Cham: Springer International; 2015. p. 39–64.

Makanya AN, Sparrow MP, Warui CN, Mwangi DK, Burri PH. Morphological analysis of the postnatally developing marsupial lung: the quokka wallaby. Anat Rec. 2001;262:253–65.

Makanya AN, Hlushchuk R, Duncker HR, Draeger A, Djonov V. Epithelial transformations in the establishment of the blood–gas barrier in the developing chick embryo lung. Dev Dyn. 2006;235:68–81.

Makanya AN, Hlushchuk R, Baum O, Velinov N, Ochs M, Djonov V. Microvascular endowment in the developing chicken embryo lung. Am J Physiol Lung Cell Mol Physiol. 2007;292:L1136–46.

Makanya AN, Hlushchuk R, Djonov VG. Intussusceptive angiogenesis and its role in vascular morphogenesis, patterning, and remodeling. Angiogenesis. 2009;12:113–23.

Makanya AN, Koller T, Hlushchuk R, Djonov V. -Pre-hatch lung development in the ostrich. Respir Physiol Neurobiol. 2012;180:183–92.

Makanya A, Anagnostopoulou A, Djonov V. Development and remodeling of the vertebrate blood-gas barrier. Biomed Res Int. 2013;2013:101597.

Makanya AN, Kavoi BM, Djonov V. Three-dimensional structure and disposition of the air conducting and gas exchange conduits of the avian lung: the domestic duck (*Cairina moschata*). ISRN Anat. 2014;2014:621982. doi:10.1155/2014/62198.

deMello DE, Reid LM. Embryonic and early fetal development of human lung vasculature and its functional implications. Pediatr Dev Pathol. 2000;3:439–49.

deMello DE, Sawyer D, Galvin N, Reid LM. Early fetal development of lung vasculature. Am J Respir Cell Mol Biol. 1997;16:568–81.

Metzler G, Schaumburg-Lever G, Fehrenbacher B, Moller H. Ultrastructural localization of actin in normal human skin. Arch Dermatol Res. 1992;284:242–5.

Noden DM. Origins and assembly of avian embryonic blood vessels. Ann N Y Acad Sci. 1990;588:236–49.

Parera MC, van Dooren M, van Kempen M, de Krijger R, Grosveld F, Tibboel D, Rottier R. Distal angiogenesis: a new concept for lung vascular morphogenesis. Am J Physiol Lung Cell Mol Physiol. 2005;288(1):L141–9.

Sakiyama J, Yamagishi A, Kuroiwa A. Tbx4-Fgf10 system controls lung bud formation during chicken embryonic development. Development. 2003;130:1225–34.

Satoh Y, Ishikawa K, Oomori Y, Takede S, Ono K. Secretion mode of the harderian gland of rats after stimulation by cholinergic secretagogues. Acta Anat (Basel). 1992;143:7–13.

Schaumburg-Lever G, Lever WF. Secretion from human apocrine glands: an electron microscopic study. J Invest Dermatol. 1975;64:38–41.

Schittny JC, Burri PH. Morphogenesis of the mammalian lung: aspects of structure and extracellular matrix. In: Massaro JD, Massaro G, Chambon P, editors. Lung development and regeneration. New York: Mercel Dekker; 2003. p. 275–317.

Schittny JC, Miserocchi G, Sparrow MP. Spontaneous peristaltic airway contractions propel lung liquid through the bronchial tree of intact and fetal lung explants. Am J Respir Cell Mol Biol. 2000;23:11–8.

Shannon JM, Hyatt BA. Epithelial–mesenchymal interactions in the developing lung. Annu Rev Physiol. 2004;66:625–45.

Shook D, Keller R. Mechanisms, mechanics and function of epithelial-mesenchymal transitions in early development. Mech Dev. 2003;120:1351–83.

Smith JD, Hearn GW. Ultrastructure of the apocrine-sebaceous anal scent gland of the woodchuck, *Marmota monax*: evidence for apocrine and merocrine secretion by a single cell type. Anat Rec. 1979;193:269–91.

Stoeckelhuber M, Sliwa A, Welsch U. Histo-physiology of the scent-marking glands of the penile pad, anal pouch, and the forefoot in the aardwolf (*Proteles cristatus*). Anat Rec. 2000;259:312–26.

Stoeckelhuber M, Stoeckelhuber BM, Welsch U. Human glands of Moll: histochemical and ultrastructural characterization of the glands of moll in the human eyelid. J Invest Dermatol. 2003;121:28–36.

Tomanek RJ, Sandra A, Zheng W, Brock T, Bjercke RJ, Holifield JS. Vascular endothelial growth factor and basic fibroblast growth factor differentially modulate early postnatal coronary angiogenesis. Circ Res. 2001;88:1135–41.

Volberg T, Geiger B, Kartenbeck J, Franke WW. Changes in membrane-microfilament interaction in intercellular adherens junctions upon removal of extracellular Ca^{2+} ions. J Cell Biol. 1986;102:1832–42.

Warburton D, Bellusci S, De Langhe S, Del Moral PM, Fleury V, Mailleux A, Tefft D, Unbekandt M, Wang K, Shi W. Molecular mechanisms of early lung specification and branching morphogenesis. Pediatr Res. 2005;57:26R–37R.

Warburton D, El-Hashash A, Carraro G, Tiozzo C, Sala F, Rogers O, De Langhe S, Kemp PJ, Riccardi D, Torday J, Bellusci S, Shi W, Lubkin SR, Jesudason E. Lung organogenesis. Curr Top Dev Biol. 2010;90:73–158.

West NH, Bamford OS, Jones DR. A scanning electron microscope study of the microvasculature of the avian lung. Cell Tissue Res. 1977;176:553–64.

Zeller U, Richter J. The monoptychic glands of the jugulo-sternal scent gland field of Tupaia: a TEM and SEM study. J Anat. 1990;172:25–38.

Structure and Function of Avian Pulmonary Capillaries: Comparison with Mammals

7

John B. West

Abstract

The avian pulmonary circulation has many interesting features and much information can be gained by comparing the avian and mammalian systems. Two features are primarily responsible for the unique features of avian capillaries. First, the avian lung has separated the ventilation and gas exchange functions. Second, the avian lung uses a flow-through process for ventilation rather than the reciprocal pattern adopted by mammals. As a consequence, the environment of the pulmonary capillaries is very different in the avian compared with the mammalian lung. The avian capillaries are nested in a syncytium of air capillaries, whereas the capillaries of the mammalian lung are strung out along the alveolar walls. This means that the mechanical support of the categories is very different in birds compared with mammals. A consequence of this is that the walls of the capillaries are very different. In the avian lung, the blood–gas barrier is extremely thin and uniform throughout the capillaries. Contrast this with the mammalian lung where a type I collagen cable is required for the support of the capillaries, and as a result the blood–gas barrier is much thicker in places and the diffusion characteristics are therefore inferior. Another striking difference is that avian pulmonary capillaries are remarkably rigid unlike those in mammals, which undergo recruitment and distention when the cardiac output rises. The implications of this for pulmonary vascular resistance during intense exercise such as flying are still unclear.

Keywords

Gas exchange • Vascular resistance • Blood-gas barrier • Diffusion • Ventilation

Contents

7.1 Introduction

For those of us interested in pulmonary gas exchange of vertebrates, it is fascinating that

J.B. West (✉)
Department of Medicine, University of California San Diego, 9500 Gilman Drive, La Jolla, CA 92093-0623, USA
e-mail: jwest@ucsd.edu

J.N. Maina (ed.), *The Biology of the Avian Respiratory System*, DOI 10.1007/978-3-319-44153-5_7

nature has evolved two very different systems, one for the mammals and another for the birds. The divergence can be traced back to some 300 million years ago when the ancestors of the present reptiles emerged from water and made a commitment to air breathing. As is the case with modern reptiles, they were exothermic and incapable of sustained levels of high oxygen consumption. But from them developed the two great clades of vertebrates capable of high-sustained physical activity, the mammals and birds. Specifically it has been argued that the mammals were derived from a group of carnivorous reptiles, the cynodonts, while the birds descended from the theropod dinosaurs. It is a remarkable fact that although the two groups share many similarities in the physiology of their cardiovascular, renal, gastrointestinal, endocrine, and nervous systems, the respiratory systems are dramatically different.

A major difference between the mammalian and avian systems of pulmonary gas exchange is that the mammalian lung uses its delicate alveolar structure for both ventilation and blood flow, whereas the avian lung has separated the ventilation and gas exchange functions. This makes a great deal of sense. Why use delicate alveolar tissue for pumping air when robust air sacs can do this? Another important feature that is consistent with this anatomical difference is that the avian lung has not adopted the reciprocal pattern of breathing used by mammals, but instead has evolved a flow-through process. Many important physiological and anatomical characteristics flow from these differences, and this article is devoted to these.

7.2 The Environment of the Pulmonary Capillaries Is Very Different in the Avian Compared with the Mammalian Lung

Figure 7.1 shows typical microscopic appearances of the gas-exchanging regions of avian and mammalian lungs. Note that in the mammalian lung, the capillaries are strung out along the alveolar wall like a string of beads. By contrast, the capillaries in the gas-exchanging tissue of the avian lung are nested in a syncytium of small air capillaries. As we shall see, this difference in the environment has a marked effect on the ultrastructure and physiology of the capillaries.

Figure 7.1 shows that the terminal air spaces of the avian lung are very much smaller than those of the mammalian lung. For example, the diameter of the air capillaries in the avian lung is only of the order of 10–20 μm whereas in the human lung the diameter of the alveoli is about 300 μm. Why is this? The answer is that this dramatic difference in dimensions of the terminal airways is necessary because of the different airflow regimes of the two types of lung. In the bird, airflow through the gas-exchanging tissue is continuous and unidirectional. By contrast, airflow into the distal regions of the mammalian lung is reciprocating because of the sequential pattern of inspiration and expiration.

Airflow in the avian parabronchi is almost purely convective. It is true that some diffusion must take place in the short air capillaries that branch off the larger conducting airways, but the distance to be traveled is very small. By contrast, airflow into the most distal regions of the mammalian lung is a complex mixture of convection and diffusion over a substantial distance, for example, several mm in the human lung. This is a difficult area of study because of the complicated geometry of the small peripheral airways. However, extensive experimental and theoretical studies have been performed, and most experimental evidence supports the existence of some stratification. This means that there are residual differences in gas concentration along the small airways at the end of inspiration (Scheid and Piiper 1980). These studies have been carried out using gases of different molecular weights and therefore diffusivity with both single-breath and multi-breath washout maneuvers. The experimental results applied to normal resting conditions, but on exercise when the breathing rate is faster and the time available for diffusion is reduced, the stratified inhomogeneity along the peripheral airways is presumably more pronounced.

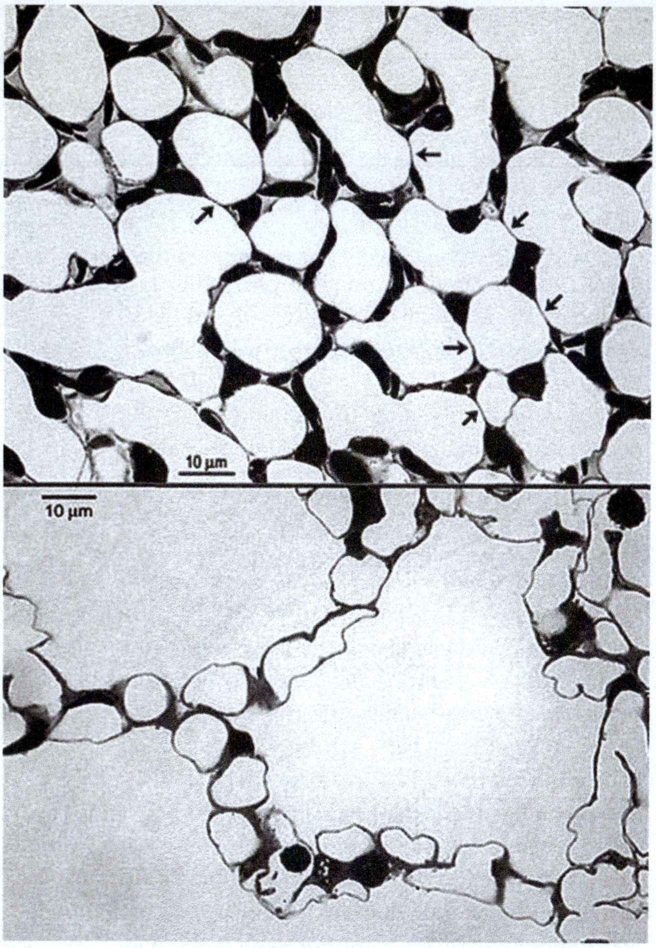

Fig. 7.1 Light micrographs of a portion of lung from chicken (*top*) and rabbit (*bottom*). Note the small size of the air capillaries in the chicken compared with the alveoli in the rabbit. Both micrographs are the same magnification. From West et al. (2006)

Another cause of stratified inhomogeneity for oxygen along the peripheral airways of mammals is known as screening (Sapoval et al. 2002). This term refers to the fact that alveoli in the proximal part of the acinus may take up oxygen as the inspired gas moves through this region. The result is that the more distal alveoli are exposed to a lower PO_2. The capture of some of the oxygen before the inspired gas reaches the most peripheral regions is another negative consequence of the reciprocating pattern of ventilation.

These studies highlight the disadvantages of reciprocating ventilation in the mammalian lung compared with the flow-through convective pattern in the bird. On one hand, in mammals, the air spaces need to be relatively large to keep the resistance of the diffusion-convection process to a minimum. On the other hand, as the dimensions are increased, the diffusion distances become greater leading to increased stratified in homogeneity along the peripheral airways and oxygen screening. The avian solution is clearly superior in this respect.

7.3 Mechanical Support of the Capillaries

Figure 7.1 immediately shows that the support of the pulmonary capillaries is very different in the avian compared with the mammalian lung. As already indicated, the alveolar wall of the mammalian lung is essentially composed of a series of capillaries adjacent to each other like a string of

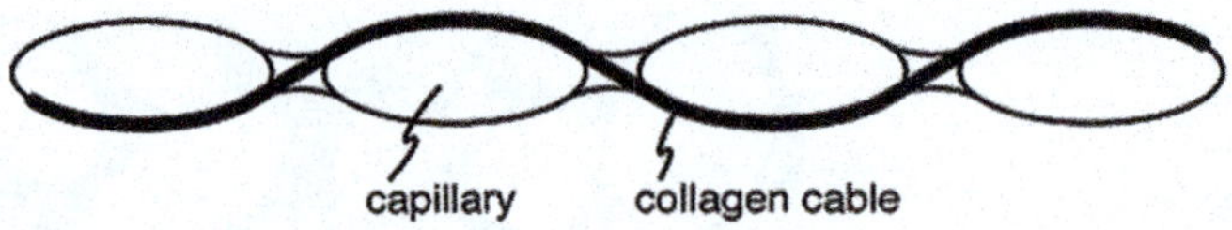

Fig. 7.2 Diagram showing the type I collagen cable that runs along the alveolar wall of the mammalian lung. The cable traverses one side of each capillary and apparently heps to maintain the mechanical integrity of both the capillaries and the alveolar wall itself. From West (2011)

beads. Although some mechanical support is provided by adjacent capillaries in the wall, there appears to be no support at right angles to the alveolar wall. Note also that the alveolar wall is remarkably long. For example, in the human lung, the diameter of the alveoli is about 300 μm, and this means that the wall can be as long as about 100 μm with no external support at all.

However, examination of the ultrastructure of the alveolar wall shows that there is a type I collagen cable that snakes its way along the wall. As Fig. 7.2 shows, this cable moves from one side of a capillary to the other side of an adjacent capillary. This collagen cable apparently supplies provides support to each capillary at right angles to the wall. It also provides stability to the alveolar wall itself which otherwise is essentially composed of only delicate tissue including the endothelium, extracellular matrix, and epithelium of the capillaries together with occasional fibroblasts. Therefore, this collagen cable seems to be critical in maintaining the mechanical integrity of both the capillaries and the alveolar wall itself.

The situation for the capillaries in the avian lung is clearly very different. As Fig. 7.1 shows in a light micrograph and as Fig. 7.3 shows in electron micrographs, the avian capillaries are connected to each other by a network of epithelial bridges that make up the walls of the air capillaries. These bridges are composed of two very thin epithelial plates, the total thickness in some parts being as small as 1 μm. This array of epithelial bridges contributes very substantial mechanical support to the blood capillaries, as we shall see shortly.

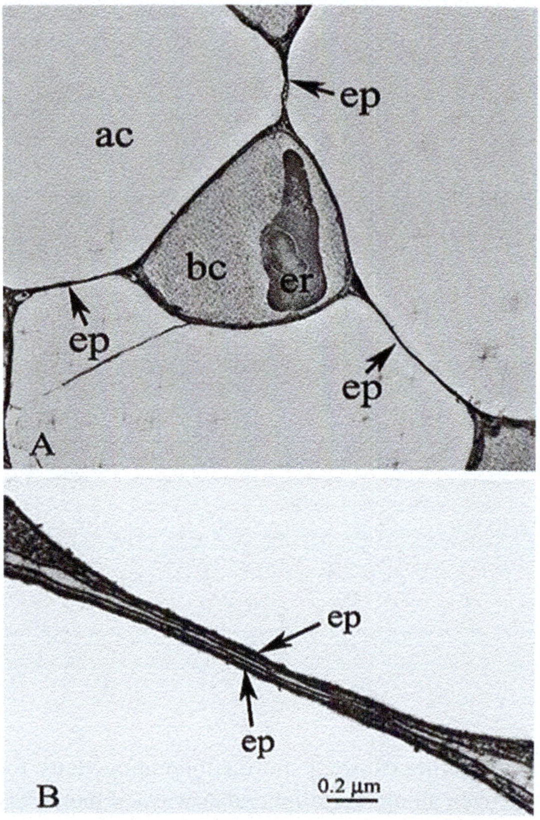

Fig. 7.3 (**a**) Shows an avian pulmonary capillary with its epithelial bridges (ep). bc is the blood capillary and ec is an erythrocyte. (**b**) Shows that the epithelial bridges are composed of two very thin epithelial plates (ep). Note that the thickness of the bridge is only about 0.1 μm. From West et al. (2010)

The structure of the junctions between the epithelial bridges and the walls of the pulmonary capillaries is particularly interesting. Figure 7.4 shows a high-power electron micrograph of one of these junctions, and it can be seen that this contains a series of small structures that are believed to be microtubules. They have a diameter of about 20 nm and, as can be seen, they form a cluster of small structures that presumably have a mechanical function. This is a provocative finding because most studies of microtubules emphasize their role in the transport of intracellular material and little emphasis has been given to their possible mechanical function.

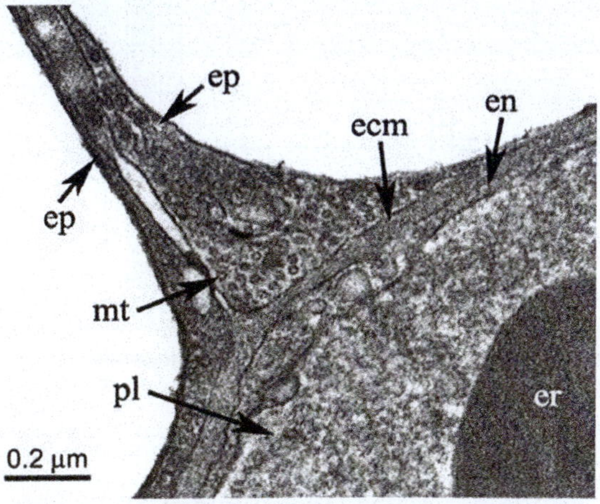

Fig. 7.4 High-power electron micrographs showing a junction of an epithelial bridge with a capillary wall. Note that the expanded part of the cell that is part of the bridge includes a group of small circular inclusions of diameter about 20 nm. These are believed to be microtubules. *ep* epithelial cell, *en* endothelial cell, *ecm* extracellular matrix, *mt* microtubules, *pl* plasma, *er* erythrocyte. From West et al. (2010)

Another factor contributing to the support of avian capillaries is the surface tension in fluid lining the air capillaries. Since these have a diameter of about 10–20 µm, calculations using the Laplace equation give a pressure of about 7 kPa (53 mm Hg) tending to reduce the lumen of the airway if the fluid has the same surface tension as water. However it is known that avian lungs, like their mammalian counterparts, contain surfactant (Pattle 1976) so the pressures are presumably less. Nevertheless this factor probably contributes to the support of the avian pulmonary capillaries.

7.4 Ultrastructure of the Walls of the Pulmonary Capillaries

Figure 7.5 shows electron micrographs of the pulmonary capillaries of the avian lung on the left and the mammalian lung on the right. The avian lung is from chicken and the mammalian lung from dog. The ultrastructure of the capillary walls is clearly seen in both instances. Note that both micrographs have the same magnification.

A striking feature of the walls of the capillaries is that those of the avian lung

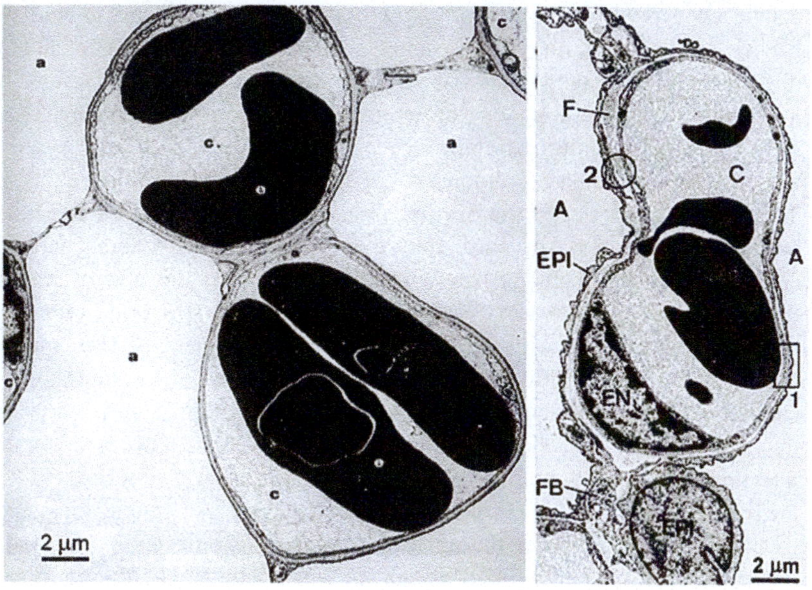

Fig. 7.5 Electron micrographs of pulmonary capillaries in chicken (*left*) and dog (*right*). Note that the blood–gas barrier in the bird is much thinner and much more uniform in thickness than in the mammal. *EPI* epithelium, *F* fibrils of type I collagen, *FB* fibroblast. *Left panel* is modified from Maina (2002). *Right panel* is modified from Weibel (1973)

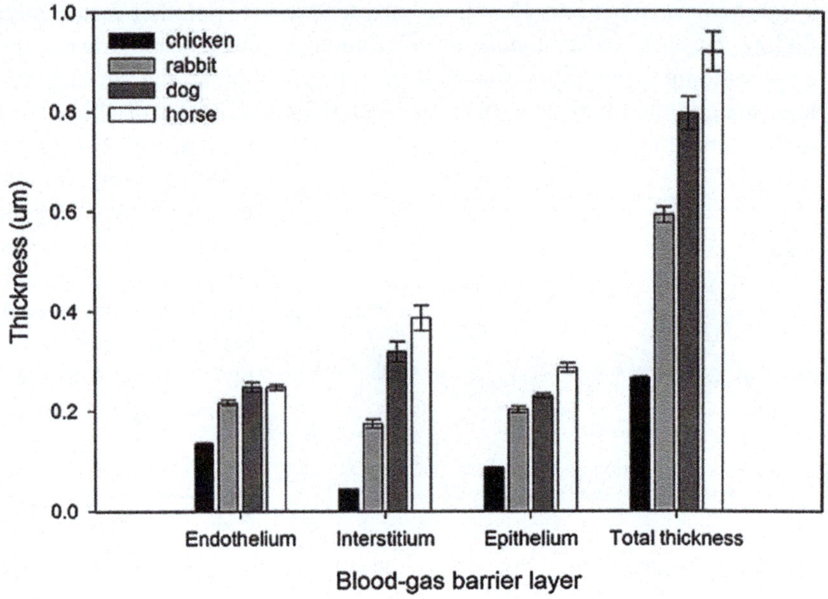

Fig. 7.6 Bar graph showing average thicknesses ± SD of all layers of the blood–gas barrier measured in the lungs of chicken, horse, dog, and rabbit. Note the small thickness of all layers in the bird. From Watson et al. (2007)

appeared to be much thinner than those of the mammalian lung. Morphometry confirms this. Figure 7.6 shows measurements of the thickness of the blood–gas barrier in chicken and three mammals, rabbit, dog, and horse (Watson et al. 2007). Note that the thicknesses of all three layers of the blood–gas barrier, that is, the endothelium, interstitium, and epithelium, are all much less in the bird than the mammals. Of course the very thin nature of the blood–gas barrier in the avian lung is advantageous for diffusive gas exchange.

Turning again to Fig. 7.5, close inspection shows another striking difference between the avian and mammalian pulmonary capillaries. The blood–gas barrier of the bird is remarkably uniform in thickness. In particular, the lower capillary in the bird image shows very clearly that the barrier thickness is extraordinarily uniform. Contrast this with the mammalian capillary shown on the right. We can see that the capillary wall is "polarized" in the sense that the thickness on the right side is considerably less than that on the left. Specifically the thickness on the right is of the order of 0.3 μm whereas that on the right, particularly near the upper part of the image, is several μm thick. Also we can see type I collagen fibrils in the interstitium of the capillary wall in this region. These are part of the type I collagen cable shown diagrammatically in Fig. 7.2. As was pointed out there, this collagen cable is apparently essential to support the capillaries in the mammalian lung because they are strung out along the long alveolar wall. However, in the avian lung where support is available from numerous epithelial bridges as shown in Fig. 7.3 and perhaps surface tension as well, this collagen cable is not required.

The collagen cable in the pulmonary capillary wall of the mammal has important implications for gas exchange. In studies of the thickness of the blood–gas barrier in mammals, for example, those carried out by Gehr et al. (1978), it was shown that the thick side of the blood–gas barrier occupies about half of the total area of the barrier. Because of its thickness, this region of the barrier contributes little to diffusive gas exchange. This means that about half of the area of the blood–gas barrier in a typical mammalian lung is very inefficient for gas exchange. Again the arrangement in the avian lung is clearly superior.

An interesting question is how the avian capillaries handle the escape of fluid from the lumen of the capillary when the transcapillary pressure rises, for example, on exercise, and the

Starling equilibrium is therefore disturbed. It is well established that in mammals, the fluid enters the thick side of the capillary where the type I collagen tissue is located, and from there, it finds its way to perivascular and peribronchial spaces that provide a path out of the lung (Fishman 1972). Since avian capillaries do not have a thick side, this mechanism is presumably unavailable. Possibly the fluid exits at the junctions between the epithelial plates and the wall such as shown in Fig. 7.4. Possibly the microtubules act as conduits. It is known that edema in the avian lung is associated with thickening of the epithelial bridges (Weidner and Kinnison 2002).

7.5 Rigidity of Avian Pulmonary Capillaries

As Figs. 7.1 and 7.3 indicate, the support offered to the pulmonary capillaries by the surrounding tissue appears to be very different in birds compared with mammals. This led us to study the mechanical properties of the capillaries. It is well known that in the mammalian lung, the capillaries can exhibit marked recruitment and distention. Recruitment means that when the capillary hydrostatic pressure rises, some capillaries open up, or at least, they exhibit flow, whereas previously they did not. It is also well established that mammalian pulmonary capillaries are capable of distention. This refers to the fact that the caliber of the capillary increases when the pressure inside it is raised compared with the pressure outside.

Figures 7.7 and 7.8 show evidence for both recruitment and distention in mammalian pulmonary capillaries. These measurements were made in isolated perfused dog lungs where the pulmonary arterial, alveolar, and the venous pressures were accurately known. After the appropriate conditions were established, the peripheral part of the lung was rapidly frozen with liquid Freon gas, and the microscopic appearances were carefully studied (Glazier et al. 1969; Warrell et al. 1972). In the studies on distension, the difference between pulmonary arterial and venous pressures was kept very small so that the capillary pressure that lay between these two values could be accurately estimated. Figure 7.7 shows that the number of open capillaries increased markedly as the capillary transmural pressure was raised indicating recruitment. Figure 7.8 shows that

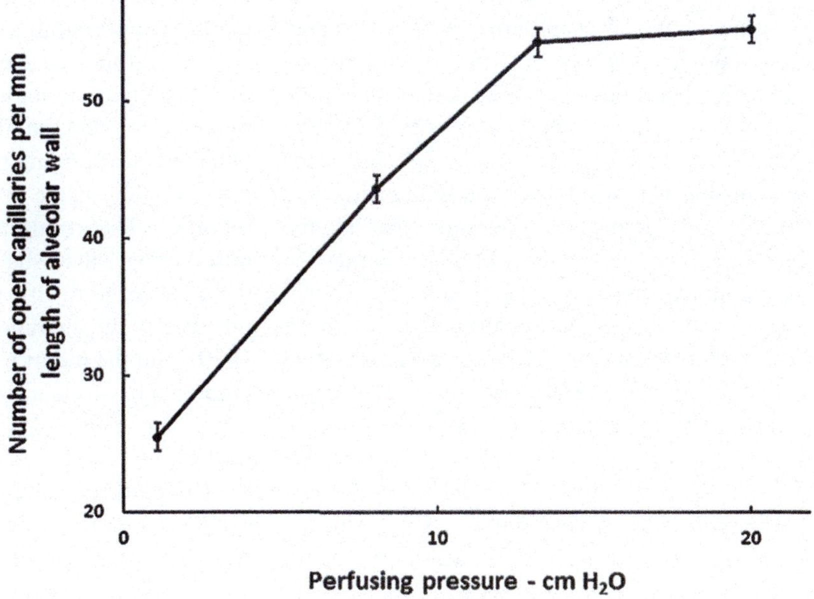

Fig. 7.7 Number of open pulmonary capillaries in dog lung as the capillary transmural pressure is increased. This is evidence of recruitment. From Warrell et al. (1972)

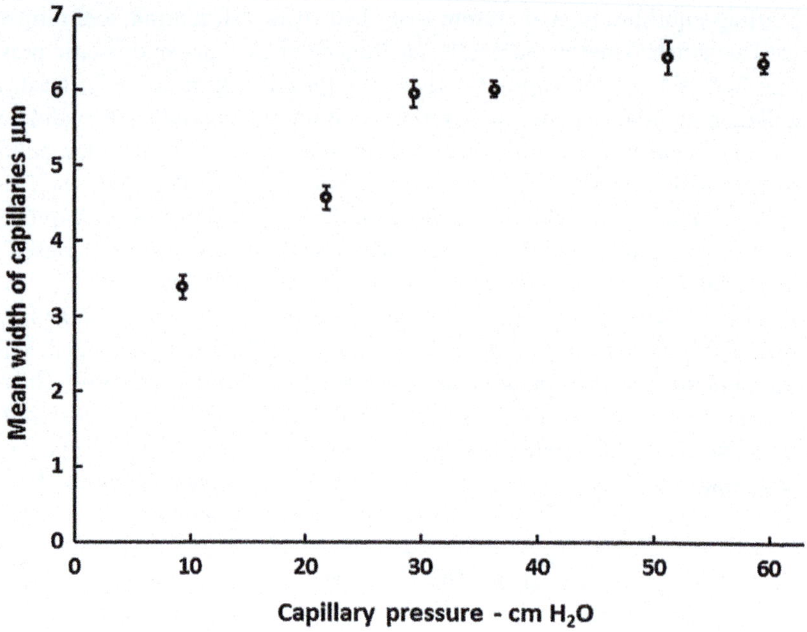

Fig. 7.8 Increase in the width of pulmonary capillaries in dog lung as the capillary transmural pressure is increased. This is evidence of distention. Modified from Glazier et al. (1969)

the diameter of the capillaries measured at right angles to the alveolar wall also increased markedly as the capillary transmural pressure was increased. This was evidence of distension.

We carried out similar studies in chicken lung after the animals had been anesthetized, the chest opened, and the pulmonary artery and left atrium had been cannulated (Watson et al. 2008). The pressure in the air capillaries was determined from the tracheal pressure. After the appropriate pressure conditions had been established, the lungs were rapidly fixed by instilling fixative solution into the trachea. Figure 7.9 shows the relationship between the capillary diameter and the capillary transmural pressure, that is, the blood capillary hydrostatic pressure minus the air capillary pressure. The results are shown over a wide range of pressures, and it can be seen that the data fell close to a straight line and that the slope was very small. The right-hand part of the figure shows that the increase in diameter as the capillary pressure was raised from about 5 to about 25 cm water above the pressure in the air capillaries. Note that the resulting increase in diameter was very small. Even more striking was that when the air capillary pressure exceeded the blood capillary pressure as shown in the left-hand part of the figure, the change in capillary diameter was also very small.

Figure 7.10 shows the same data together with measurements made on dog lung by Mazzone (1980) and cat lung by Sobin et al. (1972). In both instances, the capillary diameter increased markedly with capillary transmural pressure. It is clear therefore that the pulmonary capillaries in the chicken are remarkably rigid compared with those in dog and cat. These morphological data obtained by changing the transcapillary pressure over a wide range have not previously been reported. The result is consistent with earlier observations that the parabronchial tissue in ducks undergoes little volume change when the pressure in the airways is raised (Macklem et al. 1979), and the report by Powell et al. (1985) that pulmonary vascular resistance nearly doubles when the pulmonary artery to one lung is occluded.

It is particularly surprising that the diameter of the capillaries in the bird changes so little when the air capillary pressure exceeds the blood capillary pressure. This result is in marked

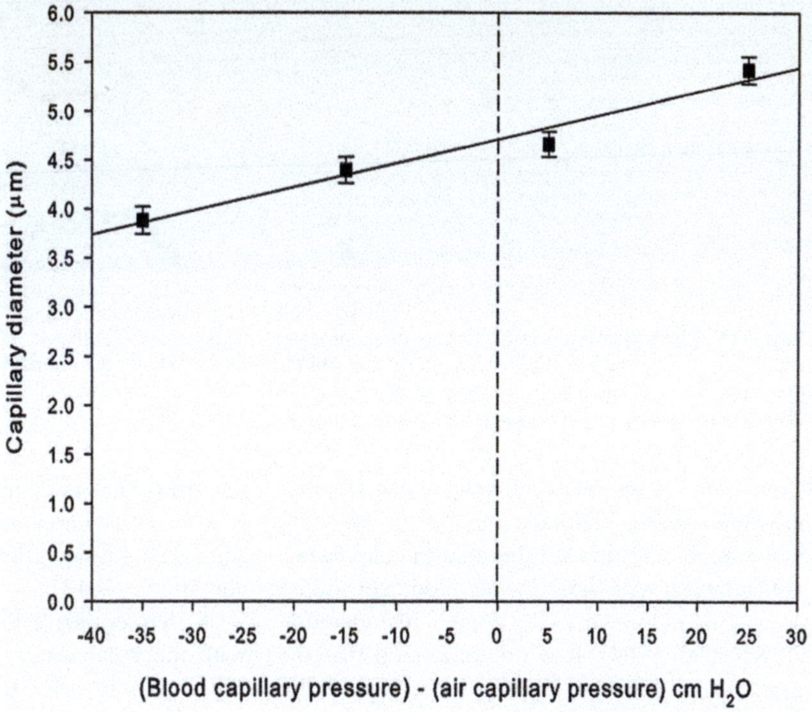

Fig. 7.9 Increase in the diameter of pulmonary capillaries in chicken as the transmural pressure of the capillaries is increased. Note that the capillary diameter changes little over a very large range of transcapillary pressures. From Watson et al. (2008)

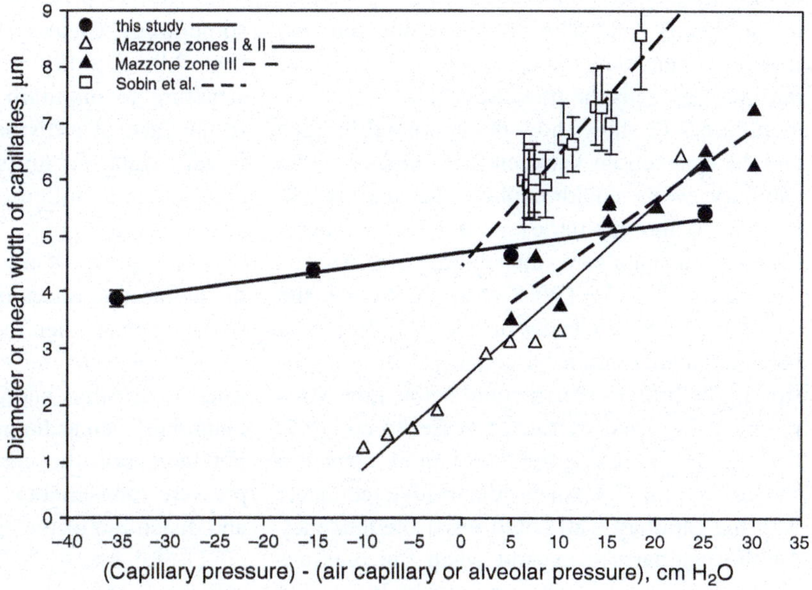

Fig. 7.10 Comparison of the distensibility of avian (chicken) and mammalian (dog and cat) pulmonary capillaries as the capillary pressure is raised. Dog data from Mazzone (1980). Cat data from Sobin et al. (1972)

contrast to what happens in the mammalian lung when alveolar pressure exceeds the pressure in the blood capillaries (Glazier et al. 1969). Figure 7.11 shows light micrographs of two conditions. On the right we see the microscopic appearances when the capillary pressure exceeds alveolar pressure in the dog. As expected the vessels are well filled under these conditions.

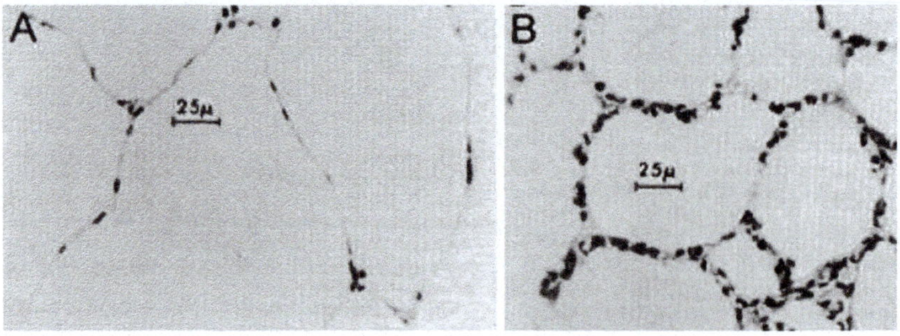

Fig. 7.11 Light microscope appearances of pulmonary capillaries in dog lung at different capillary transmural pressures. The *left image* shows the result of raising the alveolar pressure above the capillary hydrostatic pressure. This causes collapse of the capillaries. The *right image* shows the normal situation when capillary hydrostatic pressure exceeds alveolar pressure. From Glazier et al. (1969)

On the left we see the situation when the alveolar pressure exceeds capillary pressure in the same preparation, and it is striking that the capillaries are collapsed with very few red blood cells to be seen. This is known as the zone 1 phenomenon (West et al. 1964). It is not surprising that the capillaries behave in this way if we look at the electron micrograph of the mammalian lung capillary shown in Fig. 7.5. The extremely thin wall of the capillary would not be expected to provide much rigidity, and it is not surprising that the capillary completely collapses when the pressure outside it exceeds the pressure inside.

Figure 7.12 shows how the epithelial bridges and the connections between the bridges and the capillary walls could explain the remarkable rigidity of the capillaries in the face of large changes in capillary transmural pressure as depicted in Fig. 7.9 (West et al. 2010). On the one hand, the tendency of the capillaries to become smaller could be countered by the tension in the bridges which could act as guy wires resisting the tendency for the vessel to collapse. On the other hand, the substantial structure shown in Fig. 7.4 where the bridge connects with the capillary wall could resist the tendency for the capillary to expand when the pressure inside it exceeds the pressure outside. Figure 7.4 shows the junction in two dimensions only but of course this continues around the capillary as shown in Fig. 7.12. As such it could behave like the iron hoop around a barrel of beer maintaining its mechanical integrity. In fact, the stress that is countered by such a structure is known as hoop stress. Other authors have suggested similar schemes, for example, Maina and Jimoh (2013).

A frequent error that is made in connection with the behavior of the capillaries shown in Fig. 7.9 is to infer the great strength of their walls. In fact, the figure gives no information about strength but only about rigidity. Strength refers to the forces required to break or disrupt something. Clearly this is not relevant to Fig. 7.9. Rigidity refers to the ability of a structure to maintain its dimensions or shape in the face of a distorting force. A wine glass is very rigid but not strong. It easily breaks if dropped. A bulletproof vest is immensely strong but not rigid.

The rigidity of the pulmonary capillaries in the avian lung raises interesting questions about what happens when the cardiac output increases during heavy exercise as occurs during flying. This is a very energetic activity indeed. In mammals such as humans, when the cardiac output increases, the consequent rise in capillary pressure presumably causes both recruitment and distention of the capillaries, and this has the effect of markedly reducing pulmonary vascular resistance. As a result, the rise in pulmonary artery pressure during exercise is relatively small. But in the avian lung where all the evidence suggests that the pulmonary capillaries behave as essentially rigid tubes, we would expect the pulmonary artery pressure to rise

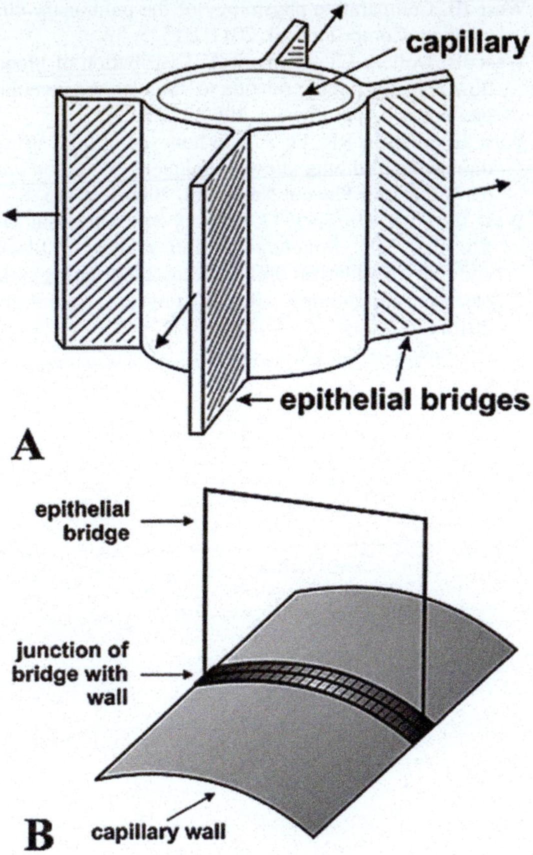

Fig. 7.12 Diagram showing how the epithelial bridges could be responsible for the rigidity of avian pulmonary capillaries. (**a**) Shows how the tension in bridges could hold the vessels open when the pressure outside the capillary exceeds the pressure inside. (**b**) Shows how the junction between the bridge and the capillary wall could provide a hoop stress preventing expansion of the vessel when the pressure inside the capillary exceeded the pressure outside. From West et al. (2010)

very substantially as the cardiac output increases during vigorous exercise.

Surprisingly nobody has apparently yet measured the pulmonary artery pressure in an exercising bird. Indeed it seems counterintuitive that evolution would have provided a solution in which the pulmonary artery pressure rises greatly during strenuous exercise such as flying. We need to wait until the appropriate measurements are made of the pulmonary artery pressure in exercising birds to get a better understanding of the situation.

7.6 Summary

It is fascinating that nature has evolved two very different types of vertebrate lungs. Although as mammals, we are likely to be biased toward our own design, many features of the avian lung suggest that it is a superior design. Further comparison of the two types of gas-exchanging systems is a rich area for further research.

References

Fishman AP. Pulmonary edema: the water-exchanging function of the lung. Circulation. 1972;46:390–408.

Gehr P, Bachofen M, Weibel ER. The normal human lung: ultrastructure and morphometric estimation of diffusion capacity. Respir Physiol. 1978;32:121–40.

Glazier JB, Hughes JMB, Maloney JE, West JB. Measurements of capillary dimensions and blood volume in rapidly frozen lungs. J Appl Physiol. 1969;26:65–76.

Macklem PT, Bouverot P, Scheid P. Measurement of the distensibility of the parabronchi in duck lungs. Respir Physiol. 1979;38:23–35.

Maina JN. Functional morphology of the vertebrate respiratory systems. Enfield: Science Publishers; 2002.

Maina JN, Jimoh SA. Study of stress induced failure of the blood-gas barrier and the epithelial-epithelial cells connections of the lung of the domestic fowl, *Gallus gallus* variant *domesticus* after vascular perfusion. Biomed Eng Comput Biol. 2013;5:77–88.

Mazzone RW. Influence of vascular and transpulmonary pressures on the functional morphology of the pulmonary microcirculation. Microvasc Res. 1980;20:295–306.

Pattle RE. The lung surfactant in the evolutionary tree. In: Hughes GM, editor. Respiration of amphibious vertebrates. London: Academic; 1976. p. 233–55.

Powell FL, Hastings RH, Mazzone RW. Pulmonary vascular resistance during unilateral pulmonary artery occlusion in ducks. Am J Physiol. 1985;249:R34–43.

Sapoval B, Filoche M, Weibel ER. Smaller is better-but not too small: a physical scale for the design of the mammalian pulmonary acinus. Proc Natl Acad Sci USA. 2002;99:10411–6.

Scheid P, Piiper J. Intrapulmonary gas mixing and stratification. In: West JB, editor. Pulmonary gas exchange. New York: Academic; 1980.

Sobin SS, Fung YC, Tremer HM, Rosenquist TH. Elasticity of the pulmonary alveolar microvascular sheet in the cat. Circ Res. 1972;30:440–50.

Warrell DA, Evans JW, Clarke RO, Kingaby GP, West JB. Pattern of filling in the pulmonary capillary bed. J Appl Physiol. 1972;32:346–56.

Watson RR, Fu Z, West JB. Morphometry of the extremely thin pulmonary blood-gas barrier in the chicken lung. Am J Physiol Lung Cell Mol Physiol. 2007;292:L769–77.

Watson RR, Fu Z, West JB. Minimal distensibility of pulmonary capillaries in avian lungs compared with mammalian lungs. Respir Physiol Neurobiol. 2008;160:208–14.

Weibel ER. Morphological basis of alveoli-capillary gas exchange. Physiol Rev. 1973;53:419–95.

Weidner WJ, Kinnison JR. Effect of hydrostatic pulmonary edema on the interbronchial septum of the chicken lung. Poult Sci. 2002;81:1563–6.

West JB. Comparative physiology of the pulmonary circulation. Compr Physiol. 2011;1:1525–39.

West JB, Dollery CT, Naimark A. Distribution of blood flow in isolated lung; relation to vascular and alveolar pressures. J Appl Physiol. 1964;9:713–24.

West JB, Watson RR, Fu Z. The honeycomb-like structure of the bird lung allows a uniquely thin blood-gas barrier. Respir Physiol Neurobiol. 2006;152:115–8.

West JB, Fu Z, Deerinck TJ, Mackey MR, Obayashi JT, Ellisman MH. Structure-function studies of blood and air capillaries in chicken lung using 3D electron microscopy. Respir Physiol Neurobiol. 2010;170:202–9.

Functional Design of the Mature Avian Respiratory System

8

John N. Maina

Abstract

The avian respiratory system is structurally exceptionally complex and functionally remarkably efficient. It comprises a lung that serves as the gas exchanger and air sacs that function as the ventilators. Topographically, the lung is located between two sets of air sacs, namely, a cranial and a caudal group. They continuously ventilate the lung in a craniocaudal direction by synchronized activities. The air sacs are delicate, transparent and compliant structures. Since they are avascular, they play no part in gas exchange. The lung is attached to the ribs and the vertebrae on the dorsolateral aspects and to the horizontal septum on the ventral one. It renders the lung practically rigid. This allows the exchange tissue (parenchyma) to be very intensely subdivided, resulting in minuscular terminal respiratory units, the air capillaries. That way, the respiratory surface area is increased. The bronchial system of the avian lung forms a continuous loop that consists of a three-tier system of airways. These are a primary bronchus, secondary bronchi and tertiary bronchi (parabronchi). The arrangement of the airways and the blood vessels in the lung determines where and how air and blood are distributed and the respiratory media exposed to each other for gas exchange. A cross-current system is formed by the essentially perpendicular disposition between the flow of air in the parabronchial lumen and that of venous blood in the exchange tissue; an auxiliary counter-current arrangement is formed by the relationship between the direction of the flow of air in the air capillaries and that of the blood in the blood capillaries, and a multicapillary serial arterialization system consists of the sequential interaction between the blood in the blood capillaries and the air in the capillaries in the parabronchus. In addition to these features, the bird lung has large capillary blood volume, extensive surface respiratory surface area and particularly thin blood-gas (tissue) barrier. The morphological specializations and the adaptive physiological features such as large tidal volume and cardiac output all together explain the exceptional gas exchange efficiency of the avian respiratory system, supporting the high metabolic capacity and energetic lifestyles of birds.

Keywords

Birds • Lung • Air sacs • Gas exchange • Parabronchus • Air capillaries • Blood capillaries

J.N. Maina (✉)
Department of Zoology, University of Johannesburg, Auckland Park, Campus, Kingsway, Johannesburg 2006, South Africa
e-mail: jmaina@uj.ac.za

J.N. Maina (ed.), *The Biology of the Avian Respiratory System*, DOI 10.1007/978-3-319-44153-5_8

Contents

8.1 Introduction

> 'Muscle powered flight requires a high metabolic rate and a very efficient respiratory system'.
> Constable (1990)

From the recorded literature, the avian respiratory apparatus, i.e. the lung-air sac system, has been studied since Coiter (1573). In spite of this, some important aspects of its functional design still remain unknown (Maina 2005, 2015a, b, c, 2016a, b; Powell 2000; Kardong 2009). Based on existing data, Donald Farner (1970) ranked the avian respiratory system among the most controversial organ systems. Subsequently, regarding the structure and function of the avian respiratory system, significant advances have been made (Scheid 1979; Abdalla 1989; McLelland 1989; Powell 2000; Maina 2005, 2006, 2015a, b, c, 2016a, b; Miura et al. 2009; Moura et al. 2011, 2016). Among the air-breathing vertebrates, the avian respiratory system is structurally the most complex gas exchanger (King 1966; Duncker 1971; Fedde 1980, 1998; Abdalla 1989; McLelland 1989; Maina 2005, 2006, 2015a, b, c, 2016a, b) and functionally the most efficient (Scheid 1979; Scheid and Piiper 1989; Powell 2000).

The present-day non-volant birds evolved from volant progenitors (Welty 1979) and for various reasons lost capacity of powered flight. For example, the penguins (Spheniscidae), a group of flightless birds that for the most part dwell in water, use their paddle-shaped wings for propulsion (King and King 1979). The domestic fowl (*Gallus gallus* variant *domesticus*) is a group of flightless birds that was domesticated from the wild jungle fowl (*Gallus gallus*) of South East Asia ~8000 years ago (West and Zhou 1988): by intense genetic breeding for growth and productivity, ~40 different breeds of birds of commercial value exist today. The phylogeny and the lifestyle birds lead of powered (active) flight fundamentally explain the evolution of the exceptionally efficient respiratory system in birds (Duncker 1979, 2004; Perry 1989a, b, 1992; Maina 2000a). After they evolved independently from reptiles much later after mammals (de Beer 1954; Ostrom 1975; Pough et al. 1989) and then accomplished endothermic-homeothermy (Koteja 2004; Grigg et al. 2004; Geiser 2008; Tzschentke and Rumpf 2011), birds accomplished metabolic levels between resting and maximal rate of exercise- or cold-induced thermogenesis that are 4–15 times greater than those of their progenitors at same body temperature (Lasiewski 1962). A significant metabolic barrier separates ectothermic and endothermic animals and volant from non-volant ones. The daily energy expenditure of birds is higher than that of a mammal of equivalent body mass (King 1974). Compared to the relatively lower general body temperature of mammals of 38°C (Lasiewski and Dawson 1967; Aschoff and Pohl 1970), birds maintain higher body temperatures of between 40 and 42 °C. According to Powell (1983), a 380 g bird consumes ~74% more energy daily compared to a mammal of corresponding body mass. The mass-specific aerobic capacities of flying birds and bats are 2.5–3 times higher than those of a non-flying mammal of the same body mass running hard on the ground (Thomas 1987). In order to generate and sustain large energetic needs, particularly during flight, a 4–5 g body mass hummingbird consumes ~2 g of sucrose per day (Powers and Naggy 1988), and before it migrates, the body fat content increases enormously (Carpenter et al. 1983). Before migration, the ruby-throated hummingbird (*Archilochus colubris*) stores ~0.15 g of triacylglycerols per day per gram body weight, a value which in a human being is equivalent to a weight gain of 10 kg per day (Hochachka 1973). Karasov et al. (1986) and Martinez del Rio (1990) observed that to promote absorptive rate, the gastrointestinal system of the hummingbird has the highest sucrase activity per cm^2 of surface area and higher densities of intestinal glucose transporters compared to other vertebrate species.

Energetically, powered flight is a very costly form of locomotion, and consequently considerable metabolic capacity and its capacity to acquire substantially large quantities of O_2 are required for it (Tucker 1974; Wells 1993; Nudds and Bryant 2000; Ward et al. 2002; Tobalske et al. 2003; Videler 2006; Hunter 2007). Some birds, e.g. the bar-headed goose (*Anser indicus*), are excellent endurance and high-altitude flyers (Köppen et al. 2010; Hawkes et al. 2011; Scott et al. 2014; Bishop et al. 2015). Showing how extremely restrictive it has been, powered flight has evolved only in two animal phyla, the Chordata and the Arthropoda. Chronologically, insects attained it ~350 million years ago (mya) (Wigglesworth 1972), the now extinct pterodactyls ~220 mya (Bramwell 1971), birds ~150 mya (de Beer 1954) and bats ~50 mya (Yalden and Morris 1975) (Plate 8.1). The various animals that are said to fly like the freshwater butterflyfish (*Pantodon buchholzi*) of the West African Rivers, the parachuting frog of Borneo (*Rhacophorus dulitensis*), the flying snakes of the jungles of Borneo (*Chrysopelea* sp.), the flying squirrel (*Glaucomys volans*) of North America, the flying lemur (*Cynocephalus volans*) and the East Indian gliding lizard (*Draco volans*) are strictly acrobatic passive gliders or parachutists that use certain modifications of their body to delay a fall by using drag and lift: such animals have not had to contend with the strict aerodynamic requirements and high metabolic demands that are needed for powered flight. In birds, the aerodynamic requirements for flight have been so exacting that in the entire class Aves that consists of ~9000 species (Morony et al. 1975; Gruson 1976), the external anatomical homogeneity is remarkable (Marshall 1962, p. 555). Yapp (1970, p. 40) observed that there is less variation in the external body form of birds than that which exists in only 90 species of primates and 290 species of carnivora. Hovering, i.e. motionless flight relative to the surrounding air, when power is generated by the beating of wings that supports the body weight by generating downward displacement of air (Plate 8.2a, b), is energetically a highly demanding mode of flight (Wells 1993). With each bout lasting for less than a minute, hummingbirds generally hover more than 100 times per day (Krebs and Harvey 1986). According to Suarez et al. (1991), Mathieu-Costello et al. (1992) and Suarez (1992), the hummingbird's flight muscle is the most metabolically active vertebrate skeletal muscle, and during flight the maximum enzyme activities are incomparable to those of other birds (Suarez et al. 1986). The oxygen consumption of a hovering bird is 2.5 times greater than that during forward flapping flight (Wells 1993). In hummingbirds, the O_2 consumption (VO_2) ranges from 40 to 85 $mLO_2\ g^{-1}\ h^{-1}$ (Bartholomew and Lighton 1986; Wells 1993). For mammals, the VO_{2s} are considerably lower, with those of a 7 g running pygmy mouse (*Mus minutoides*), a 1.1 kg kangaroo rat (*Bettongia penicillata*) and a 21 kg dog (*Canis familiaris*), respectively, being 15.7, 10.6 and 9.5 $mLO_2\ g^{-1}\ h^{-1}$ (Seeherman et al. 1981). In level flight, a budgerigar (*Melopsittacus undulatus*) increases its VO_2 13 times its standard metabolic rate, a value that is ~1.5 times greater than that of a similar-sized mouse (*Mus musculus*) running hard on a treadmill (Tucker 1968). The VO_2 of a domestic pigeon (*Columba livia*) running on the ground is 27.4 mL min^{-1}, but in a flying bird (at a speed of 10 m s^{-1}), it is 77.8 mL min^{-1}, a factorial increase of ~3 (Grubb 1982). In the herring gull (*Larus argentatus*) and the grey-headed albatross (*Diomedea chrysostoma*), during flight, the metabolic rate is, respectively, two and three times greater than the basal one (Costa and Prince 1987). Among birds, some examples of exceptional endurance- and champion high-altitude flying are: a Rüppell's griffon vulture (*Gyps rueppellii*) struck a jet craft at an altitude of 11.3 km (Laybourne 1974); during its annual migration, the Arctic tern (*Sterna paradisaea*) covers a return distance of ~71,000 km (Egevang et al. 2010); the bar-headed geese (*Anser indicus*)

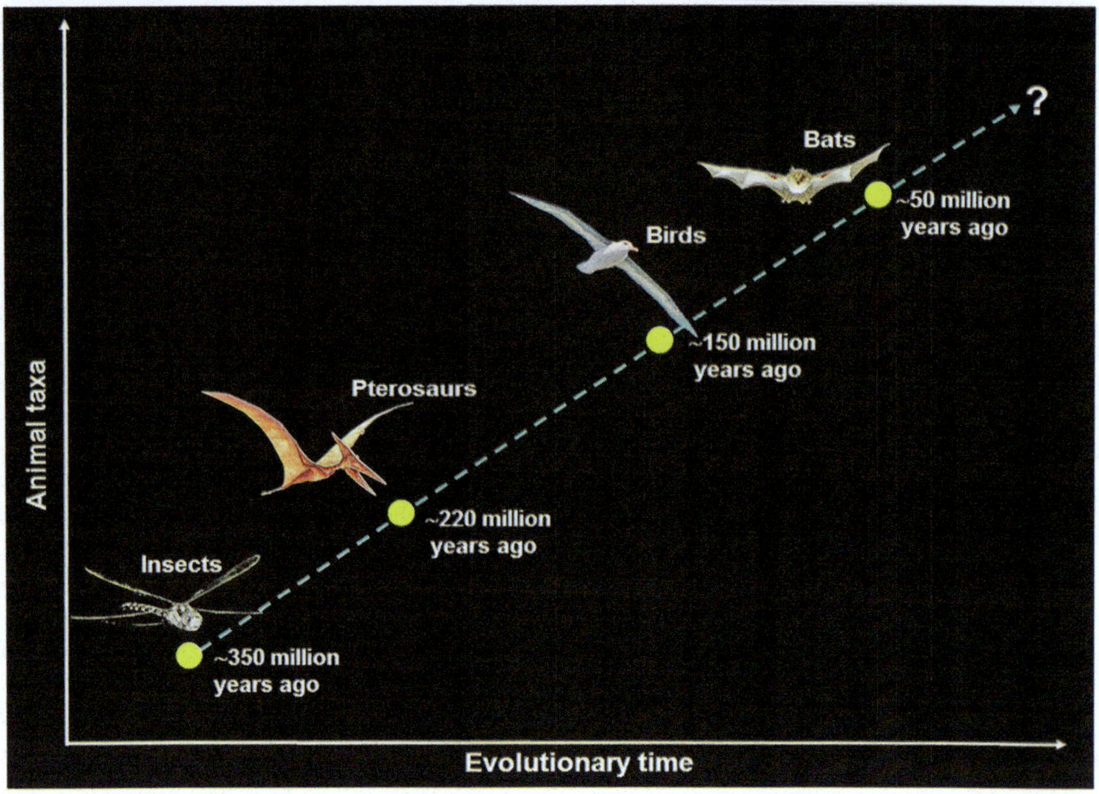

Plate 8.1 Evolution of powered (active) flight in the animal kingdom. The *question mark* points out to an interesting uncertainty as to which animal taxon is likely, if at all, to evolve powered flight in the future

fly from practically sea level and without acclimatizing rise to cross the Himalayas (Köppen et al. 2010; Hawkes et al. 2011; Scott et al. 2014; Bishop et al. 2015); and Powell and Scheid (1989) calculated that operating on arterial blood gases equivalent to those of a bird, a human being at the peak of Mt. Everest can ascend by ~780 m, if the mammalian (bronchioalveolar) lung was replaced with the avian one, i.e. the parabronchial lung.

Here, a succinct account of the structure and function of the avian respiratory system is given.

8.2 Structure of the Mature Avian Respiratory System

8.2.1 Lung

The fundamental structural distinctiveness of the avian respiratory system is that the gas-exchanging part (the lung) is completely separated from the ventilatory part (the air sacs) (King 1966; Duncker 1971; McLelland 1989; Maina 2005) (Plate 8.3a–d). Furthermore, the lung is practically rigid: it is firmly held by the ribs and the vertebrae on the dorsolateral aspects, and it attaches to the horizontal septum on the ventral aspect (Plate 8.4a). From one-fifth to as much as one-third of the volume of the lung is held between the ribs and the vertebrae (King and Molony 1971) (Plate 8.4b–d). Between inspiration and expiration, the avian lung changes in volume by only 1.4% (Jones et al. 1985). The rigidity (non-compliance) of the avian lung greatly contributed to the functional efficiency of the respiratory system. It allowed extreme subdivision of the exchange tissue which conferred large respiratory surface area to a relatively small lung (Maina 2005, 2015a, b, c). Furthermore, because the force of surface tension does not have to be overcome to ventilate the lung, as occurs in the

Plate 8.2 (**a**) During hovering the body weight of bird (*red arrow*) is supported by the power generated by the beating of its wings (*yellow arrow*). (**b**) In order to remain stationary, the wing rotates at the pectoral girdle in a figure of eight, producing and maintaining lift. Energetically, hovering is the most costly form of activity. Only a few species of birds can perform it

mammalian lung (Hlastala and Berger 1996, 2001), the work of breathing has been reduced. It is because of the rigidity of the avian lung that stable very small terminal respiratory units, the air capillaries, exist in the avian lung (Plate 8.5a–h): the air capillaries range in diameter from 3 to 20 μm (Duncker 1971, 1974; West et al. 1977; Dubach 1981; Abdalla 1989; Maina and Nathaniel 2001; Woodward and Maina 2005, 2008; Maina and Woodward 2009). In the compliant mammalian lung that changes in shape and volume during breathing, the ultimate size of the alveoli is determined by surface tension force in the alveoli: they collapse when they are too small. While some early investigators claimed that the respiratory surface of the avian lung lacks surfactant (Miller and Bondurant 1961), the complex mixture of lipids and phospholipoproteins that lowers surface tension (Daniels et al. 2001; Li 2005; Zuo et al. 2008), it has been confirmed that the surface of the air capillaries is lined by the material (Fujiwara et al. 1970; Pattle 1978; Bernhard et al. 2001). Laplace's law (Laplace 1902; Kellog 1987) predicts that surface tension is greater in smaller air spaces compared to the larger ones. All other conditions the same, higher surface tension should exist in the air capillaries of the avian lung compared to the alveoli of the mammalian lung (Hlastala and Berger 2001; Bernhard et al. 2004). The surfactant performs other roles such as prevention of adhesion of the respiratory surfaces (Daniels and Orgeig 2001; Foot et al. 2006), protection of the pulmonary surface from pathogens (Spragg 2007) and prevention of exudation of blood plasma onto the respiratory surface (Daniels et al. 1998; Alonso et al. 2005).

8.2.2 Bronchial (Airway) System

The bronchial system of the avian lung consists of a three-tier assembly of passageways. Their locations and arrangement have been described in detail by King (1966), Duncker (1971, 1972, 1974), McLelland (1989) and Maina (2005, 2015a, b, c). More recently, Makanya and Djonov (2008) and Makanya et al. (2011, 2014) have described a rather different configuration of

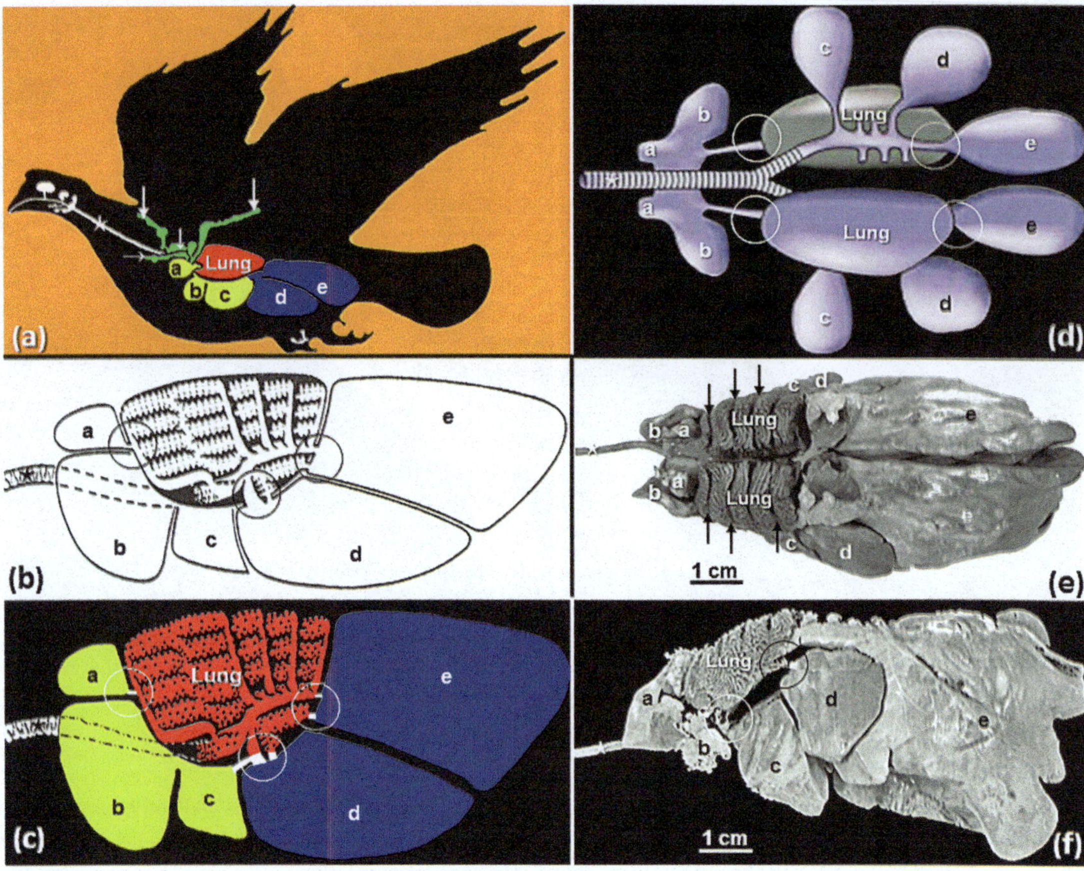

Plate 8.3 Diagrams showing the topographical locations of the lungs and the air sacs of birds. (**a**) Location of the air sacs (**a–d**) in a bird. *Arrows*, extensions of the air sacs that pneumatize adjacent bones; *star*, trachea. (**b**, **c**) Layout of the lungs and the air sacs (**a–d**). *Star*, trachea; *circles*, ostia. (**d–f**) Schematic diagram (**d**) and dorsal and lateral views of a latex cast (**e**, **f**) of the lung and the air sacs (**a–d**) of the domestic fowl (*Gallus gallus* variant *domesticus*). The air sacs are: a, cervical air sac; b, interclavicular air sac; c, craniothoracic air sac; d, caudothoracic air sac; e, abdominal air sac. Tr, trachea; *circles*, ostia; *arrows*, costal sulci

the airways in the lungs of the domestic fowl and the domestic duck (*Cairina moschata*). Because this description has not been confirmed by another investigator(s), the following summary synthesizes the earlier accounts. The principal airway, i.e. the intrapulmonary primary bronchus, passes through the lung giving rise to four setas of secondary bronchi. These are the medioventral, the mediodorsal, the laterodorsal and the lateroventral secondary bronchi (Plate 8.6a, b). In the lung of the domestic fowl, there are four medioventral and seven mediodorsal secondary bronchi, while the numbers of lateroventral and laterodorsal secondary bronchi vary even between specimens. The laterodorsal secondary bronchi range in diameter from 1 to 2 mm. They originate from the lateral side of the caudal part of the intrapulmonary primary bronchus between the origins of the mediodorsal and the lateroventral secondary bronchi. The secondary bronchi give rise to the parabronchi that consist of a parabronchial lumen that is surrounded by a gas exchange tissue mantle. From the parabronchial lumen, the atria give rise to the infundibulae (Plates 8.6c, d and 8.7a–h) which in turn give rise to air capillaries. Duncker (1971) indicated that in avian lung two phylogenetically different groups of parabronchi, namely, the paleopulmonic (ancient or old) and the neopulmonic (modern or new), exist.

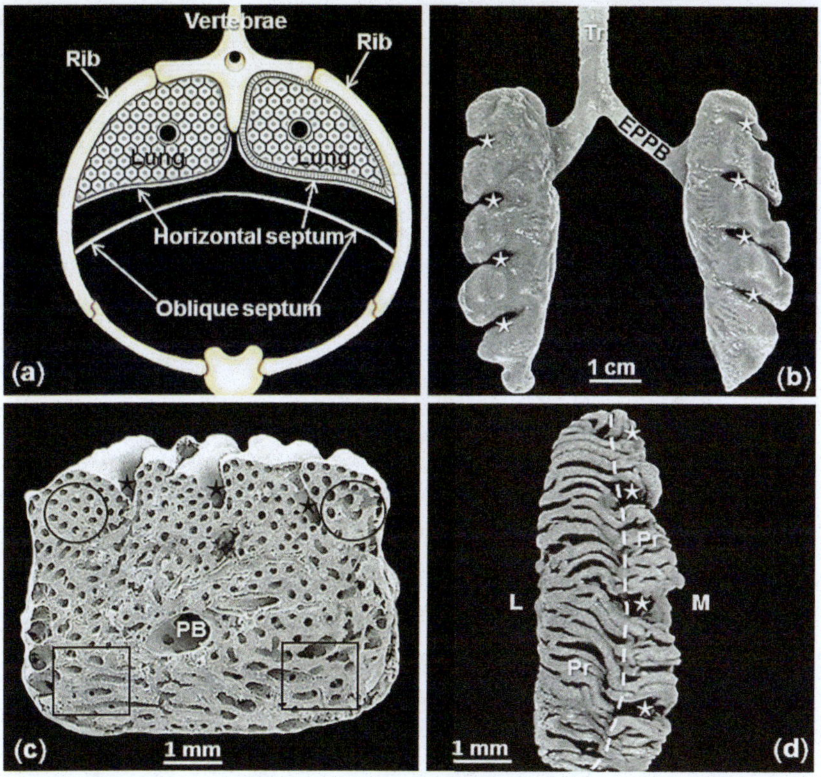

Plate 8.4 (**a**) Location of the avian lung. It is held between the vertebrae and the ribs on the dorsolateral aspect and is ventrally attached to the horizontal septum. (**b**) Dorsal view of the lung of the ostrich (*Struthio camelus*) showing the deep impressions made by the vertebrae and the ribs. *Stars*, costal sulcae; Tr, trachea; EPPB, extrapulmonary primary bronchus. (**c**) Longitudinal section of a developing lung of the domestic fowl (*Gallus gallus* variant *domesticus*) at day 15 of incubation showing deep costal sulci (*stars*). PB, primary bronchus. *Circles*, location of the paleopulmonic parabronchi; *squares*, location of the neopulmonic parabronchi. (**d**) Dorsal view of the lung of the domestic fowl showing the linea anastomotica (*dashed line*) where parabronchi (Pr) from the mediodorsal secondary bronchi on the lateral side of the lung (L) and the medioventral secondary bronchi from the medial side (M) anastomose. *Stars*, costal sulci

The former are mainly located dorsal to the intrapulmonary primary bronchus, while the latter exist ventral to the airway (Plates 8.4c and 8.6a). The paleopulmonic parabronchi connect the medioventral secondary bronchi to the mediodorsal secondary ones, whereas the neopulmonic ones connect the laterodorsal, the lateroventral, the medioventral and the mediodorsal secondary bronchi and the caudal air sacs (Duncker 1971; McLelland 1989). In basal species of birds such as the kiwi (*Family Apteryx*), the neopulmo is totally lacking, whereas in the derived species, the neopulmo may form as much as one-third of the volume of the lung (King and Molony 1971; McLelland 1989). The neopulmo is poorly developed in birds such as the storks (*Ciconiidae*) and the emu (*Dromaius novaehollandiae*) (McLelland 1989). The paleopulmonic region consists of stacks of parabronchi that interconnect irregularly (Plate 8.7c), whereas the neopulmonic ones anastomose profusely to form the so-called pulmo reteformis (Plate 8.7d). The area where the paleopulmonic parabronchi connect the mediodorsal and the medioventral secondary bronchi on the dorsal longitudinal central plane of the lung is called 'planum anastomoticum' (Duncker 1971; King 1966, 1979; McLelland 1989; Maina 2005) (Plate 8.4d). In the mature lung of the domestic

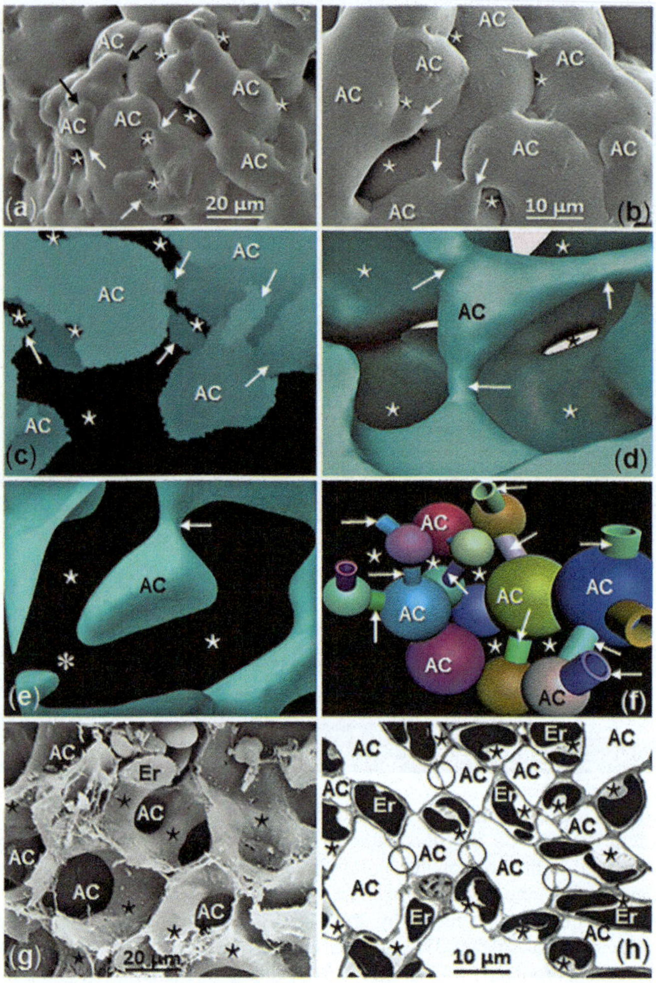

Plate 8.5 (**a**, **b**) Scanning electron micrographs of latex rubber casts of the air capillaries (AC) and three-dimensional computer reconstructions (**c–e**) showing air capillaries (AC) which are connected narrow passageways (*arrows*). *Stars*, location of the blood capillaries; *asterisk* (**c**), site where part of an interconnecting passageway is missing because it was not included in the reconstruction. (**f**) Schematic diagram showing the rotund shape of the air capillaries (AC) that are connected by narrow passageways (*arrows*). *Stars*, location of the blood capillaries which entwine closely with the air capillaries. (**g**, **h**) Critical-point dried preparation and a transmission electron micrograph showing air capillaries (AC) which network with the blood capillaries (*stars*) in the lungs of, respectively, the domestic fowl (*Gallus gallus* variant *domesticus*) (**g**) and the house sparrow (*Passer domesticus*) (**h**). *Circles*, epithelial-epithelial cell connections. Er, erythrocyte

fowl, Maina et al. (1982, 1983) determined that morphometric differences do not exist between the gas exchange tissues of the paleopulmonic and the neopulmonic parabronchi. In some avian lungs, e.g. those of the galliform birds, the parabronchi are well separated by interparabronchial septa (Plate 8.7e), while in others, e.g. the passerine birds, the septa are lacking (Plate 8.7f).

The patterns of airflow in the paleopulmonic and the neopulmonic parabronchi of the avian lung are different (Brackenbury 1971; Scheid 1979, Fedde 1980, 1998). In the former, the flow is caudocranial and continuous, while in the latter it is tidal, i.e. the direction changes with the respiratory cycles. Identification of the topographical location of the neopulmonic parabronchi in the avian lung helped resolve a

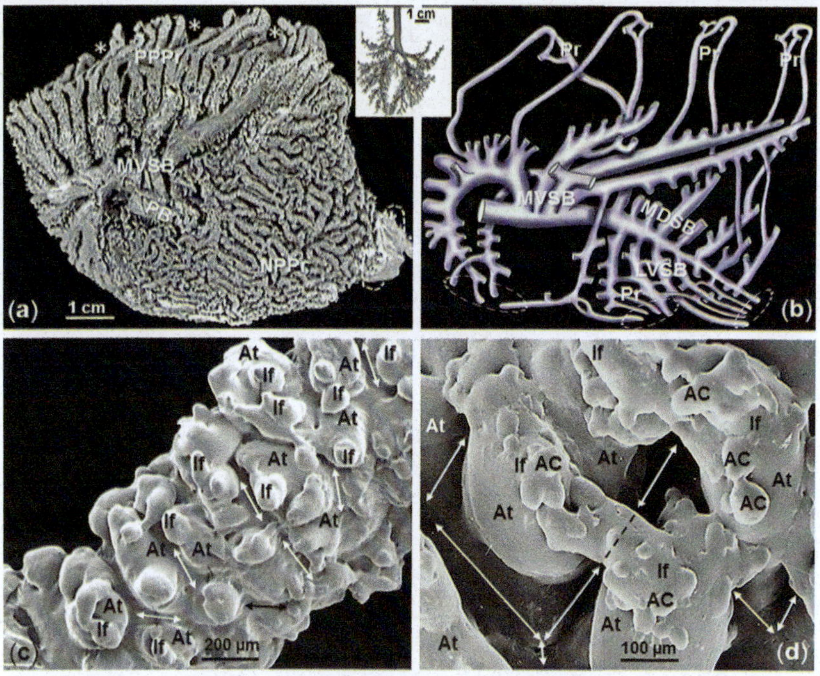

Plate 8.6 (**a**) Medial view of a cast of the lung of the domestic fowl (*Gallus gallus* variant *domesticus*) showing airways that include the medioventral secondary bronchi (MVSB) and a primary bronchus (PB). The other secondary bronchi are covered by the paleopulmonic parabronchi (PPPr) and the neopulmonic parabronchi NPPr). *Encircled (dashed) area*, ostium; *asterisks*, costal sulci. (**b**) Avian lung drawn as if transparent to show the arrangement of the airways. MVSB, medioventral secondary bronchi; MDSB, mediodorsal secondary bronchi; LVSB, lateroventral secondary bronchi; Pr, parabronchi; encircled (*dashed*) areas, ostia. The insert between (**a**) and (**b**) is a cast of the lung of the pig (*Sus scrofa*) showing regular dichotomous branching the airways in a mammalian lung. (**c, d**) Latex cast of a parabronchus of the lung of the domestic fowl showing atria (At) originating from the parabronchial lumen. They in turn give rise to infundibulae (If) and then air capillaries (AC). *Double-sided arrows*, locations of the interatrial septa

long-term dispute of explaining why the concentration of CO_2 in the air in the caudal air sacs is greater than that in the inspired air. The explanation is that while much of the inspired air flows through the intrapulmonary primary bronchus to enter the abdominal and the caudothoracic air sacs directly, in the lungs of the derived birds where neopulmonic parabronchi exist, some of the air passes through the airways where it collects CO_2 which is produced in the exchange tissue (Piiper 1978). Without empirical evidence, Duncker (1972) suggested that the neopulmonic parabronchi form the main site of gas exchange during rest, while the paleopulmonic parabronchi only become involved during exercise. After observing that in a resting duck the neopulmo and the paleopulmo are equally well perfused, Holle et al. (1978) challenged the suggestion on the basis that if both parts are well ventilated, they should contribute to gas exchange equally. Jammes and Bouverot (1975) reached the same conclusion after observing that in awake domestic Pekin duck (*Anas platyrhynchos domesticus*), differences in respired gas partial pressures existed between the caudal air sacs and the mediodorsal secondary bronchi.

The thickness of the parabronchial gas exchange mantle ranges from 200 to 500 μm (Duncker 1974; Maina et al. 1982). In the galliform species, the atria that are deep and are surrounded by prominent bands of smooth muscle and connective tissue bands are called interatrial septa (Plate 8.7b, e, g). In the lungs of the domestic pigeon and the Pekin duck, the atria are, respectively, ~80 and ~125 μm in diameter (West et al. 1977). The interatrial septa are

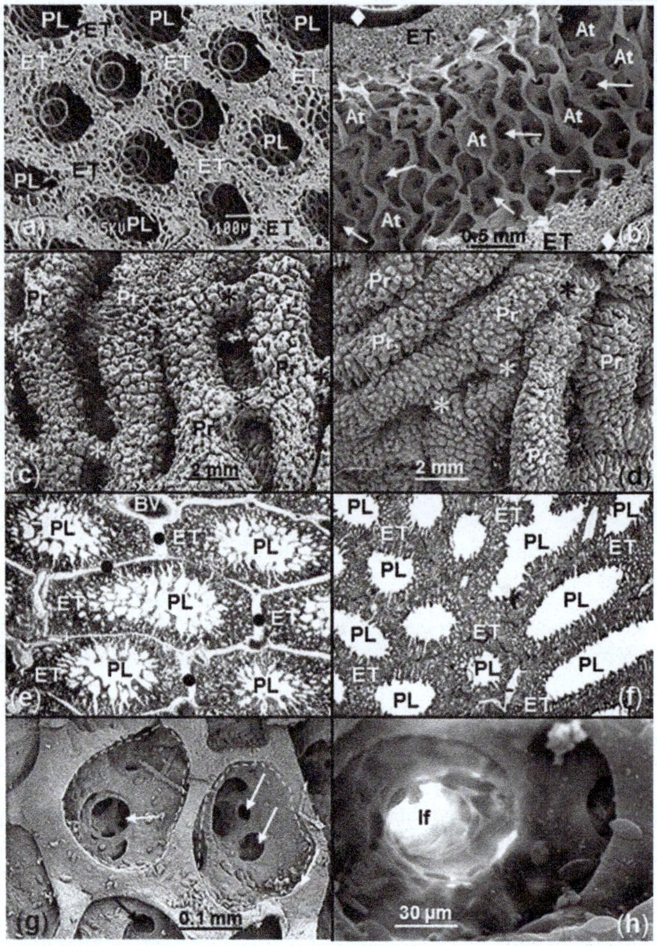

Plate 8.7 (**a**, **b**) Transverse (**a**) and longitudinal (**b**) views of the parabronchi of the lung of the domestic fowl (*Gallus gallus* variant *domesticus*). PL, parabronchial lumen; ET, exchange tissue; *filled diamond*, interparabronchial septa; *circles* (**a**), atria; At (**b**), atria; *arrows* infundibulae. (**c**, **d**) Casts of parabronchi from lungs of the domestic fowl, respectively, showing stacks of paleopulmonic parabronchi (Pr) that periodically anastomose (*asterisk*) and the neopulmonic parabronchi that are irregularly arranged. (**e**, **f**) Histological sections of lungs of the domestic fowl and the house sparrow (*Passer domesticus*), respectively, showing the presence of interparabronchial septa (*dots*) in the lung of the former and their absence in the later. ET, exchange tissue; PL, parabronchial lumen. (**g**) Scanning electron micrographs of the lung of the domestic fowl showing prominent atria (*dashed spaces*) giving rise to infundibulae (*arrows*). (**h**) Close-up of an infundibulum (If)

less prominent and the atria very shallow in the lungs of the small and highly metabolically active species of birds (Duncker 1974; Maina et al. 1982) (Plate 8.7f). The atria project outwards from the parabronchial lumen into the exchange tissue, giving rise to 3–8 narrower air passages, the infundibulae (McLelland 1989; Maina 2005) (Plate 8.7g, h). In the domestic pigeon and mallard duck, the diameters of the infundibulae range from 25 to 40 μm and are, respectively, 100–150 μm long (West et al. 1977). The infundibulae give rise to the air capillaries that are ~3 μm in diameter in the lungs of the songbirds; ~10 μm in those of the penguins, swans (Anatidae) and coot (*Fulica atra*); and ~20 μm in that of the ostrich (*Struthio camelus*) (Maina and Nathaniel 2001). Recently, using three-dimensional computer reconstruction, the air capillaries were shown to be rotund structures that were interconnected by short and narrow passageways (Plate 8.5a–f). Since the lungs of only two species have so far been investigated, it cannot be ruled out that the shapes of air capillaries may differ between birds. The air capillaries intertwine closely with the blood capillaries (Plate 8.5g, h). The blood capillaries consist of conspicuous segments that are about as long as they are wide (Woodward and Maina 2005, 2008; Maina and Woodward 2009) (Plate 8.8a–d). Previously, the air capillaries were claimed to be straight, non-branching, blind-ending tubules that traversed the exchange tissue from the parabronchial lumen and the blood capillaries to be corresponding structures that run in the opposite direction, i.e. from the periphery of the parabronchus in direct contact with the air capillaries (Piiper and Scheid 1973; Brackenbury and Akester 1978; Scheid 1987, 1979; West et al. 1977; Scheid and Piiper 1989; Powell and Scheid 1989).

8.2.3 Pulmonary Vasculature

The avian pulmonary vasculature has been well investigated (Radu and Radu 1971; Abdalla and King 1975, 1976a, b, 1977; West et al. 1977; Maina 1988). Abdalla (1989) reviewed aspects of the distribution and the topographical relationships between the branches of the pulmonary artery and vein and the airways in the lung of the domestic fowl. Contrary to the mammalian lung where the arterial system closely tracks the airways (Maina and van Gils 2001), in the avian lung, the pulmonary artery and its branches and the vein and its branches do not follow each other, nor do they follow the bronchial system. The pulmonary artery enters the lung at the hilum ventral to the first medioventral secondary bronchus and then divides into four branches (Plate 8.9a–f): the accessory branch that supplies blood to a small area of the lung ventral to the hilum is the smallest blood vessel and is reported to be lacking in lungs of species such as the guinea fowl (*Numida meleagris*), the turkey (*Meleagris gallopavo*) and the domestic duck (Abdalla and King 1977); the cranial branch runs dorsocaudally almost parallel to and lateral to the origins of the medioventral secondary bronchi, delivering blood to the craniodorsal part of the lung; because it is the most direct continuation of the pulmonary artery, the caudomedial branch supplies blood to most of the lung; and the caudolateral branch supplies blood to the caudolateral, ventral and caudoventral parts of the lung. The supply of blood to the avian lung can essentially be divided into two regions: the cranial part is supplied by the cranial branch and the accessory branch of the pulmonary artery, while the caudal part is supplied by the caudomedial branch and the caudolateral branch of the pulmonary artery. The four branches of the pulmonary artery do not anastomose (Abdalla and King 1976a, b). The first-order branches of the pulmonary artery are the interparabronchial arteries that give rise to a series of the smaller intraparabronchial arteries that enter exchange tissue of the parabronchus where the intraparabronchial arterioles terminate in blood capillaries (Plate 8.10a–f). The blood capillaries interdigitate very closely with the air capillaries (Plate 8.11a–f).

In the avian lung, the topographical arrangement between the airways and the blood vessels importantly determines the delivery and exposure of air to blood and therefore determines gas exchange efficiency. The cross-current gas exchange system that consists of the orthogonal disposition between the mass (convective) airflow in the parabronchial lumen and the centripetal (inward) flow of the venous blood from the periphery of the exchange tissue (Plate 8.12) is certainly the most important functional design of the avian lung (Scheid and Piiper 1972; Scheid 1979). Superimposed on it is an auxiliary counter-current system that comprises the centrifugal (outward) flow of air from the parabronchial lumen into the air capillaries and that of the centripetal (inward) flow of blood in the blood capillaries from periphery of the parabronchus (Hsia et al. 2013; Maina 2015b). In addition, a multicapillary serial arterialization system consists of a succession of the infinitely many sites where the blood capillaries and the air

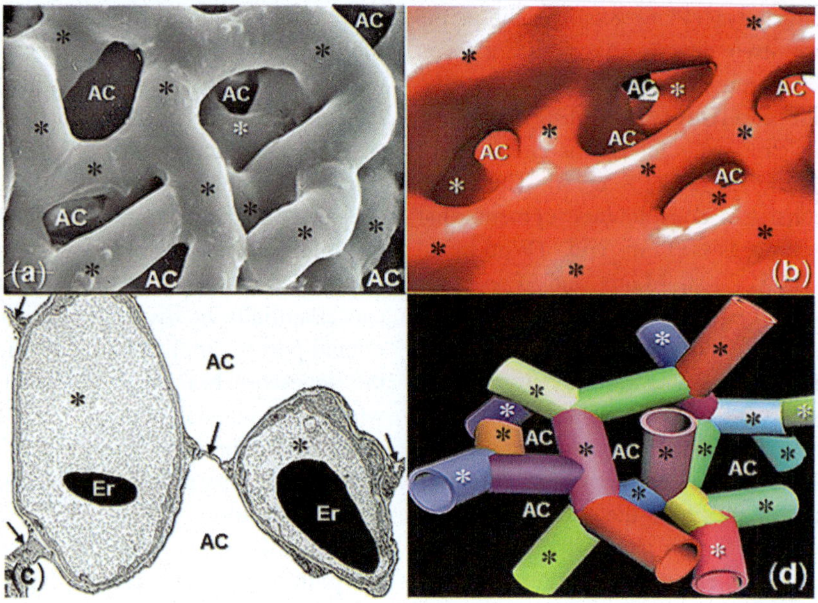

Plate 8.8 Latex cast (**a**) and three-dimensional computer reconstruction (**b**) of the blood capillaries (*asterisks*) of the avian lung. AC, air capillaries. (**c**) Transmission electron micrograph showing transverse views of the blood capillaries (*asterisks*) of the lung of the domestic fowl (*Gallus gallus* variant *domesticus*). AC, air capillaries; Er, erythrocytes; *arrows*, epithelial-epithelial cell connections. (**d**) Schematic illustration of the blood capillaries (*asterisks*) of the lung of the domestic fowl. The blood capillaries are about as wide as they are long and interconnect in three dimensions. AC, air capillaries

capillaries contact and exchange gases occurs in the lung (Hsia et al. 2013; Maina 2015b, 2016a).

8.2.4 Blood-Gas (Tissue) Barrier

In the gas exchangers that have evolved in the air-breathing vertebrates, the structure of the blood-gas (tissue) barrier has been greatly conserved (Maina 1998). It consists of a thin trilaminar tissue barrier that separates blood in the blood capillaries from air in the terminal air spaces (Weibel 1973; Meban 1980; Maina and King 1982; Maina and West 2005). Respiratory gases, i.e. O_2 and CO_2, diffuse across the blood-gas (tissue) barrier (BGB) (Plate 8.13a–f) which consists of a very thin (squamous) epithelial cell that is separated from an endothelial cell by an extracellular matrix space or by a common basement lamina (Plate 8.13c–f). The epithelial cell has scanty organelles, while the endothelial one is endowed with organelles such as the mitochondria, Golgi bodies, rough endoplasmic reticulum and numerous micropinocytotic vesicles (Plate 8.13c, d). For the lung of the domestic fowl, the epithelium forms 12.28%, the basement lamina 21% and the endothelium 66.73% of the BGB (Maina and King 1982). Birds have BGBs that are 56–67% thinner than those in the lungs of mammals of equivalent body mass (Maina et al. 1989) (Plate 8.14). From the very low slopes (scaling factors) that are evident in the regression lines (Plate 8.14), the thickness of the BGB appears to have been optimized within and between the different air-breathing vertebrate taxa (Maina 1998, 2005). For the non-flying mammals on which data are available, the harmonic mean thickness of the BGB in a 2.2 g shrew (*Suncus etruscus*) is 0.230 μm (Gehr et al. 1980), a value which is only 1.5 times smaller than that of the lung of the immensely large bowhead whale (*Balaena mysticetus*) of 0.35 μm (Henk and Haldman 1990); in bats, the thickness of the BGB in the lung of 5 g *Pipistrellus pipistrellus* is 0.206 μm (Maina and King 1984), a value which is only 1.5

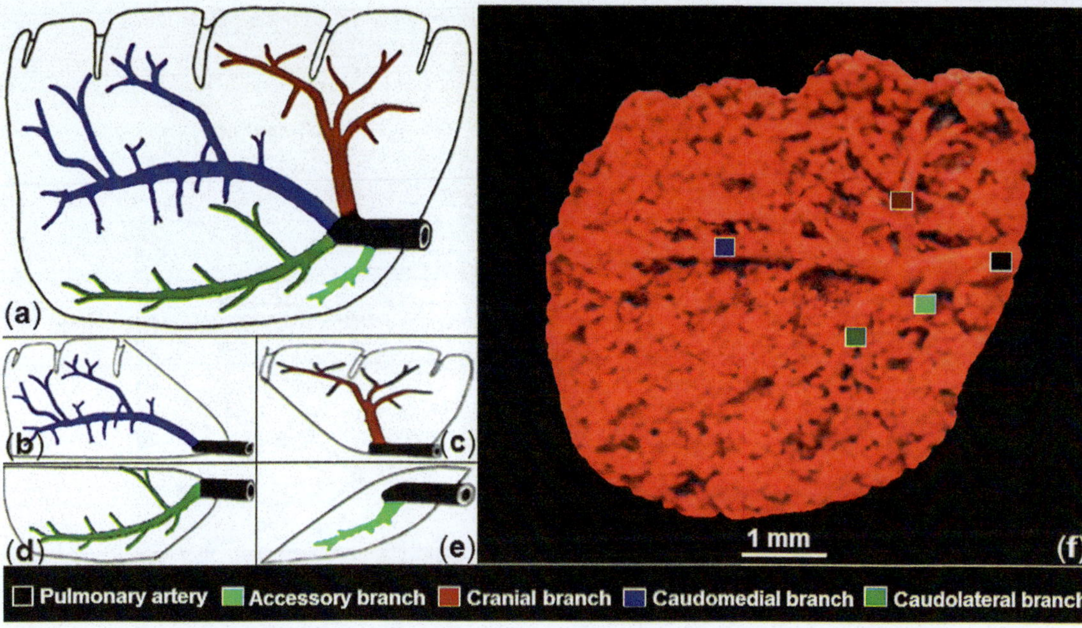

Plate 8.9 (**a–e**) Diagrams showing the branching of the pulmonary artery and the distribution of blood to various parts of the lung of the domestic fowl (*Gallus gallus* variant *domesticus*). (**f**) Latex cast preparation of the arterial vasculature of the lung of the domestic fowl showing the main branches of the pulmonary artery and the regions of the lung they supply blood to

times thinner that of a 900 g flying fox (*Pteropus poliocephalus*) of 0.303 μm (Maina et al. 1991) ; and in birds, the thickness of the BGB of the lung of a 7.3 g violet-eared hummingbird (*Colibri coruscans*) is 0.09 μm (Dubach 1981), a value which is only 6 times thinner than that of 0.56 μm in the lung of a 45 kg ostrich (Maina and Nathaniel 2001). The optimization of the thickness of the BGB suggests it may be the least adaptable structural parameter in the lung. Among the species of birds on which data are available, the violet-eared hummingbird (Dubach 1981) and the African rock martin (*Hirundo fuligula*) (13.7 g) (Maina 1984, 1989) have the thinnest BGBs that are ~0.09 μm thin. The thickest BGBs exist in the ostrich lung (0.56 μm) (Maina and Nathaniel 2001) and the Humboldt penguin (*Spheniscus humboldti*) (0.53 μm) (Maina and King 1987). In the lung of the emperor penguin (*Aptenodytes forsteri*), the inordinate thickness of the BGB and the abundance of connective tissue elements in its basement membrane may maintain the strength of the barrier, allowing it to tolerate the high hydrostatic pressures that the animals encounter during dives.

While the BGB of the avian lung is rather uniform in thickness (Plates 8.8c and 8.13a), on closer scrutiny, the barrier is seen to exhibit sporadic attenuations (Plate 8.13c–f). According to Weibel and Knight (1964), periodic thinning of the BGB promotes the flux of respiratory gases across the barrier without sacrificing its structural integrity. The degree of sporadic attenuation of the BGB is expressed as the ratio of the arithmetic mean thickness of the BGB ($^{-}\tau t$) to its harmonic mean thickness (τ_{ht}). From the data that are available, in birds, the highest ratio occurs in the lung of a house sparrow (*Passer domesticus*) of 10.8 (Maina 1984, 1989), and the lowest is that of 1.2 in the lung of the ostrich (*Struthio camelus*) (Maina and Nathaniel 2001). The minimum harmonic mean thickness of the BGB ($\tau_{htmin.}$) marks the extent to which a barrier can thin (Maina and King 1982). In the lung of the greylag goose (*Anser anser*), the $^{-}\tau t$, the τht and the τt_{min} are, respectively, 0.887, 0.112 and 0.05 μm.

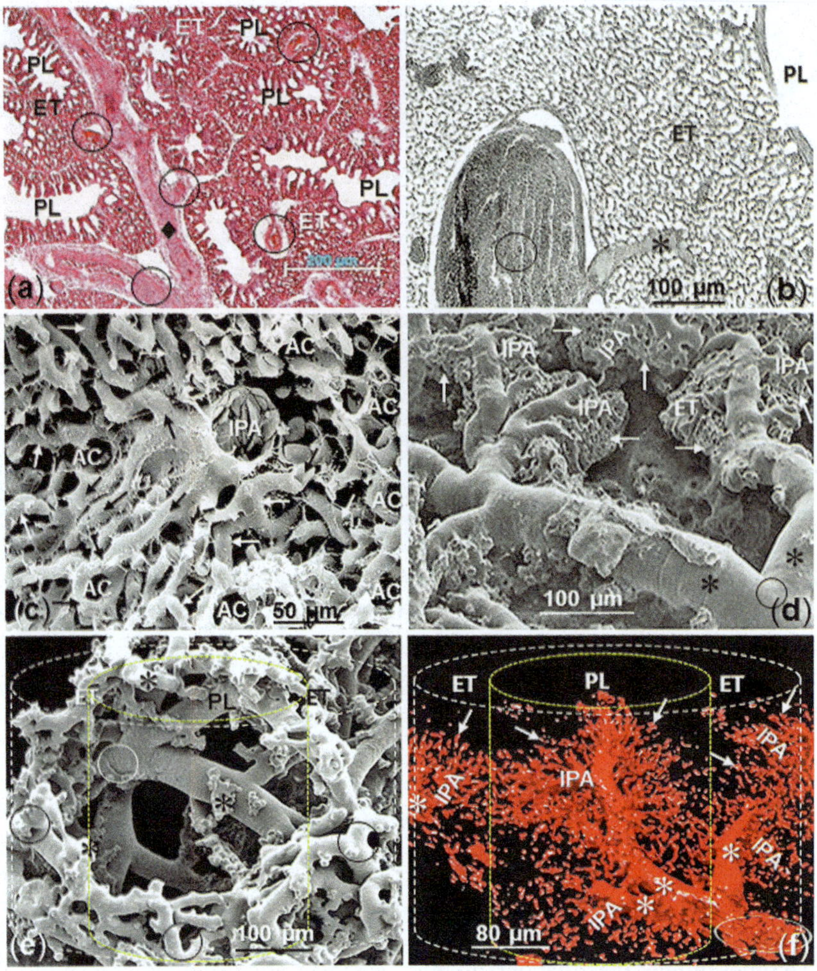

Plate 8.10 (**a**) Histological section of the lung of the domestic fowl (*Gallus gallus* variant *domesticus*) showing a branch of the pulmonary artery (*filled diamond*) giving rise to interparabronchial arteries (*circles*). ET, exchange tissue; PL, parabronchial lumen. (**b**) Histological section of the lung of the ostrich (*Struthio camelus*) showing an interparabronchial artery (*circle*) giving rise to an intraparabronchial artery (*asterisk*). PL, parabronchial lumen; ET, exchange tissue. (**c**) Scanning electron micrograph of the lung of the domestic fowl showing an intraparabronchial arteriole (IPA) giving rise to blood capillaries (*arrows*). AC, air capillaries. (**d**) An interparabronchial artery (*circle*) giving rise to intraparabronchial arteries (*asterisks*) that in turn branch into intraparabronchial arterioles (IPA) which terminate in blood capillaries (*arrows*). (**e**) Scanning electron micrograph of a cast of a parabronchus of the lung of the domestic fowl showing interparabronchial arteries (*circles*) that give rise to intraparabronchial arteries (*asterisks*). The outlines show the extents of the parabronchus which is shown by the *white dashed figure* and that of the parabronchial lumen by the *yellow dashed outline*. ET, exchange tissue; PL, parabronchial lumen. (**f**) A three-dimensional computer reconstruction of the interparabronchial artery (*dashed circle*) giving rise to intraparabronchial arteries (*asterisks*) and intraparabronchial arterioles (IPA) that terminate in blood capillaries (*arrows*). The outline of the bronchus is shown by the *white dashed figure* and that of the parabronchial lumen by the *yellow dashed one*

Nearly four decades ago, Macklem et al. (1979) observed that the air capillaries did not collapse after the parabronchial exchange tissue was subjected to a pressure of 20 cm H_2O (2 kPa). Thereafter, Powell et al. (1985) determined that the blood capillaries were practically rigid on observing that doubling of the flow of blood into a lung resulted in the doubling of

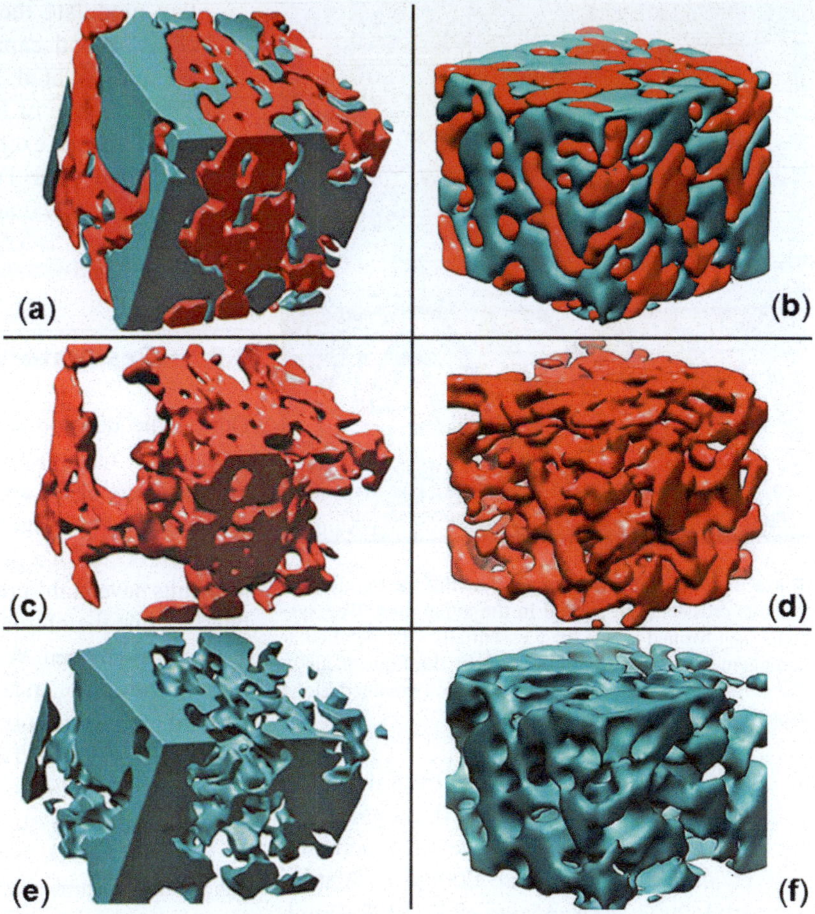

Plate 8.11 (**a**, **b**) Three-dimensional computer reconstructions of the exchange tissue of the lung of the ostrich (*Struthio camelus*) showing the air capillaries (*cyan*) interdigitating with the blood capillaries (*red*). (**c**, **d**) Blood capillaries extracted from the reconstructions shown in (**a** and **b**). (**e**, **f**) Air capillaries extracted from the reconstructions shown in (**a** and **b**). The respiratory units entwine very closely but are not mirror images

pulmonary vascular resistance. More recently, West et al. (2007a, b) and Watson et al. (2008) noted that increasing the pulmonary arterial blood pressure from 10 to 30 cm H_2O (1–3 kPa) or the pulmonary venous pressure from 5 to 15 cm H_2O (0.5–1.5 kPa) did not change the pulmonary vascular resistance and that the blood capillaries remained open even after they were subjected to an external pressure of 35 cm H_2O (3.5 kPa). In an experiment in which chicken lungs were perfused at different pressures (Maina and Sikiru 2013), it was observed that epithelial-epithelial cell connections failed (broke) at an intramural pressure of 2.90 kPa, whereas the BGB failed at a higher one of 3.39 kPa. Understanding the basis of the strengths of the air capillaries and blood capillaries of the avian lung has elicited interest, controversy and debate (Maina 2007a, b; West et al. 2007a, b, c; Maina and Jimoh 2013; Maina and Sikiru 2013). Interestingly, while the BGB of the avian lung is the thinnest among the lungs of the air-breathing vertebrates (Meban 1980; Maina and King 1982; Abdalla et al. 1982; Maina 1989, 2005; Maina et al. 1989; Maina and West 2005), it lacks a dedicated supporting system similar to that in the interalveolar septum of the mammalian lung (Weibel 1973, 1984) (Plate 8.13d—insert). Although the conundrum has not yet been fully resolved, the most important observations that may explain the sources of the strengths of the air capillaries and blood capillaries are the following:

(a) The presence of epithelial-epithelial cell connections (Plates 8.5g, 8.8c and 8.13a, f)

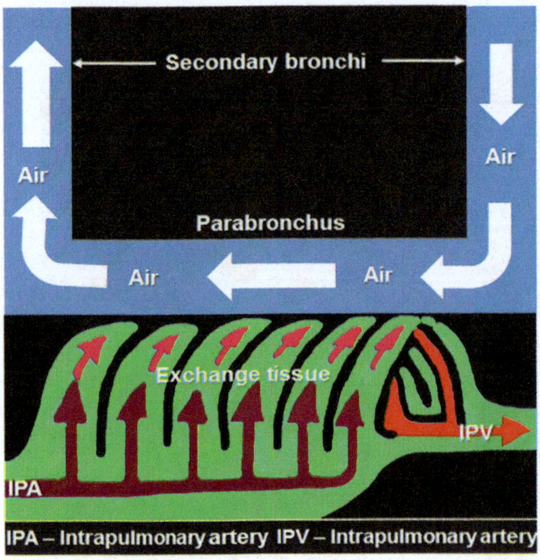

Plate 8.12 Simplified diagram showing the highly efficacious cross-current system in the avian lung. The system is formed by the essentially perpendicular arrangement between the direction of the flow of air in the parabronchial lumen and that of deoxygenated blood from the periphery of the parabronchus

that work as suspensory cables of the blood capillaries, distributing and dissipating tension in the exchange tissue (Scheuermann et al. 1997; West et al. 2007b, c; Watson et al. 2007, 2008; Maina et al. 2010; Jimoh and Maina 2013)

(b) Accumulation of a proteinaceous substance at sites where the epithelial-epithelial cell connections attach to the blood capillaries, providing extra support (Scheuermann et al. 2000)

(c) The presence of a trilaminar substance in the cytoplasm of the squamous epithelial cells that serves as an intracellular scaffold (Klika et al. 1997)

(d) The interdependence between the air capillaries and the blood capillaries where individual respiratory units are supported by the adjoining ones (Hlastala and Berger 1996; West et al. 2006) stemming from their close entwining (Plates 8.5, 8.8 and 8.11)

(e) Thickenings of the epithelial-epithelial cell connections at the triangular areas, parts that modulate the hoop stress in the walls of the blood capillaries (Maina and West 2005; West et al. 2010; Maina 2016a)

(f) The presence of large amounts of collagen fibres in the exchange tissue and particularly the type IV collagen in the BGB (Jimoh and Maina 2013)

8.3 Respiratory Surface Area

During the development of the invaginated gas exchangers, respiratory surface area is created by the branching of the airways which leads to subdivision of the exchange tissue (Maina 1998). This, however, occurs at a cost. Small terminal respiratory units have high surface tension at the air-tissue interface and therefore need more energy to inflate and keep them open. While in the mammalian lung most of the compliance is confined to the terminal parts of the respiratory tree, mainly the alveolar spaces (Dubois et al. 1956), in the avian lung, it is practically rigid (Jones et al. 1985): the compliance of the respiratory system has been relegated to the air sacs (Scheid and Piiper 1989). Although a bird has a lung volume that is smaller compared to mammals of equivalent body mass (Maina 1989, 2005; Maina et al. 1989) and the volume density of the exchange tissue in the lung is ~50% less compared to the that of ~90% in that of a mammal (Gehr et al. 1981; Maina et al. 1982), the surface area of the BGB is on average 15% greater compared to that of a mammal (Plate 8.15). This is explained by the intense subdivision of the exchange tissue of a bird lung (Plates 8.10a–f and 8.11a–f). In the rigid avian lung, surface tension is not a severely limiting factor to the reduction of the size of the terminal respiratory air spaces. The intensity of the subdivision of the exchange tissue of the lung is expressed as the surface density of the respiratory surface area, i.e. the surface area of the BGB divided by the volume of the exchange tissue (Maina 1989, 1993, 2002, 2005; Maina et al. 1989). The values of birds are considerably greater compared to those of non-flying mammals and bats (Maina 2005, p. 148). The surface density of

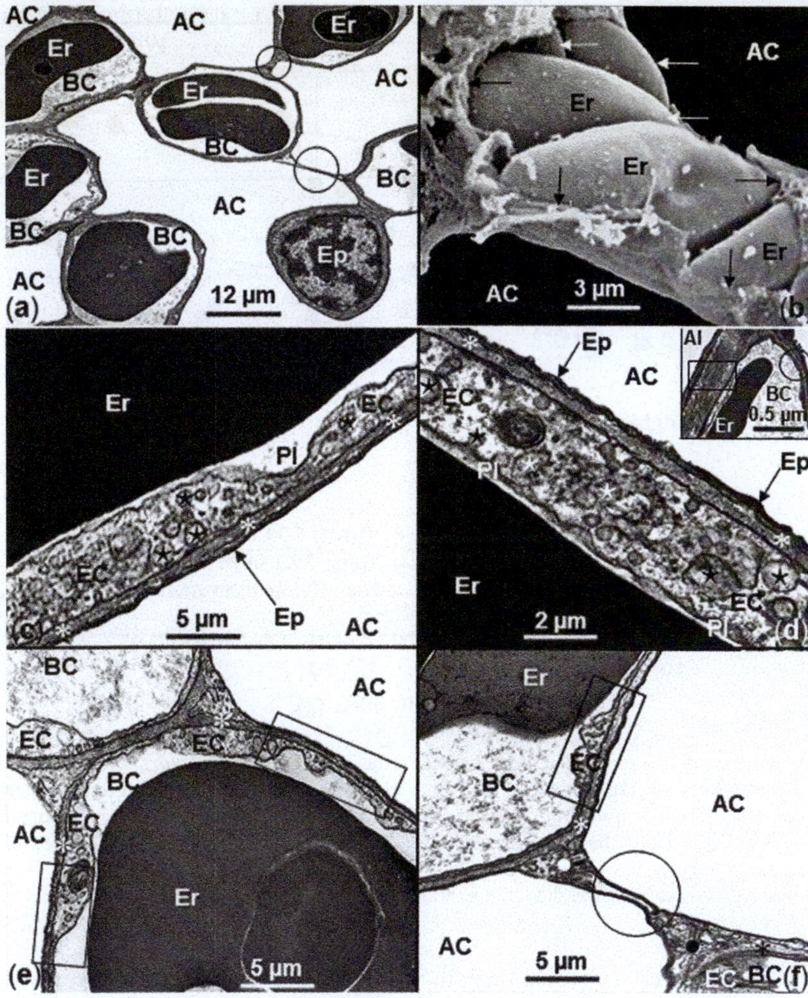

Plate 8.13 (**a**) Transmission electron micrograph of the exchange tissue of the lung of the domestic fowl (*Gallus gallus* variant *domesticus*) showing air capillaries (AC), blood capillaries (BC) and erythrocytes (Er). *Circles*, epithelial-epithelial cell connections; Ep, epithelial cell. (**b**) Scanning electron micrograph from the lung of the domestic fowl showing erythrocytes (Er) passing through a blood capillary in a file. A blood-gas barrier (*arrows*) separates the erythrocytes from air in the air capillaries (AC). (**c**) Transmission electron micrograph of the blood-gas barrier of the lung of the black-headed gull (*Larus ridibundus*) showing its components, namely, the epithelial cell (Ep), the basement membrane (*asterisk*) and the endothelial cell (Ec). *Stars*, micropinocytotic vesicles; Er, erythrocyte; Pl, plasma layer; AC, air capillary. (**d**) Close-up of a transmission electron micrograph showing the structure of the blood-gas barrier of the lung of the black-headed gull. Ep, epithelial cell; *asterisk*, basement membrane; Ec, endothelial cell; *stars*, micropinocytotic vesicles; Er, erythrocyte, plasma layer (Pl); air capillary (AC). d—insert: Interalveolar septum of the lung of the lesser bushbaby (*Galago senegalensis*) showing a thick supporting side of the septum (*boxed area*) containing plentiful collagen fibres and a relatively thinner gas-exchange side (*circled*). Al, alveolus; Er, erythrocyte; BC, blood capillary. (**e**, **f**) Transmission electron micrographs of the lung of the domestic fowl showing the periodic attenuation of the blood-gas barrier (*boxed areas*). BC, blood capillaries; AC, air capillaries; Er, erythrocytes; EC, endothelial cell; *asterisk*, basement membrane; *dots* (**f**), the triangular areas; *circle*, epithelial-epithelial cell connection

the BGB ranges from 82 $mm^2\ mm^{-3}$ in the emu (Maina and King 1989) to 389 $mm^2\ mm^{-3}$ in the violet-eared hummingbird (Dubach 1981), and the highest mass-specific respiratory surface area of ~90 $cm^2\ g^{-1}$ occurs in the violet-eared hummingbird (Dubach 1981) and the African rock martin

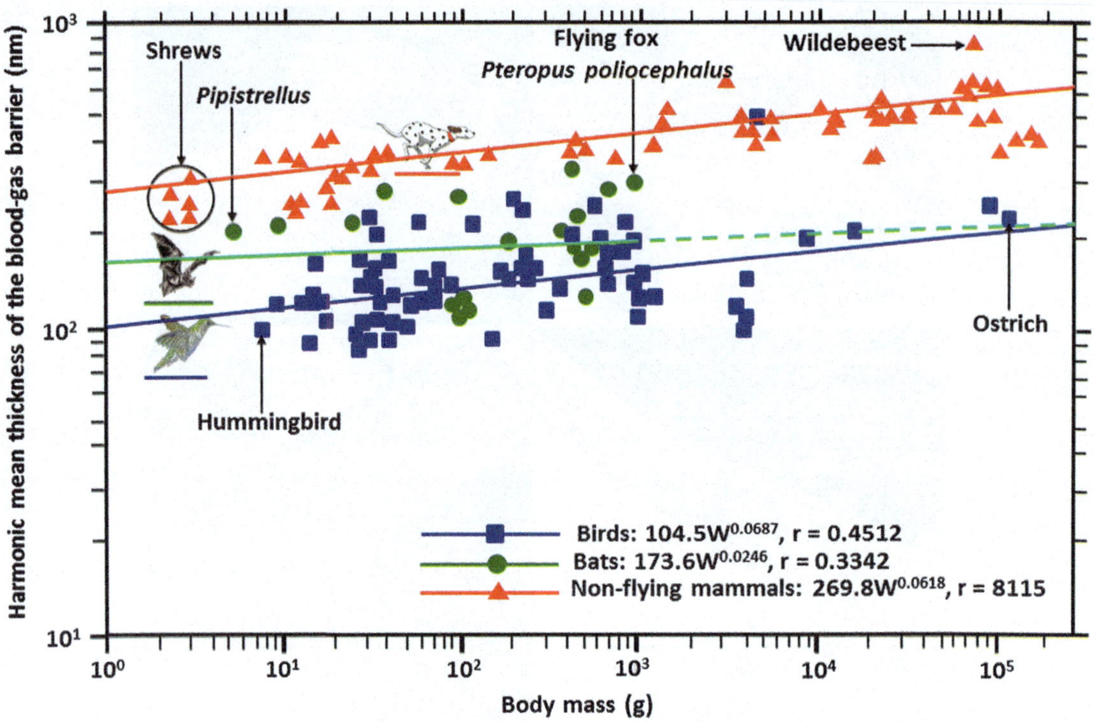

Plate 8.14 Comparison of the regression lines of the harmonic mean thicknesses of the blood-gas barriers against body mass (W) of the lungs of birds, bats and non-flying mammals. Birds have the thinnest barriers followed by bats and non-flying mammals. The heaviest bat (*Pteropus poliocephalus*) studied weighed 928 g, and therefore the *dashed part* of the regression line is an extrapolation. Data for non-flying mammals are from Gehr et al. (1981), bats are from Maina and King (1984) and Maina et al. (1991), and birds are from Maina (1989, 2005) and Maina et al. (1989)

(Maina 1984, 1989). The very large mass-specific respiratory surface area of 800 $cm^2\ g^{-1}$ reported in an unnamed species of hummingbird by Stanislaus (1937) should be treated with caution since it was not explained how it was determined. The lowest mass-specific respiratory surface area of 5.4 $cm^2\ g^{-1}$ has been reported in the lung of the emu by Maina and King (1989). The emu is a large bird native to Australia. It evolved in a habitat with low level of predation when placental mammals were absent.

8.4 Pulmonary Capillary Blood Volume

The volume of the pulmonary capillary blood and the way blood is exposed to air at the gas exchange level significantly determine respiratory efficiency. The exchange tissue of the avian lung is highly vascularized (Plates 8.5g, h, 8.10c and 8.11a–d). In the avian lung, the pulmonary capillary blood volume is 2.5–3 times greater than that in the parenchyma of the lung of a mammal of equivalent body mass. In the mammalian lung, 20% of the blood is found in the alveolar capillaries (Weibel 1963). In the avian lung, however, blood forms as much as 36% of the volume of the lung with 58–80% of it in the blood capillaries (Duncker and Güntert 1985a, b; Maina et al. 1989). In the African rock martin, 29% of the volume of the lung consists of blood with 79% of it in the blood capillaries (Maina 1984). If a sheet-flow design (Fung 1993) exists in the gas exchange tissue of the avian lung as suggested by (Maina 2000b), in the African rock martin, 0.075 cm^3 of blood is literally spread over a respiratory surface area of

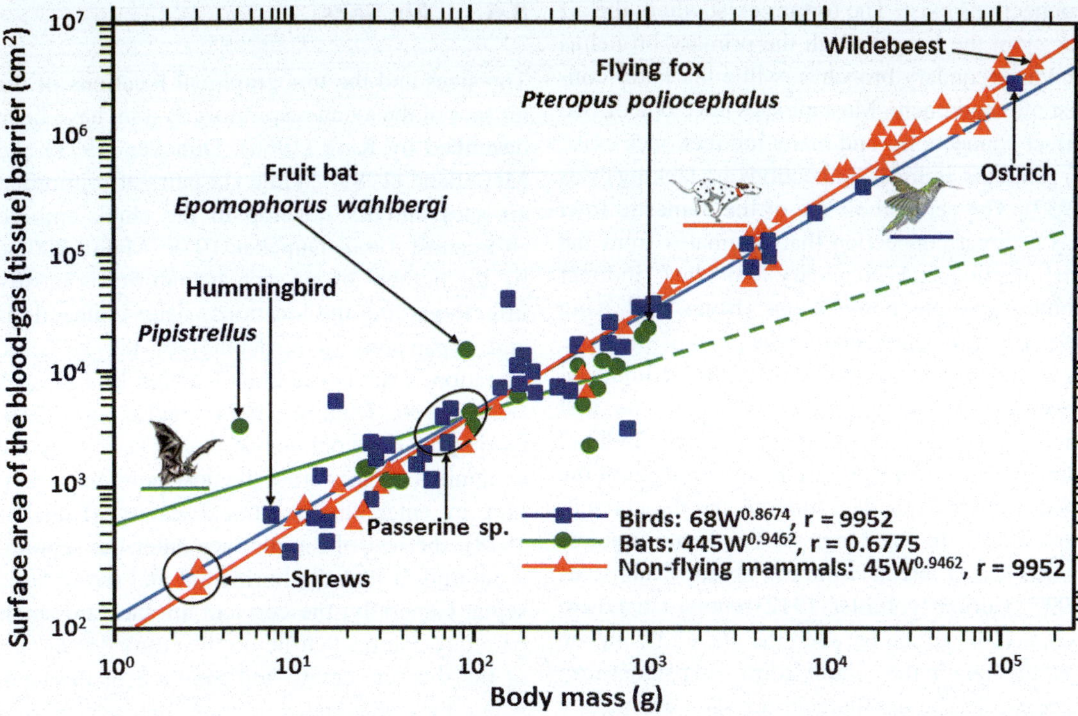

Plate 8.15 Comparison of the regression lines of the surface area of the blood-gas barriers against body mass (W) of the lungs of birds, bats and non-flying mammals. Bats have more extensive respiratory surface areas than non-flying mammals and birds. Small birds have a more extensive respiratory surface area compared to small non-flying mammals, whereas for larger animals, the relationship is reversed. The mean respiratory surface area in birds is 15% greater in birds compared to non-flying mammals. The heaviest bat (*Pteropus poliocephalus*) weighed 928 g, and therefore the dashed part of the regression line is an extrapolation. Data for non-flying mammals are from Gehr et al. (1981), bats are from Maina and King (1984) and Maina et al. (1991) and birds are from Maina (1989, 2005) and Maina et al. (1989)

0.12 m^2. Erythrocytes squeeze through the blood capillaries in a single file (Plate 8.13b).

Capillary loading is the ratio of the pulmonary capillary blood volume to the respiratory surface area. It depicts the degree of exposure of the pulmonary capillary blood to air. In birds, the values range from 0.7 $cm^3\ m^{-2}$ in the African rock martin (Maina 1984) to 4.4 $cm^3\ m^{-2}$ in the Humboldt penguin (Maina and King 1987): low values are indicative of more optimal exposure of pulmonary capillary blood to air. In lungs where a 'double capillary system' exists, e.g. the amphibian, the reptilian and the dipnoan lungs, (Maina 2015c) and capillary blood is exposed to air only on one side of a blood capillary, capillary loading as high as 13 $cm^3\ m^{-2}$ has been reported in the lung of the turtle (*Pseudemys scripta elegans*) (Perry 1978) and 12–19 $cm^3\ m^{-2}$ in that of the lungfish (*Lepidosiren paradoxa*) (Hughes and Weibel 1976).

8.5 Ostia

Because the diaphragm is lacking in birds, the air sacs spread extensively in the continuous coelomic cavity, with some of them pneumatizing certain bones and others even extending out to lie subcutaneously (King 1966; Duncker 1971; McLelland 1989; Maina 2005; Bezuidenhout et al. 1999). The account below encapsulates the summary given by McLelland (1989). The air sacs connect to the lungs at sites called ostia (Plates 8.3 and 8.6a, b). Direct and indirect connections exist. The former exist where the air sacs join the lung through the primary bronchus and/or secondary bronchus, while the latter consist of parabronchi. Most air sacs have one or two direct connections, and many indirect ones exist. In penguins, indirect connections are lacking (Vos 1937). The cervical air sac of the domestic fowl has a direct connection that originates from the first mediodorsal secondary bronchus: indirect connections are non-existent (Biggs and King 1957). For the clavicular air sac, two ostia where one is connected directly to the third mediodorsal secondary bronchus and the other to the first mediodorsal secondary bronchus occur. Indirect connections occur between the parabronchi of the medioventral secondary bronchi and the clavicular air sac. In the lungs of the hummingbirds (trochlids), a lateral ostium is absent (Stanislaus 1937). Curiously, Juillet (1912) reported that there was no connection between the clavicular air sac and the lung in the columbiform and psittaciform species. The craniothoracic air sacs are mostly connected to the lung via two ostia with the medial one joined to the third mediodorsal secondary bronchus, while the lateral ostium is indirectly connected by means of the parabronchi of one or more mediodorsal secondary bronchi. In the hummingbirds, the craniothoracic air sacs are indirectly connected to the lung (Stanislaus 1937), and in many species of birds, the craniothoracic air sac has one ostium that is located at or near the caudal part of the costal septal border of the lung: direct and indirect connections occur. In species of penguin where the neopulmo is lacking, e.g. the kiwi, indirect connections do not exist between the lungs and the caudothoracic air sacs. In most birds, each of the abdominal air sac has one ostium that is located on the lateral aspect of the caudal border of the lung, and direct and indirect connections occur at the ostium, with the direct one joining the primary bronchus and the indirect ones consisting of the parabronchi, laterodorsal secondary bronchi, lateroventral secondary bronchi and last mediodorsal secondary bronchus. In hummingbirds, Stanislaus (1937) observed that only a single indirect connection occurs between the lung and the abdominal air sacs.

8.6 Air Sacs

The sizes and the topographical locations of the air sacs of the avian respiratory system have been described by King (1966), Duncker (1971) and McLelland (1989). While six pairs of primordial air sacs initially develop in the chick embryo (Romanoff 1960; Duncker 1978; Maina 2003a, b), the number of air sacs at maturity is mostly less. Regarding the locations, sizes, connections and expansions (diverticulae) of the air sacs, variations exist between and within bird species (King 1966; Duncker 1971; McLelland 1989. During the development of the air sacs, only the abdominal air sacs cross the post-pulmonary septum to enter the peritoneal cavity (Duncker 1978): the rest of the air sacs enter the septum, separating it into the horizontal and the oblique septa. Generally, the cervical air sacs are small paired structures that lie on the craniodorsal part of the thoracic cavity and the base of the neck where they pneumatize (aerate) the cervical and the thoracic vertebrae and the vertebral ribs (Duncker 1971; McLelland 1989). In the loons (Gaviidae) and the grebes (Podicipedidae), the cervical air sacs are lacking. In the turkey, the left and the right cervical air sacs connect on day 17 of incubation (King and Atherton 1970). In bird species such as the gannet (*Morus bassanus*), the ostrich and the pigeon, the air sacs extend out of the coelomic cavity to lie subcutaneously (McLelland 1989; Bezuidenhout et al. 1999). Formed by fusion of the right and left primordia and coming to occupy the cranioventral part of the thorax, the base of the neck and most of the right and left axillary space, generally, the clavicular air sac is large and unpaired. In birds such as gulls (Laridae), the primordia of the air sacs remain unconnected (Locy and Larsell 1916a, b). In some bird species, e.g. the common pigeon (*Columba livia*) (Müller 1908) and in the house sparrow (*Passer domesticus*) (Wetherbee 1951), the clavicular air sac and the cervical air sacs join to form a single, large cervicoclavicular air sac. In the passeriform and the trochilid species, the clavicular air sac connects to the craniothoracic air sacs (Wetherbee 1951; Duncker 1971). Because the clavicular air sacs that are compliant closely relate to the syrinx, changes in pressure and movements of the air sacs may modulate sound and airflow in the syrinx (Maina 2005). The clavicular air sacs extend to the heart and the surface of the body and extensively pneumatizing bordering bones such as the sternum, the coracoid, the humerus and the ribs (Duncker 1971, McLelland 1989). The craniothoracic and the caudothoracic air sacs which are paired are located in the subpulmonary cavity, i.e. in the space ventral to the lung and the horizontal septum: normally, the craniothoracic air sacs are smaller than the caudothoracic air sacs. In species such as the coot, the craniothoracic air sac is particularly large (Duncker 1971), while in penguins and passerine species, the air sacs are conspicuously small. In the hummingbirds, the caudothoracic air sacs are the largest (Stanislaus 1937). King and Atherton (1970) reported that in the turkey, the caudothoracic air sacs are absent. Usually, the craniothoracic and the caudothoracic air sacs have few, if any, diverticulae (King 1966, 1979). In the kiwi, the abdominal air sacs are located in the subpulmonary cavity (Duncker 1979). Owing to the asymmetry of the viscera, generally, the left abdominal air sac is usually smaller than the right one. However, in species such as the common loon (*Gavia immer*) and the herring gull (*Larus argentatus*), the left abdominal air sac is larger than the right one (Gier 1952). In some species of bird, to different extents, the left and right abdominal air sacs connect (Duncker 1971). While exceptionally large in some species such as the flamingo (*Phoenicopterus ruber*) (Groebbels 1932), the abdominal air sacs are very small in birds such as the penguins and the rhea (Duncker 1971) and the hummingbirds (Groebbels 1932; Stanislaus 1937).

The delicate and transparent walls of the air sacs are lined on the inside by a simple epithelium that covers a thin layer of connective tissue space (Walsh and McLelland 1974). The epithelial cells are connected by junctional complexes at the luminal aspect and laterally by interdigitation. The epithelium consists of squamous cells, but close to the ostia, ciliated cuboidal and columnar cells exist (Fletcher 1980). In the lung of the domestic fowl, a broad band of pseudostratified ciliated columnar epithelium with goblet cells extends from the primary bronchus into the abdominal air sac (Cook and King 1970). On the surface of the caudothoracic air sac, Cook et al. (1987) observed a pseudostratified, ciliated, cuboidal-to-columnar epithelium. In the penguins, the epithelium of the air sacs is generally high and nearly cuboidal. Microvilli project into the luminal space, and electron-dense lysosome-like granules occur in the cytoplasm (Carlson and Beggs 1973; Walsh and McLelland 1974). In some species of birds, Fletcher (1980) reported that muscle cells and collections of fat cells were sparingly dispersed in the walls of the air sacs. According to Trampel and Fletcher (1980), the smooth muscle cells in the wall of the air sac are extensions of a layer that surrounds the parabronchial lumen. Rawal (1976) and Cook et al. (1987) observed cholinergic nerve plexuses in the walls of the air sacs, while vasoimmunopressive (VIP), substance P-, somatostatin- and enkephalin-immunoreactive fibres were demonstrated in the wall of the air sacs of the domestic fowl by Cook et al. (1987). Because the air sacs are typically avascular (Fletcher 1980), they do not play a role in gas exchange (Magnussen et al. 1976).

8.7 Summary

(a) The avian respiratory system consists of a lung (the gas exchanger) that is completely separated from the air sacs (the ventilators). The lung is held firmly by the ribs and the vertebrae on its dorsolateral aspect, and on the ventral aspect, it is attached to the horizontal septum. This has rendered the lung practically rigid. Between respiratory cycles, the volume of the lung does not change significantly. The air sacs, which disseminate extensively in the coelomic cavity pneumatizing certain bones, connect to the lung at sites called ostia and ventilate the lung by synchronized bellows-like activities.

(b) Three hierarchically arranged groups of airways, namely, the intrapulmonary primary bronchus, the secondary bronchi and the tertiary bronchi (parabronchi), exist in the avian lung. Due to the complex arrangement of the airways, the path followed by the inspired air cannot be discerned by simple examination of the arrangement of the airways. While the valves or sphincters were hitherto presumed to control airflow in the avian lung, it has now been shown that the process is regulated by the process of aerodynamic valving. To a large extent, shape, size and orientation of the airways determine the process.

(c) The parabronchi connect the secondary bronchi. They contain most of the gas exchange tissue. In the lumina of the paleopulmonic parabronchi, the flow of air is continuous and unidirectional in a caudocranial direction, while in the neopulmo the flow is tidal, i.e. it changes with the respiratory cycles. A cross-current system which is formed by the essentially perpendicular arrangement between the mass (convective) airflow in the parabronchial lumen and the centripetal (inward) flow of the venous blood in the exchange tissue, an auxiliary counter-current system which is formed by the centrifugal (outward) flow of air in the air capillaries and the centripetal (inward) flow of blood in the blood capillaries and a multicapillary serial arterialization system which is formed by the sequential interaction between the blood and the air capillaries at infinitely many points in the parabronchial exchange tissue greatly contribute to the gas exchange efficiency of the avian lung. Together with structural refinements such as large pulmonary capillary blood volume, extensive respiratory surface area and thin blood-gas (tissue) barrier, the avian lung transfers large amounts of O_2.

(d) The air capillaries are rotund units that connect across narrow passageways, whereas the blood capillaries consist of conspicuous segments that are about equal in length and width. Although very thin, the BGB lacks a dedicated support system. Unexpectedly, the air and the blood capillaries are remarkably strong. In addition to other features, the strengths arise from the arrangement of the respiratory units and the presence of type IV collagen in the basement membrane of the BGB.

Acknowledgements The preparation of this work was supported by the National Research Foundation (NRF) of South Africa. The views and opinions expressed here are, however, mine and not those of the NRF. To the many people that have collaborated with me over the years, I am eternally grateful for their ideas shared with me, giving me their time unreservedly, and particularly for their continued friendship.

References

Abdalla MA. The blood supply to the lung. In: King AS, McLelland J, editors. Form and function in birds, vol. 4. London: Academic; 1989. p. 281–306.

Abdalla MA, King AS. The functional anatomy of the pulmonary circulation of the domestic fowl. Respir Physiol. 1975;23:267–90.

Abdalla MA, King AS. Pulmonary arteriovenous anastomoses in the avian lung: do they exist? Respir Physiol. 1976a;27:187–91.

Abdalla MA, King AS. The functional anatomy of the bronchial circulation of the domestic fowl. J Anat. 1976b;121:537–50.

Abdalla MA, King AS. The avian bronchial arteries: species variations. J Anat. 1977;123:697–704.

Abdalla MA, Maina JN, King AS, King DZ, Henry J. Morphometrics of the avian lung. 1. The domestic fowl, *Gallus domesticus*. Respir Physiol. 1982;47:267–78.

Alonso C, Waring A, Zasadzinki JA. Keeping lung surfactant where it belongs: protein regulation of two-dimensional viscosity. Biophys J. 2005;89:266–73.

Aschoff J, Pohl H. Rhythmic variations in energy metabolism. Fed Proc. 1970;29:1541–52.

Bartholomew GA, Lighton JRB. Oxygen consumption during hover-feeding in free-ranging Anna hummingbirds. J Exp Biol. 1986;123:191–9.

Bernhard W, Gerbert A, Vieten G, Rau G, Hollfield JM, Postle AD, Frenhorst J. Pulmonary surfactant in birds:

coping with surface tension in a tubular lung. Am J Physiol Regul Intergr Comp Physiol. 2001;281: R327–37.

Bernhard W, Haslam PL, Floros J. From birds to humans: new concepts on airways relative to alveolar surfactant. Am J Respir Cell Mol Biol. 2004;30:6–11.

Bezuidenhout AJ, Groenewald HB, Soley JT. An anatomical study of the respiratory air sacs in ostriches. Onderstepoort J Vet Res. 1999;66:317–25.

Biggs PM, King AS. A new experimental approach to the problem of the air pathway within the avian lung. J Physiol Lond. 1957;138:282–99.

Bishop CM, Spivey RJ, Hawkes LA, Batbayar N, Chua B, Frappell PB, Milsom WK, et al. The roller coaster flight strategy of bar-headed geese conserves energy during Himalayan migrations. Science. 2015;347:250–4.

Brackenbury JH. Air flow dynamics in the avian lung as determined by direct and indirect methods. Respir Physiol. 1971;13:319–29.

Brackenbury JH, Akester AR. A model of the capillary zone of the avian tertiary bronchus. In: Piiper J, editor. Respiratory function in birds, adult and embryonic. Berlin: Springer; 1978. p. 109–63.

Bramwell CD. Aerodynamics of *Pteranodon*. J Linn Soc Biol. 1971;3:313–28.

Carlson CW, Beggs EC. Ultrastructure of the abdominal air sac of the fowl. Res Vet Sci. 1973;14:148–50.

Carpenter FL, Paton DC, Hixon MA. Weight gain and adjustment of feeding territory size in migrant hummingbirds. Proc Natl Acad Sci USA. 1983;80:7259–63.

Coiter V. Anatomie Avium. In: Externarum et internarum praecipalium humani corporis partium tabulae atque anatomicae exercitationes observationesque varieae. Norimbergae; 1573. p. 130–3

Constable G. How things work: flight. Alexandria, Virginia: Time Life Books; 1990.

Cook RD, King AS. Observations on the ultrastructure of the smooth muscle and its innervation in the avian lung. J Anat. 1970;106:273–83.

Cook RD, Vaillant CR, King AS. The structure and innervation of the saccopleural membrane of the domestic fowl, *Gallus gallus*: an ultrastructural and immunohistochemical study. J Anat. 1987;150:1–9.

Costa DP, Prince PA. Foraging energetics of grey-headed albatrosses, *Diomedea chrysostoma* at Bird Island, South Georgia. Ibis. 1987; 129:149–58.

Daniels CB, Lopatkov OV, Orgeig S. Evolution of surface activity related functions of vertebrate pulmonary surfactant. Clin Exp Pharmacol Physiol. 2001;25:716–21.

Daniels CB, Orgeig S. The comparative biology of pulmonary surfactant: past, present and future. Comp Biochem Physiol A. 2001;129:9–36.

Daniels CB, Orgeig S, Wood PG, Sullivan LC, Lopatkov AK, Smits AW. The changing state of surfactant lipids: new insights from ancient animals. Am Zool. 1998;38:305–20.

de Beer G. Archeopteryx lithographica. London: British Museum (Natural History); 1954.

Dubach M. Quantitative analysis of the respiratory system of the house sparrow, budgerigar, and violet-eared hummingbird. Respir Physiol. 1981;46:43–60.

Dubois A, Brody AW, Lewis DH, Burgess F. Oscillation mechanics of lungs and chest in man. J Appl Physiol. 1956;8:587–94.

Duncker HR. The lung-air sac system of birds. A contribution to the functional anatomy of the respiratory apparatus. Ergeb Anat Entwicklungsgesch. 1971;45:1–171.

Duncker HR. Structure of the avian lung. Respir Physiol. 1972;14:4–63.

Duncker HR. Structure of the avian respiratory tract. Respir Physiol. 1974;22:1–34.

Duncker HR. Development of the avian respiratory and circulatory systems. In: Piiper J, editor. Respiratory function in birds, adult and embryonic. Berlin: Springer; 1978. p. 260–73.

Duncker HR. General morphological principles of amniotic lungs. In: Piiper J, editor. Respiratory function in birds, adult and embryonic. Heidelberg: Springer; 1979. p. 1–15.

Duncker HR. Vertebrate lungs: structure, topography and mechanics: a comparative perspective of the progressive integration of respiratory system, locomotor apparatus and ontogenetic development. Respir Physiol Neurobiol. 2004;144:111–24.

Duncker HR, Guntert M. The quantitative design of the avian respiratory system: from hummingbird to the mute swan. In: Nachtigall W, editor. BIONA Report No. 3. Stuttgart: Gustav-Fischer; 1985a. p. 361–78.

Duncker HR, Guntert M. Morphometric analysis of the avian respiratory system. In: Duncker HR, Fleischer G, editors. Vertebrate morphology. Stuttgart: Gustav-Fischer; 1985b. p. 383–7.

Egevang C, Stenhouse IJ, Phillips RA, Petersen A, Fox JW, Silk JRD. Tracking of Arctic terns *Sterna paradisaea* reveals longest animal migration. Proc Natl Acad Sci USA. 2010;107:2078–81.

Farner DS. Some glimpses of comparative avian physiology. Fed Proc. 1970;29:1649–63.

Fedde MR. The structure and gas flow pattern in the avian lung. Poult Sci. 1980;59:2642–53.

Fedde MR. Relationship of structure and function of the avian respiratory system to disease susceptibility. Poult Sci. 1998;77:1130–8.

Fletcher OJ. Pathology of the avian respiratory system. Poult Sci. 1980;59:2666–79.

Foot NJ, Orgeig S, Daniels CB. The evolution of a physiological system: the pulmonary surfactant system in diving mammals. Respir Physiol Neurobiol. 2006;154:118–38.

Fujiwara T, Adams FH, Nozaki M, Dermer GB. Pulmonary surfactant phospholipids from Turkey lung: comparison with rabbit lung. Am J Phys. 1970;218:218–25.

Fung YC. Biomechanics: mechanical properties of living tissues. 2nd ed. Berlin: Springer; 1993.

Gehr P, Mwangi DK, Amman A, Maloiy GMO, Taylor CR, Weibel ER. Design of the mammalian respiratory system. V. Scaling morphometric diffusing capacity to body mass: wild and domestic animals. Respir Physiol. 1981;44:41–86.

Gehr P, Sehovic S, Burri PH, Classen H, Weibel ER. The lung of shrews: morphometric estimation of diffusion capacity. Respir Physiol. 1980;44:61–86.

Geiser F. Ontogeny and phylogeny of endothermy and torpor in mammals and birds. Comp Biochem Physiol A Mol Integr Physiol. 2008;150:176–80.

Gier HT. The air sacs of the loon. Auk. 1952;69:40–9.

Grigg GC, Beard LA, Augee ML. The evolution of endothermy and its diversity in mammals and birds. Physiol Biochem Zool. 2004;77:982–97.

Groebbels FDV. Bau, Funktion, Lebenserscheinung, Einpassung, vol. 1. Berlin: Gebriider Borntraeger; 1932.

Grubb BR. Cardiac output and stroke volume in exercising ducks and pigeons. J Appl Physiol. 1982;53:203–11.

Gruson ES. Checklist of birds of the world. London: Collins; 1976.

Hawkes LA, Balachancran S, Batbayar N, Butler PJ, Frappell PB, Milsom WK, Tseveenmyadag S, Newman SH, et al. The trans-Himalayan flights of bar-headed geese (*Anser indicus*). Proc Natl Acad Sci USA. 2011;108:9516–9.

Henk WG, Haldman JT. Microanatomy of the lung of the bowhead whale *Balaena mysticetus*. Anat Rec. 1990;226:187–97.

Hlastala MP, Berger AJ. Physiology of respiration. 1st ed. New York: Oxford University Press; 1996.

Hlastala MP, Berger AJ. Physiology of respiration. 2nd ed. New York: Oxford University Press; 2001.

Hochachka PW. Comparative intermediary metabolism. In: Prosser CL, editor. Comparative animal physiology. 3rd ed. Philadelphia: Saunders; 1973. p. 212–78.

Holle JP, Heisler N, Scheid P. Blood flow distribution in the bird lung and its controls by respiratory gases. Am J Phys. 1978;234:R146–54.

Hsia CW, Schimtz A, Lambertz M, Perry SF, Maina JN. Evolution of air breathing: oxygen homeostatsis and the transitions from water- to-land and sky. Compr Physiol. 2013;3:849–915.

Hughes GM, Weibel ER. Morphometry of fish lungs. In: Hughes GM, editor. Respiration of amphibious vertebrates. London: Academic; 1976. p. 213–32.

Hunter P. The nature of flight: the molecules and mechanics of flight in animals. EMBO Rep. 2007;8:813. doi:10.1038/sj.embor.7401050.

Jammes Y, Bouverot P. Direct PCO_2 measurement in the dorsobronchial gas of wake Pekin ducks: evidence for a physiological role of the neopulmo in respiratory gas exchanges. Comp Biochem Physiol. 1975;52A:635–7.

Jimoh SA, Maina JN. Immuno-localization of type-IV collagen in the blood-gas barrier and the epithelial cell connections of the avian lung. Biol Lett. 2013;9 doi:10.1098/rsbl.2012.0951.

Jones JH, Effmann EL, Schmidt-Nielsen K. Lung volume changes during respiration in ducks. Respir Physiol. 1985;59:15–25.

Juillet A. Recherches anatomiques, embryologiques, histologiques et comparatives sur le poumon des oiseaux. Arch Zool Exp Gen. 1912;IX:207–371.

Karasov WH, Phan D, Diamond JM, Carpenter FL. Food passage and intestinal nutrient absorption in hummingbirds. Auk. 1986;103:453–64.

Kardong KV. Vertebrates: comparative anatomy, function, and evolution. 5th ed. New York: McGraw-Hill; 2009.

Kellog RH. Laws of physics pertaining to gas exchange. In: Farhi LE, Tenney SM, editors. Handbook of physiology, section 3, the respiratory system, vol. IV: gas exchange. Bethesda, MD: American Physiological Society; 1987. P. 13–21.

King AS. Structural and functional aspects of the avian lung and its air sacs. Intern Rev Gen Exp Zool. 1966;2:171–267.

King JR. Seasonal allocation of time and energy resources in birds. In: Paynter RA, editor. Avian energetics. Cambridge, MA: Nuttall Ornithological Club; 1974. p. 4–85.

King AS. Systema respiratorium. In: Baumel JJ, King AS, Lucas AM, Breazile JE, Evans HE, editors. Nomina anatomica avium. London: Academic; 1979. p. 227–65.

King AS, Atherton JD. The identity of the air sacs of the Turkey (*Melleagris gallopavo*). Acta Anat. 1970;77:78–91.

King AS, King DZ. Avian morphology. In: King AS, McLelland J, editors. Form and function in birds, vol. 1. London: Academic; 1979. p. 1–38.

King AS, Molony V. The anatomy of respiration. In: Bell DF, Freeman BM, editors. Physiology and biochemistry of the domestic fowl, vol. 1. London: Academic; 1971. p. 347–84.

Klika E, Scheuermann DW, De Groodt-Lassel MHA, Bazantova I, Switka A. Anchoring and support system of pulmonary gas exchange tissue in four bird species. Acta Anat. 1997;159:30–41.

Köppen U, Yakoview A, Barth R, Kaatz M, Berthold P. Seasonal migrations of four individual bar-headed geese, *Anser indicus* from Kyrgyzstan followed by satellite telemetry. J Ornithol. 2010;151:703–12.

Koteja P. The evolution of concepts on the evolution of endothermy in birds and mammals. Physiol Biochem Zool. 2004;77:1043–50.

Krebs JR, Harvey PH. Busy doing nothing—efficiently. Nature. 1986;320:18–9.

Laplace PS. A philosophical essay on probabilities (translated from the 6th French edition by Truscott FW, Emory FL). New York: Wiley; 1902.

Lasiewski RC. The energetics of migrating hummingbirds. Condor. 1962;64:324.

Lasiewski RC, Dawson WR. A re-examination of the relation between standard metabolic rate and body weight in birds. Condor. 1967;69:13–23.

Laybourne RC. Collision between a vulture and an aircraft at an altitude of 37,000 ft. Wilson Bull. 1974;86:461–2.

Li J. Processing, stability and interactions of lung surfactant protein C. Stockholm: Karolinska Institutet University Press; 2005.

Locy WA, Larsell O. The embryology of the bird's lung based on observations of the bronchial tree. Part I. Am J Anat. 1916a;19:447–504.

Locy WA, Larsell O. The embryology of the bird's lung based on observations of the domestic fowl. Part II. Am J Anat. 1916b;20:1–44.

Macklem P, Bouverot P, Scheid P. Measurement of the distensibility of the parabronchi in duck lungs. Respir Physiol. 1979;33:23–35.

Magnussen H, Willmer H, Scheid P. Gas exchange in the air sacs: contribution to respiratory gas exchange in ducks. Respir Physiol. 1976;26:129–46.

Maina JN. Morphometrics of the avian lung. 3. The structural design of the passerine lung. Respir Physiol. 1984;55:291–309.

Maina JN. Scanning electron microscopic study of the spatial organization of the air- and blood conducting components of the avian lung *(Gallus gallus domesticus)*. Anat Rec. 1988;222:145–53.

Maina JN. The morphometry of the avian lung. In: King AS, McLelland J, editors. Form and function in birds, vol. 4. London: Academic; 1989. p. 307–68.

Maina JN. Morphometries of the avian lung: the structural-functional correlations in the design of lungs of birds. Comp Biochem Physiol. 1993;105:397–410.

Maina JN. The gas exchangers: structure, function, and evolution of the respiratory processes. Berlin: Springer; 1998.

Maina JN. What it takes to fly: the novel respiratory structural and functional adaptations in birds and bats. J Exp Biol. 2000a;203:3045–64.

Maina JN. Is the sheet-flow design a 'frozen core' (a Bauplan) of the gas exchangers? Comparative functional morphology of the respiratory microvascular systems: illustration of the geometry and rationalization of the fractal properties. Comp Biochem Physiol. 2000b;126A:491–515.

Maina JN. Some recent advances of the study and understanding of the functional design of the avian lung: morphological and morphometric perspectives. Biol Rev. 2002;77:97–152.

Maina JN. A systematic study of the development of the airway (bronchial) system of the avian lung from days 3 to 26 of embryogenesis: a transmission electron microscopic study on the domestic fowl, *Gallus gallus* variant *domesticus*. Tissue Cell. 2003a;35:375–91.

Maina JN. Developmental dynamics of the bronchial (airway)- and air sac systems of the avian respiratory system from days 3 to 26 of life: a scanning electron microscopic study of the domestic fowl, *Gallus gallus* variant *domesticus*. Anat Embryol. 2003b;1207:119–34.

Maina JN. The design of the lung-air sac system of birds: development, structure, and function. Heidelberg: Springer; 2005.

Maina JN. Development, structure, and function of a novel respiratory organ, the lung-air sac system of birds: to go where no other vertebrate has gone. Biol Rev. 2006;81:545–79.

Maina JN. Minutialization at its extreme best! The underpinnings of the remarkable strengths of the air and the blood capillaries of the avian lung: a conundrum. Respir Physiol Neurobiol. 2007a;159:141–5.

Maina JN. Spectacularly robust! Tensegrity principle explains the mechanical strength of the avian lung. Respir Physiol Neurobiol. 2007b;155:1–10.

Maina JN. Structural and biomechanical properties of the exchange tissue of the avian lung. Anat Rec. 2015a;298:1673–88.

Maina JN. The design of the avian respiratory system: development, morphology and function. J Ornithol. 2015b;156:41–63.

Maina JN. Morphological and morphometric properties of the blood-gas barrier: comparative perspectives. In: Makanya AN, editor. The vertebrate blood-gas barrier in health and disease. New York: Springer; 2015c. p. 15–38.

Maina JN. Pivotal debates and controversies on the structure and function of the avian respiratory system: setting the record straight. Biol Rev. 2016a; doi:10.1111/brv.12292.

Maina JN. Critical appraisal of some factors pertinent to the functional designs of the gas exchangers. Cell Tissue Res. 2016b; doi:10.1007/s00441-016-2549-9.

Maina JN, Abdalla MA, King AS. Light microscopic morphometry of the lungs of 19 avian species. Acta Anat. 1982;112:264–70.

Maina JN, Howard CV, Scales L. Length densities and maximum diameter distribution of the air capillaries of the paleopulmo and neopulmo region of the avian lung. Acta Stereol. 1983;2:101–7.

Maina JN, Jimoh SA. Structural failures of the blood-gas barrier and the epithelial-epithelial cell connections in the different vascular regions of the lung of the domestic fowl, *Gallus gallus* variant *domesticus*, at rest and during exercise. Biol Open. 2013;2:267–76.

Maina JN, Jimoh SA, Hosie M. Implicit mechanistic role of the collagen-, smooth muscle, and elastic tissue components in strengthening the air- and the blood capillaries of the avian lung. J Anat. 2010;217:597–608.

Maina JN, King AS. The thickness of the avian blood-gas barrier: qualitative and quantitative observations. J Anat. 1982;134:553–62.

Maina JN, King AS. The structural functional correlation in the design of the bat lung. A morphometric study. J Exp Biol. 1984;111:43–63.

Maina JN, King AS. A morphometric study of the lung of a Humboldt penguin *(Spheniscus humboldti)*. Zentralb Vet Med C Anat Histol Embryol. 1987;16:293–7.

Maina JN, King AS. The lung of the emu, *Dromaius novaehollandiae*: a microscopic and morphometric study. J Anat. 1989;163:67–74.

Maina JN, King AS, Settle G. An allometric study of the pulmonary morphometric parameters in birds, with mammalian comparison. Phil Trans R Soc. 1989;326B:1–57.

Maina JN, Nathaniel C. A qualitative and quantitative study of the lung of an ostrich, *Struthio camelus*. J Exp Biol. 2001;204:2313–30.

Maina JN, Sikiru AJ. Study of stress induced failure of the blood-gas barrier and the epithelial-epithelial cells connections of the lung of the domestic fowl, *Gallus gallus* variant *domesticus* after vascular perfusion. Biomed Eng Comput Biol. 2013;5:77–88.

Maina JN, Thomas SP, Dallas DM. A morphometric study of bats of different size: correlations between structure and function of the chiropteran lung. Phil Trans R Soc Lond B. 1991;333:31–50.

Maina JN, van Gils P. Morphometric characterization of the airway and vascular systems of the lung of the domestic pig, *Sus scrofa*: comparison of the airway, arterial, and venous systems. Comp Biochem Physiol. 2001;130A:781–98.

Maina JN, West JB. Thin and strong! The bioengineering dilemma in the structural and functional design of the blood-gas barrier. Physiol Rev. 2005;85:811–44.

Maina JN, Woodward JD. Three-dimensional serial section computer reconstruction of the arrangement of the structural components of the parabronchus of the ostrich, *Struthio camelus* lung. Anat Rec. 2009;291:1685–98.

Makanya AN, Djonov V. Development and spatial organization of the air conduits in the lung of the domestic fowl, *Gallus gallus* variant *domesticus*. Microsc Res Tech. 2008;71:689–702.

Makanya AN, El-Darawish Y, Kavoi BM, Djonov V. Spatial and functional relationships between air conduits and blood capillaries in the pulmonary gas exchange tissue of adult and developing chickens. Microsc Res Tech. 2011;74:159–69.

Makanya AN, Kavoi BM, Djonov V. Three dimensional structure and disposition of the air conducting and gas exchange conduits of the avian lung: the domestic duck (*Cairina moschata*). ISRN Anat. 2014. Doi:org/10.1155/2014/621982.

Marshall AJ. Cited in King and King 1979. Avian morphology. In: King AS, McLelland J, editors. Form and function in birds, vol. 1. London: Academic; 1962. P. 1–38.

Martinez del Rio C. Dietary, phylogenetic and ecological correlates of intestinal sucrase and maltase activity in birds. Physiol Zool. 1990;63:987–1011.

Mathieu-Costello O, Szewczak JM, Logermann RB, Agey PJ. Geometry of blood-tissue exchange in bat flight muscle compared with bat hindlimb and rat soleus muscle. Am J Phys. 1992;262:R955–65.

McLelland J. Anatomy of the lungs and air sacs. In: King AS, McLelland J, editors. Form and function in birds, vol. 4. London: Academic; 1989. p. 221–79.

Meban C. Thicknesses of the air-blood barriers in vertebrate lungs. J Anat. 1980;131:299–307.

Miller DN, Bondurant S. Surface characteristics of vertebrate lung extracts. J Appl Physiol. 1961;16:1075–7.

Miura T, Hartmann D, Kinboshi M, Komada M, Ishibashi M, Shiota K. The cyst-branch difference in developing chick lung results from a different morphogen diffusion coefficient. Mech Dev. 2009;126:160–72.

Morony JJ, Bock WJ, Farrand J. Reference list of the birds of the world. New York: American Museum of Natural History, Department of Ornithology; 1975.

Moura RS, Coutinho-Borges JP, Pacheco AP, daMota PO, Correia-Pinto J. FGF signaling pathway in the developing chick lung: expression and inhibition sites. PLoS One. 2011;6(3):e17660. doi:10.1371/journal.pone.0017660.

Moura RS, Silva-Gonçalves C, Vaz-Cunha P. Expression analysis of Shh signaling members in early stages of chick lung development. Histochem Cell Biol. 2016; doi:10.1007/s00418-016-1448-1.

Müller B. The air sacs of the pigeon. Smithson misc Colls. 1908;50:365–414.

Nudds RL, Bryant DM. The energy cost of short flights in birds. J Exp Biol. 2000;203:1561–82.

Ostrom JH. The origin of birds. Annu Rev Earth Planet Sci. 1975;3:55–77.

Pattle RE. Lung surfactant and lung lining in birds. In: Piiper J, editor. Respiratory function in birds, adult and embryonic. Berlin: Springer; 1978. p. 23–32.

Perry SF. Quantitative anatomy of the lungs of the red-eared turtle, *Pseudemys script elegans*. Respir Physiol. 1978;35:245–62.

Perry SF. Mainstreams in the evolution of vertebrate respiratory structures. In: King AS, McLelland J, editors. Form and function in birds, vol. V. London: Academic; 1989a. p. 1–67.

Perry SF. Structure and function of the reptilian respiratory system. In: Wood SC, editor. Comparative pulmonary physiology: current concepts. New York: Marcel Dekker; 1989b. p. 193–236.

Perry SF. Gas exchange strategies in reptiles and the origin of the avian lung. In: Wood SC, Weber RE, Hargens AR, Millard RW, editors. Physiological adaptations in vertebrates: respiration, circulation, and metabolism. New York: Marcel Dekker; 1992. p. 149–67.

Piiper J. Origin of carbon dioxide in caudal air sacs of birds. In: Piiper J, editor. Respiratory function in birds, adult and embryonic. Berlin: Springer; 1978. p. 221–48.

Piiper J, Scheid P. Gas exchange in the avian lung: model and experimental evidence. In: Bolis L, Schmidt-Nielsen K, Maddrell SHP, editors. Comparative physiology. Amsterdam: Elsevier; 1973. p. 161–85.

Pough FH, Heiser JB, McFarland WN. Vertebrate life. 3rd ed. New York: Macmillan; 1989.

Powell FL. Respiration. In: Abs M, editor. Physiology and behaviours of the pigeon. New York: Academic; 1983. p. 73–95.

Powell FL. Respiration. In: Whittow GC, editor. Sturkie's Avian Physiology. 5th ed. San Diego, CA: Academic; 2000. p. 233–64.

Powell FL, Hastings RH, Mazzone RW. Pulmonary vascular resistance during unilateral pulmonary artery occlusion in ducks. Am J Phys. 1985;249: R34–43.

Powell FL, Scheid P. Physiology of gas exchange in the avian respiratory system. In: King AS, McLelland J, editors. Form and function of the avian lung, vol. 4. London: Academic; 1989. p. 393–437.

Powers DR, Naggy KA. Field metabolic rate and food consumption by free-living Anna's hummingbirds (*Calypte anna*). Physiol Zool. 1988;61:500–6.

Radu C, Radu L. Le dispositif vasculaire du poumon chez les oiseaux domestiques (coq, dindon, oie, canard). Rev Med Vet. 1971;122:1219–26.

Rawal UM. Nerves in the avian air sacs. Pavo. 1976;14:57–60.

Romanoff AL. The avian embryo. New York: Macmillan; 1960.

Scheid P. Mechanisms of gas exchange in bird lungs. Rev Physiol Biochem Pharmacol. 1979;86:137–86.

Scheid P. The use of models in physiological studies. In: Feder ME, Bennett AF, Burggrenn WW, Huey RB, editors. New direction in ecological physiology. Cambridge: Cambridge University Press; 1987. p. 275–88.

Scheid P, Piiper J. Cross-currrent gas exchange in the avian lungs: effects of reversed parabronchial air flow in ducks. Respir Physiol. 1972;16:304–12.

Scheid P, Piiper J. Respiratory mechanics and air flow in birds. In: King AS, McLelland J, editors. Form and function in birds, vol. 4. London: Academic; 1989. p. 364–91.

Scheuermann DW, Klika E, Lasseel DG, Bazantova I, Switka A. An electron microscopic study of the parabronchial epithelium in the mature lung of four bird species. Anat Rec. 1997;249:213–25.

Scheuermann DW, Klika E, Lasseel DG, Bazantova I, Switka A. Lamellar inclusions and trilaminar substance in the parabronchial epithelium of the quail (*Coturnix coturnix*). Ann Anat. 2000;182:221–33.

Scott GR, Hawkes LA, Frappel PB, Butler PJ, Bishop CM, Milsom WK. How bar-headed geese fly over the Himalayas. Physiology. 2014;30:107–15.

Seeherman HJ, Taylor CR, Maloiy GMO, Armstrong RB. Design of the mammalian respiratory system. II. Measuring maximum aerobic capacity. Respir Physiol. 1981;44:11–23.

Spragg RS. Surfactant for acute lung injury. Am J Respir Cell Mol Biol. 2007;37:377–8.

Stanislaus M. Untersuchungen an der Kolibrilunge. Z Morphol Tiere. 1937;33:261–89.

Suarez RK. Hummingbird flight: sustaining the highest mass-specific metabolic rates among vertebrates. Experientia. 1992;48:565–70.

Suarez RK, Brown GS, Hochachka PW. Mitochondrial respiration in hummingbird muscles. Am J Phys. 1986;251:R537–42.

Suarez RK, Lighton JRB, Brown GS, Mathieu-Costello O. Mitochondrial respiration in hummingbird flight muscles. Proc Natl Acad Sci USA. 1991;87:9207–10.

Thomas SP. The physiology of bat flight. In: Fenton MB, Racey P, Rayner JMV, editors. Recent advances in the study of bats. Cambridge: Cambridge University Press; 1987. p. 75–99.

Tobalske BW, Hedrik TL, Dial KP, Biewener AA. Comparative power curves in bird flight. Nature. 2003;421:363–6.

Trampel DW, Fletcher OJ. Ring-stabilizing technique for collection of avian air sacs. Am J Vet Res. 1980;14:1730–4.

Tucker V. Respiratory physiology of house sparrows in relation to high altitude flight. J Exp Biol. 1968;48:55–66.

Tucker VA. Energetics of natural avian flight. In: Paynter RA, editor. Avian energetics. Cambridge, MA: Nuttall Ornithological Club; 1974. p. 298–333.

Tzschentke B, Rumpf M. Embryonic development of endothermy. Respir Physiol Neurobiol. 2011;178:97–107.

Videler JJ. Avian flight. Oxford: Oxford University Press; 2006.

Vos HJ. Über das Fehlen der rekurrenten Bronchien beim Pinguin und bei den Reptilien. Zool Anz. 1937;117:176–81.

Walsh C, McLelland J. The ultrastructure of the avian extrapulmonary respiratory epithelium. Acta Anat. 1974;89:412–22.

Ward S, Bishop CM, Woakes AJ, Butler PJ. Heart rate and the rate of oxygen consumption of flying and walking barnacle geese (*Branta leucopsis*) and bar-headed geese (*Anser indicus*). J Exp Biol. 2002;205:47–3356.

Watson RR, Fu Z, West JB. Morphometry of the extremely thin pulmonary blood-gas barrier in the chicken lung. Am J Physiol Lung Cell Mol Physiol. 2007;292:L769–77.

Watson RR, Fu Z, West JB. Minimal distensibility of pulmonary capillaries in avian lungs compared with mammalian lungs. Respir Physiol Neurobiol. 2008;160:208–14.

Weibel ER. Morphometry of the human lung. Berlin: Springer; 1963.

Weibel ER. Morphological basis of the alveolar-capillary gas exchange. Physiol Rev. 1973;53:419–95.

Weibel ER. The pathways for oxygen. Harvard, MA: Harvard University Press; 1984.

Weibel ER, Knight BW. A morphometric study on the thickness of the pulmonary air-blood barrier. J Cell Biol. 1964;21:367–84.

Wells DJ. Muscle performance in hovering hummingbirds. J Exp Biol. 1993;178:39–57.

Welty JC. The life of birds. 2nd ed. Philadelphia: Saunders; 1979.
West NH, Bamford OS, Jones DR. A scanning electron microscope study of the microvasculature of the avian lung. Cell Tissue Res. 1977;176:553–64.
West JB, Fu Z, Deerinck TJ, Mackey MR, Obayashi JT, Ellsman MH. Structure-function studies of blood- and air capillaries in chicken lung using 3-D electron microscopy. Respir Physiol Neurobiol. 2010;170:202–9.
West JB, Watson RR, Fu Z. The honeycomb-like structure of the bird lung allows a uniquely thin blood-gas barrier. Respir Physiol Neurobiol. 2006;152:115–8.
West JB, Watson RR, Fu Z. Major differences in the pulmonary circulation between birds and mammals. Respir Physiol Neurobiol. 2007a;157:382–90.
West JB, Watson RR, Fu Z. Support of pulmonary capillaries in avian lung: letter to the editor. Respir Physiol Neurobiol. 2007b;159:146.
West JB, Watson RR, Fu Z. The human lung: did evolution get it wrong? Eur Respir J. 2007c;29:11–7.
West B, Zhou BW. Did chickens go North? New evidence for domestication. J Archeol Sci. 1988;15:515–33.
Wetherbee DK. Air sacs in the English sparrow. Auk. 1951;68:242–4.
Wigglesworth VB. The principles of insect physiology. 7th ed. London: Chapman & Hall; 1972.
Woodward JD, Maina JN. A 3-D digital reconstruction of the components of the gas exchange tissue of the lung of the Muscovy duck, *Cairina moschata*. J Anat. 2005;206:477–92.
Woodward JD, Maina JN. Study of the structure of the air- and the blood capillaries of the gas exchange tissue of the avian lung by serial section three-dimensional reconstruction. J Microsc. 2008;230:84–93.
Yalden DW, Morris PA. The lives of bats. New York: The New York Times; 1975.
Yapp WB. The life and organization of birds. London: Edward Arnold; 1970.
Zuo YY, Veldhuizen RA, Neumann AW, Ptersen NO, Possmeyer F. Current perspectives in pulmonary surfactant inhibition, enhancement and evaluation. Biochim Biophys Acta. 2008;1778:1947–77.

Structure and Function of the Shell and the Chorioallantoic Membrane of the Avian Egg: Embryonic Respiration

9

John N. Maina

Abstract

The evolution of a cleidoic (self-supporting) egg in the amniotes was pivotal to the transition of animal life from water to land: embryonic development could occur without the direct presence of water. For birds, remarkable adaptive radiation occurred after the achievement of volancy. It has culminated in ~10,000 species. Among the air-breathing vertebrates, the taxon Aves is the most speciose. In the developing avian egg, exchange of respiratory gases (O_2 and CO_2) and water vapour occurs entirely by passive diffusion across the shell and the chorioallantoic membrane (CAM) along existing concentration gradients. A multifunctional structure, the shell displays compromise design. On one hand it has to be thin enough to allow optimal flux of O_2 and CO_2 and loss of water which is generated by metabolism of the developing embryo. On the other hand, the shell must not be too thin to admit pathogens and injurious substances and susceptible to failure (breaking). Also, the shell must not be too strong for the chick to be unable to break out of at hatching. The thickness of the shell and the numbers, shapes and sizes of the pores determine the hatchability of the egg and probably the incubation period of the egg. The CAM of the developing chick embryo is analogous to the placenta of the viviparous animals. While the number of pores may be fixed at the formation of the eggshell, during incubation, the surface area and vascularization of the CAM increase considerably. Here, the structure and function of the avian eggshell and that of the CAM are succinctly outlined. Particular consideration is given to the recent observations made using X-ray microcomputer tomography, a highly instructive technique for studying biological structures. The impact of the evolution of the amniotic egg on the diversification of the animal life especially that of birds, is highlighted.

Keywords

Egg • Shell • Birds • Chorioallantoic membrane • Structure • Function

Contents

J.N. Maina (✉)
Department of Zoology, University of Johannesburg, Auckland Park Campus, Kingsway, Johannesburg 2006, South Africa
e-mail: jmaina@uj.ac.za

J.N. Maina (ed.), *The Biology of the Avian Respiratory System*, DOI 10.1007/978-3-319-44153-5_9

9.1 Introduction

Between 300 and 400 million years ago, a cleidoic (from Greek word kleistos or closed) egg, a self-supporting entity that could exist out of water, formed (Schmalhausen 1968; Luckett 1976; Szarski 1968 1976; Pough et al. 1989; Little 1983, 1989, 1990). Embryonic development occurs because the egg contains nutriments (minerals and proteins) and water (Mao et al. 2007). The evolution of the amniotic egg was a pivotal event in the transition of animal life from water (sea) to land and subsequently air which lead to remarkable diversification of animal life (Randall et al. 1981; Carroll 1988; Dejours 1988; Mao et al. 2007; Browder et al. 1991; Packard and Seymour 1997). For the avian egg, during incubation, the prenatal period prior to internal pipping when the chick punctures the chorioallantoic membrane (CAM) and the inner shell membrane into the air cell, the embryo acquires vital nutrients to build new tissues and maintain existing ones from the albumen and the yolk that are stored in the egg (Vleck et al. 1980; Mortola and Cooney 2008; Uni et al. 2012; Mueller et al. 2015). Intervening between the internal and the external environments, three extraembryonic membranes, namely, the amnion, the chorion and the allantois, surround the embryo (Lusimbo et al. 2000; Makanya et al. 2016). The allantoic membrane fuses with the chorion to form the CAM which is highly vascularized. It transports calcium from the eggshell with which it borders, delivering the mineral to the embryo through the circulatory system mainly for bone formation. The CAM performs other important functions such as gas exchange, acid–base balance and water and electrolyte reabsorption from the allantoic cavity (Ar et al. 1974; Rahn et al. 1979; Liao et al. 2013; Mueller et al. 2015).

The largest avian egg that has ever formed is that of the giant elephant bird (*Aepyornis maximus*) of Madagascar that became extinct within historical times (Flacourt 1658; Davies 2003). The egg weighed ~10 kg. Among the extant birds, weighing ~0.8 gram, the egg of the bee hummingbird (*Mellisuga helenae*) is the smallest, while the largest one which weighs as much as 2 kg is that of the ostrich (*Struthio camelus*) (Deeming et al. 1993; Şahan et al. 2003; Mueller et al. 2015). However, constituting only 1.5% of its adult body mass, among birds, the ostrich egg is one of the smallest (Badley 1997). Forming the physical barrier between the external/nest aerial environment and that of the contents in the egg, the shell is formed utilizing the highly economical dome principle (Fuller 1961; Fuller and Applewhite 1975; Fung 1993): location of structural materials, where internal supporting columns are avoided, confers external support to the egg.

The calcareous eggshell that is produced by all birds and most reptiles is a complex bioceramic material that comprises ~1–3.5% protein organic matrix infused with ~95% calcium carbonate (Baker and Balch 1962; Nys et al. 1999, 2004; Arias et al. 1993; Hunton 1995; Hincke et al. 2012; Rossi et al. 2013; Liao et al. 2013). The avian eggshell was termed 'a very special creation' by Devine (1982) and 'a marvel of nature's engineering' by Tyler and Simkiss (1959), Taylor (1970), Kaplan and Siegesmund (1973) and Hassan and Aigbodion (2013): it exhibits compromise bioengineering. On one hand, while a thin shell promotes gas exchange and water loss (McDaniel et al. 1979; Wilson 1991), such a shell is susceptible to failure (breaking) (Bain 1991; Khatkar et al. 1997) and the embryo is more susceptible to attack by pathogens and toxic substances (Sauter and Petersen 1974). On the other hand, too thick an eggshell will suffocate the embryo from lack of O_2, kill it from accumulation of CO_2, the embryo will drown from accumulation of metabolically produced water and the chick will die if it cannot break out of the shell at hatching. In the developing avian egg, most of the energy that is needed for embryonic development comes from the fat stores of the yolk: for every gram of fat burned, an almost equal mass of metabolic water is generated (Coleman and Terepka 1972; Freeman and Vince 1974; Saleuddin et al. 1976). If water is not lost in form of water vapour, the relative water content of the egg will increase excessively during incubation (Rahn et al. 1979).

An optimal incubation relative humidity is critical to allow best hatchability (El-Hanoun et al. 2012). Defects in eggshell structure ultimately lead to reduced viability of the embryos and hatchability of the eggs (Peebles and Brake 1987). To meet the conflicting functional needs and therefore optimize function, a compromise must be established between the structural parameters and components of the eggshell. The avian eggshell is an excellent biological entity for studying and understanding the evolution of innovative biological designs where conflicts, constraints, compromises and trade-offs exist.

The avian eggshell is perforated by numerous narrow passageways (pores) that provide the only means of exchange of substances between the external milieu and the developing embryo (Ar et al. 1974; Paganelli et al. 1974; Carey et al. 1980; Rahn et al. 1987) (Plates 9.1a–f and 9.2a–c). Oxygen is acquired from the external environment (atmosphere) while CO_2 and water vapour are discharged back into it. Optimal hatching is an outcome of a delicate balance between structural and functional factors of the eggshell, the prevailing environmental conditions such as humidity and the interventions made by the incubating bird, e.g. rotation of the eggs. In the megapodes, a group of galliform birds that are endemic to Australasia (Jones 1988, Jones et al. 1995; Göth and Vogel 1997), an unusual incubation strategy has evolved where heat for incubation is acquired from external environment rather than from the body of the incubating bird. In, for example, *Alectura lathami*, the surface of the eggshell displays nodes similar to those of the extinct titanosaur dinosaurs (Hechenleitner et al. 2016; Grellet-Tinner et al. 2016) which protect the surface of the egg from chemical etching in the environment (soil mounds) in which eggs are laid. The soils are high in organic acids from putrefying plant matter (Göth and Vogel 1997). A similar adaptive incubation strategy, i.e. acquisition of heat from outside, has been reported in the Cretaceous sauropods which deposited their eggs close to geothermal springs (Hechenleitner et al. 2015).

The shapes, sizes and numbers of pores determine the diffusing capacity, i.e. the conductance, of the eggshell to respiratory gases (O_2 and CO_2) and water vapour (Wangensteen 1972; Rahn et al. 1979; Hoyt et al. 1979; Wangensteen and Weibel 1982). While oxygen consumption (VO_2) increases with embryonic development and the surface area and the vascularity of the CAM also increase, the numbers, sizes and shapes of the eggshell pores are arguably prefixed during the formation of the egg (Hoyt et al. 1979). In the avian egg, the mechanisms of gas exchange differ profoundly from those that occur in the vertebrate lung: no convective processes happen between the external environment and the blood capillaries of the CAM (Wangensteen and Weibel 1982; Lusimbo et al. 2000; Gabrielli and Accili 2010; Yuan et al. 2014; Nowak-Sliwinska et al. 2014; Mîndrilă et al. 2014; Makanya et al. 2016).

The means of gas exchange in the avian eggs was last comprehensively reviewed ~3½ decades ago (Rahn and Paganelli 1981). Ever since, only succinct accounts on the evolution of the amniotic egg and the structure and function of the avian eggshell have appeared. Here, a brief treatise of the functional design of the eggshell and that of the CAM is given. The results acquired more recently with the highly instructive technique of X-ray microcomputer tomography are particularly highlighted.

9.2 Development of the Eggshell

The complex process of the formation of the avian egg has been well-documented (Gilbert 1967; Fujii 1974; Talbot and Tyler 1974; Creger et al. 1976; Stemberger et al. 1977; King and McLelland 1984; Arias et al. 1991, 1993, 1997; Meiri and Pines 2000). The following brief account is based on these studies. The eggshell is formed as the egg passes through the oviduct. The various layers of the shell are deposited sequentially as the egg passes through the different parts of the oviduct. The following events occur: after the fertilization of the ovum in the infundibulum and secretion of albumen in the magnum, the egg enters the isthmus ~2–3 h after ovulation; in the isthmus, the granular

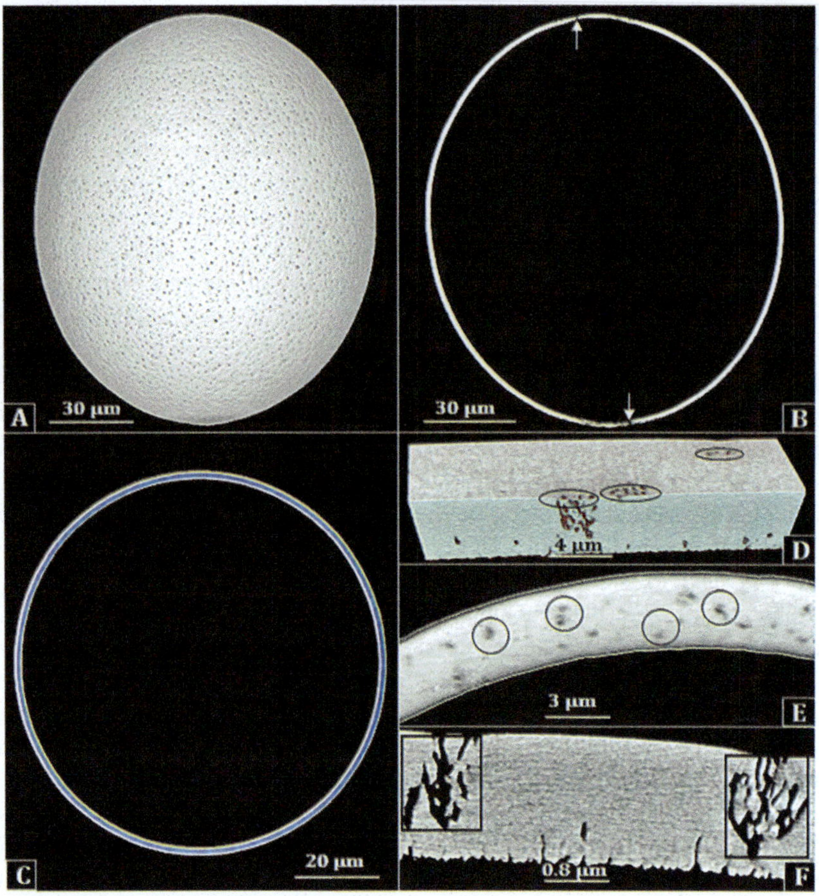

Plate 9.1 All the figures in the plate are from the ostrich (*Struthio camelus*) egg. (**a**) Gross specimen showing pores on the surface of the egg. (**b**, **c**) Long and short cross sections of the eggshell which is rather uniform in thickness. The *arrows* in (**b**) show perforations which were made to drain the contents of the eggs. (**d**, **e**): Close ups of the shell showing pores (*encircled area* in **d** and **e**) and *boxed areas* in (**f**). The branched morphology of some of the pores is seen in (**f**)

cells secrete various components of the shell membranes such as collagen type X; most of the calcium deposition in the eggshell happens in the eggshell gland; ~5–6 g of calcium carbonate is deposited into the eggshell during its formation, a process that takes 17 of the 20 h; calcium deposition starts in the isthmus around the outer shell membrane fibres at organic nodules called mammillary knobs which are spread on the membrane. The deposited crystals enlarge to form the mammillary cone layer and the palisade columns that are covered by a cuticle (Stemberger et al. 1977; Board 1982; Arias et al. 1993; Nys et al. 1999, 2004; Bušs and Keišs 2009; Hincke et al. 2012) (Plates 9.3 and 9.4a–e). The cuticle covers the external surface of the eggshell (Creger et al. 1976; Nys et al. 1999, 2004) (Plate 9.4c, d): it consists of 85–90% of proteins (mainly glycoproteins) and polysaccharides (4%), lipids (3%) and porphyrin pigments (in brown eggs) (Baker and Balch 1962; Hasiak et al. 1970; Wedral et al. 1974; Nys et al. 1999; Rose-Martel et al. 2012). Deposited by the epithelial cells that line the uterus during the last 1½ h before oviposition (Nys et al. 1999), the cuticle is also rich in phosphorus (Wedral et al. 1974; Dennis et al. 1996; Cusack et al. 2003). The cuticle varies in composition,

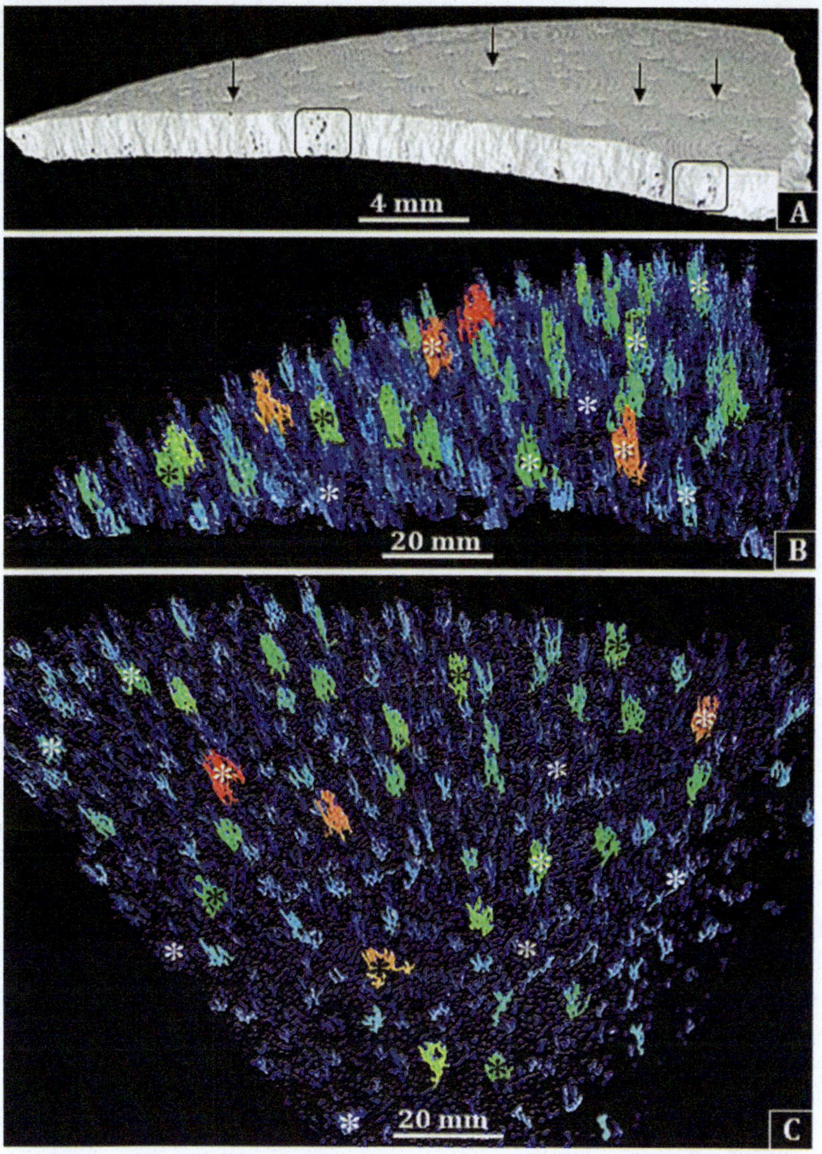

Plate 9.2 All the figures are from the ostrich (*Struthio camelus*) egg. (**a**) A piece of eggshell showing its thickness (*boxed areas*) and the pores on the outer surface (*arrows*). (**b**, **c**) Different views of the same piece of eggshell examined by X-ray microfocus computer tomography. The pores are colour-coded (*asterisks*) to show their differences in shape and size

with the outer part being richer in proteins and the inner one highly loaded with sulfated polysaccharides and phosphates (Rodríguez-Navarro et al. 2013). Moreover, the composition of the cuticle and the thickness and the extent of its coverage are greatly dependent on the age of the laying bird and the freshness of the egg (Rodríguez-Navarro et al. 2013). During the first year of laying, the thickness and the degree of glycosylation of the cuticle decreases with the bird's age; by the end of the laying cycle, it is considerably depleted of lipids, and after the eggs are laid, compositional changes of the cuticle gradually occur, with notable increase in the

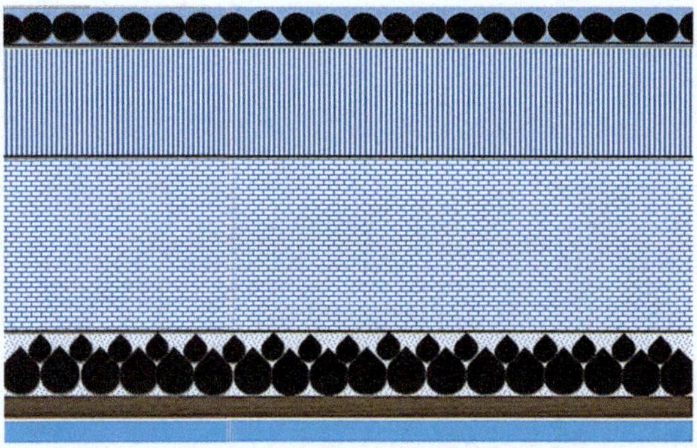

Plate 9.3 A simplified schematic diagram showing the general structural components of the shell of the egg of a bird. The development of the shell wall starts with deposition of calcite crystals on the mammillary cones next to the outer shell membrane

permeability of the eggshell taking place 24 h after laying due to the drying of the cuticle. Rodríguez-Navarro et al. (2013) and Karlson and Lilja (2008) observed that in the shells of the precocial birds (species that are at a relatively underdeveloped state), the number of mammillary knobs was greater, and calcium removal was more extensive compared to the shells of the eggs of the altricial birds (species that are at a relatively more developed state). The shells of the eggs of the precocial birds have high mammillary body density while those of the altricial birds have low mammillary density (Österström and Lilja 2012). This shows adaptive differences in the growth rate and the manner of development in these groups of birds.

The orientation of the calcite crystals determines the strength of the eggshell (Bain 1992; Hincke 1995; Rodriguez-Navarro et al. 2002; Lammie et al. 2006; Solomon 2010; Mazzuco and Bertechini 2014). According to Bain (1992), the structure of the mammillary layer determines the strength of the eggshell while the number of mammillary bodies determines the quantity of calcium available for bone formation in the developing chick embryo (Plate 9.5). The size and orientation of the calcite crystals in the palisade layer sets up the strength of the eggshell (Carnarius et al. 1996; Rodriguez-Navarro et al. 2002; Lammie et al. 2006), with smaller and less-regularly oriented crystals conferring relatively greater strength than larger and highly oriented ones. The numbers of mammillary bodies of the eggshells have been determined by Panheleux et al. (1999), Lunam and Ruiz (2000), Rodriguez-Navarro (2007) and Riley et al. (2014). About 80% of the calcium requirements of the developing chick derive from the mammillary layer before hatching (Ruiz et al. 1998). Variation in the eggshell thickness may affect the amount of calcium available to the developing embryo (Dieckert et al. 1989; Nascimento et al. 1992).

Overall, the structure of the eggshell is greatly affected by a range of factors that include the age of the laying bird, genetics, diet and level of nutrition (Rodriguez-Navarro et al. 2002). Furthermore, these factors affect the arrangement of mineral crystals in the avian eggshell (Rodriguez-Navarro et al. 2002; Fathi et al. 2007; Tumova and Gous 2012; Mazzuco and Bertechini 2014). In the ostrich eggshell, the mammillary knobs are conspicuous, the mammillary cones are set very close to each other and the boundary between the cones and the palisade layer is indistinct (Bronwyn et al. 2016) (Plate 9.4b–e). On ostrich farms in Australia and Southern Africa, over 20% of the chicks suffer trauma to their right limb (the hatching leg) because unlike other avian species

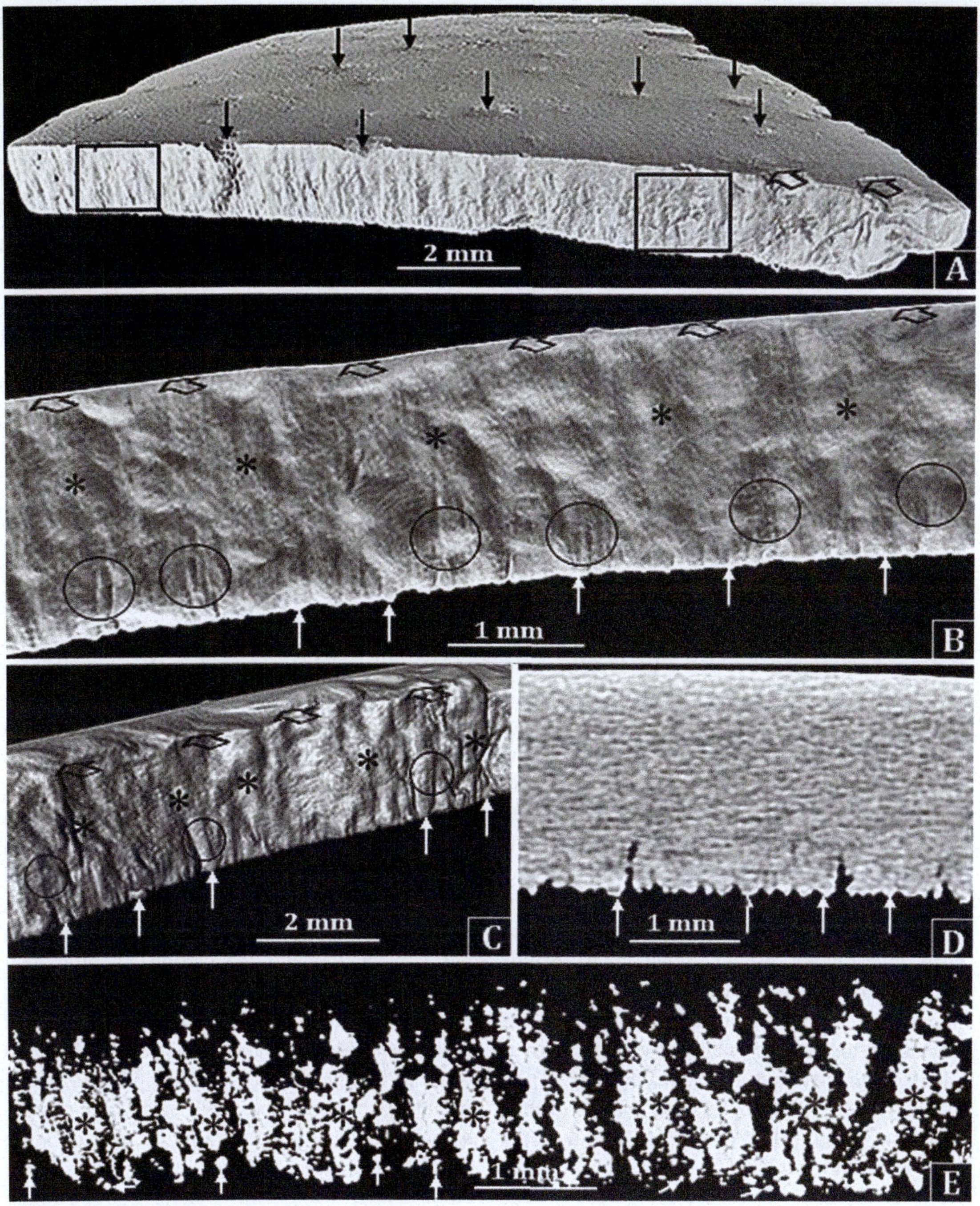

Plate 9.4 All the figures are from the ostrich (*Struthio camelus*) egg and are observations made by X-ray μ-computed tomography. (**a**) A piece of eggshell showing pores on the outer surface (*solid arrows*), the shell wall (*boxed areas*) and the cuticle (*open arrows*). (**b–e**) Structural components of the shell wall. *Solid arrows*, mammillary knobs; *circles*, mammillary cones; *asterisks*, palisade layer; *open arrows*, cuticle

that have an egg-tooth, the ostrich hatches with the assistance of its limbs (Deeming 1997). The compactness of the structural components may explain the strength of the ostrich eggshell (Bronwyn et al. 2016). Quantification and spatial and chemical analysis of the mammillary cones

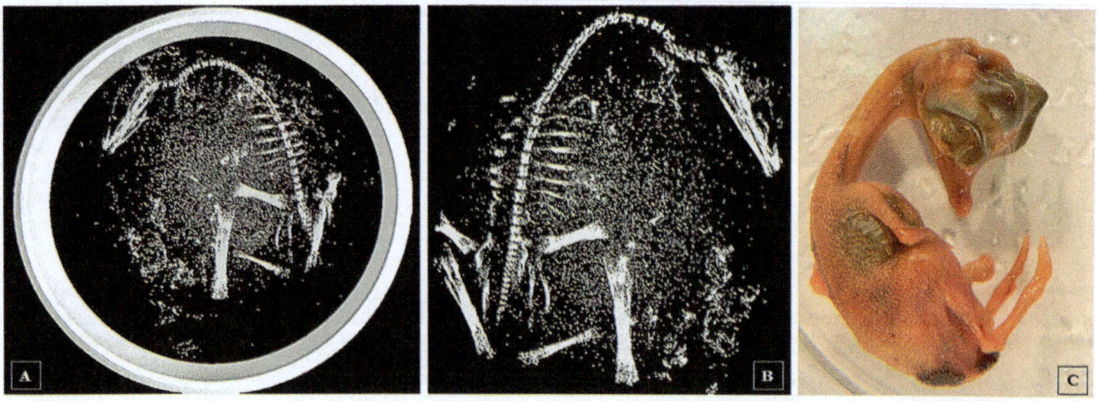

Plate 9.5 Microfocus X-ray computer tomography images (**a**, **b**) of an embryonated ostrich (*Struthio camelus*) egg at day 24 of incubation showing the bones into which calcium is deposited. The calcium is derived from the eggshell (shown in **a**) and transported to the embryo by the chorioallantoic membrane vasculature. The developing embryo is shown in (**c**)

of the ostrich eggshell and the palisade columns are needed to assess the basis of the strength of the ostrich eggshell.

9.3 Structure of the Eggshell

The bird egg has been designated as 'a reproductive cell' (Rees 2013). Varying between the nearly spherical ones of the owls, the oval to elongated ones of the megapodes (e.g. *Leipoa ocellata*) and the sharply pointed ones of the gulls (Paganelli et al. 1974; Grellet-Tinner et al. 2016), avian eggs considerably differ in shape and size. The structure of the avian eggshell has been investigated by techniques that include casting and light (optical), scanning and transmission electron microscopy as well as immunohistochemistry, mass spectrometry, mercury porosimetry and molecular biology (Tyler 1969; Taylor 1970; Tyler and Simkiss 1959; Solomon et al. 1994; La Scala et al. 2000; Rodriguez-Navarro 2007; Mann et al. 2008; Dunn et al. 2009; Jonchère et al. 2010). The main limitation that is inherent in these conventional techniques is that the structural components can only be visualized in two dimensions. What is more, for a technique like scanning electron microscopy, sputter coating the surface of the shell introduces a layer that covers or masks the underlying structures. This may lead to erroneous determination of parameters such as pore number, size and shape (Christensen et al. 1996). Recently, the structure of the eggshell has been studied using X-ray micro-computed tomography (X-ray μCT), a non-invasive technique that is based on visualization of images derived from attenuation of X-rays by materials of different physical densities (Mooney et al. 2012). Using the technique, Rasskin-Gutman et al. (2013), Riley et al. (2014) and Bronwyn et al. (2016), respectively, investigated the structure of a fossilised shell of an egg of the dinosaur, *Megaloolithus siruguei*, the domestic fowl (chicken) (*Gallus gallus* variant *domesticus*) and the ostrich. In biological studies, the technique is highly informative (Fajardo et al. 2007; Li et al. 2008; Westneat et al. 2008; Mizutani and Suzuki 2012; Hoffman et al. 2014), particularly when applied to 'hard' tissues like bone and eggshell. Three-dimensional visualization allows significant correlation between structure and function (Sato et al. 1981; Robb 1982; Wang and Vannier 2001; Ritman 2004; Hao et al. 2008; Hoffman et al. 2014). Advances in hardware and software have increased the speed and the accuracy with which biological materials can be investigated. The non-destructive nature of X-ray μCT makes the technique very attractive as an investigative method. Furthermore, sample preparation is minimal and therefore introduction of artefacts is avoided.

The outermost cuticular covering of the eggshell, the mineralized shell wall and shell membranes (Plates 9.3 and 9.4b, c) form a physical barrier that controls the movement of particles, water and bacteria through the shell pores (Board and Halls 1973; Board 1974, 1975; Board and Tranter 1986; Messens et al. 2005; De Reu et al. 2006; Wellman-Labadie et al. 2008, 2010; Rose-Martel et al. 2012). Some investigators have reported that the ostrich egg lacks a cuticle (Deeming 1995; Sparks 1995; Sparks and Deeming 1996; Sauer 1972) while others such as Christensen et al. (1996), Richards et al. (2000, 2002) and Shanawany and Dingle (1999) have reported it in different forms. Recently, using X-ray μCT, Bronwyn et al. (2016) observed that a cuticle exists on the surface of the ostrich egg (Plate 9.4b, c). In the quail egg, the cuticle was reported to thin during embryonic development (Yoshizaki and Saito 2002), a process that enhances acquisition of O_2 as the embryo develops (Kutchai and Steen 1971). The thinning of the eggshell arises from mobilization of calcium from the shell, especially for the development of bones in the embryo (Johnston and Comar 1955; Simkiss 1961; Ono and Wakasugi 1984). Yoshizaki and Saito (2002) supposed that rotation of eggs during incubation erodes the cuticle from the surface of the egg. For the ostrich eggshell, existence of a cuticular cover is necessary since the eggs are laid and incubated in the hot, desiccating African savannah. The morphological and functional differences of the eggshell that have been reported by various investigators may derive from investigation of eggs that may have been handled and/or treated differently or stored for different durations. Washing with detergent and fumigation may remove or destroy the cuticle.

Among the different species of birds, variations occur in the microstructure of the avian eggshells (Nathusius 1868; Romanoff and Romanoff 1949; Board 1982; Booth 1989; Board and Sparks 1991; Balkan et al. 2006). At low magnification, the pores of the avian eggshell appear like small pits which are randomly distributed on the surface (Plates 9.1a and 9.4a), but at the higher one, the shapes, sizes and distribution of the pores are seen to be highly complicated (Plate 9.6a–f). Single pits turn out to consist of many smaller openings, and some pores which start from the external surface of the shell do not perforate the shell (Plate 9.6c–f), while others pass through the shell wall (Plate 9.6c, e, f). Investigators such as Tyler (1955), Tyler and Simkiss (1959), Rahn et al. (1979), Riley et al. (2014) and Bronwyn et al. (2016) have reported that the sizes and the morphologies of the pores increases with the volume and weight of the egg. In the ostrich, shell thickness increases with egg volume (Plate 9.7a) and body mass (Plate 9.7b) while surface area of the shell increases with the volume of the egg (Plate 9.8). From their shapes and sizes, the pores of the shells of the avian eggs were first classified by Nathusius (1868) and afterwards reclassified by investigators such as Tyler (1956, 1964, 1969), Tyler and Simkiss (1959), Sochava (1969), Sauer (1972), Board and Tullett (1975), Becking (1975), Board et al. (1977), Erben et al. (1979), Sahni et al. (1989), Mikhailov (1991), Riley et al. (2014) and Bronwyn et al. (2016). Ranging from simple trumpet-like ones that resemble a golf tee with its wider end facing the external surface to complex multibranched ones, in the eggshells of ~70 species of birds, as many as eight different types of pores have been identified. Simple pores that perforate the shell without branching mostly occur in the small eggs that have thin shells. In the shells of larger eggs, pores vary in shape and size (Bronwyn et al. 2016). In the shell of the egg of the domestic fowl, Riley et al. (2014) observed that the pores differed in size and shape and whether they perforated the shell is not. Single pores that run straight through the shell without branching or anastomosing with others have been described in the shells of eggs of many species of birds, including the domestic fowl (Tyler and Simkiss 1959; Riley et al. 2014). Imperforate pores were reported in the shells of eggs of birds like the eagles (*Aquila*), buzzard (*Buteo buteo*), osprey (*Pandion haliaetus*) and the sparrowhawk (*Accipiter nisus*) by Board et al.

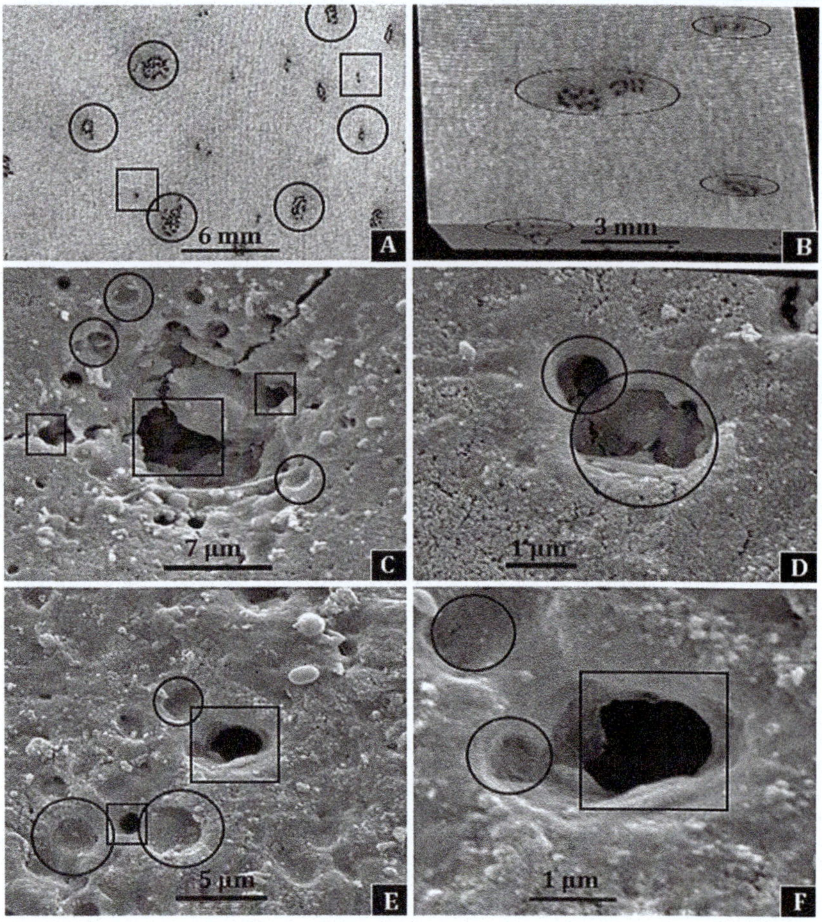

Plate 9.6 All the figures are from the ostrich (*Struthio camelus*) egg. (**a**, **b**) Microfocus X-ray computer tomography pictures of the surface of a piece of the shell showing different views of the pores some of which occur singly (*squares*) and others in clusters (*circles*). (**c**, **d**) Scanning electron micrographs of the surface of the shell showing pores, some of which appear imperforate (*circles*) while others appear to be perforate (*squares*). (**e**, **f**) Close up of perforate pores (*squares*) and inperforate ones (*circles*)

(1977) and in that of the ostrich (Tyler and Simkiss 1959; Bronwyn et al. 2016). Such pores were reported to be lacking in the shells of the eggs of the kestrel (*Falco tinnunculus*), the peregrine falcon (*Falco peregrinus*) and the secretary bird (*Sagittarius serpentarius*) (Board et al. 1977). Although the pores of the ostrich eggshell have been described as multibranched since Nathusius (1868), using plastic casting (Tyler and Simkiss 1959) and X-ray µCT (Bronwyn et al. 2016), it was shown that morphological heterogeneity exists, with multibranched pores preponderating (Bronwyn et al. 2016) (Plates 9.2, 9.9, and 9.10a–i). The multibranched pores interconnect about halfway into the thickness of the shell and closer to the outer shell membrane, and the passageways connect to form fewer trunks (Plates 9.1f, 9.2b, c, 9.9a–c, and 9.10a–e). Pores described as 'reticulated' by Board and Tullett (1975) and Sahni et al. (1989) in the eggshells of the emu (*Dromaius novaehollandiae*) and the cassowary (*Casuarius casuarius*) (Board et al. 1977) were also observed in the ostrich eggshell by Bronwyn et al. (2016) (Plate 9.10f, g). Such pores resembled an open umbrella where a large trunk (passageway) corresponds with the handle of the umbrella and the wires that support the canopy to small passageways that radiate outwards close to the outer surface of the shell (Plate 9.10f, g). Rarer stubby type of pores with few branches were reported in the shell of the egg of the ostrich by Bronwyn et al. (2016) (Plate 9.10h, i). The pores of the shell of the egg of the

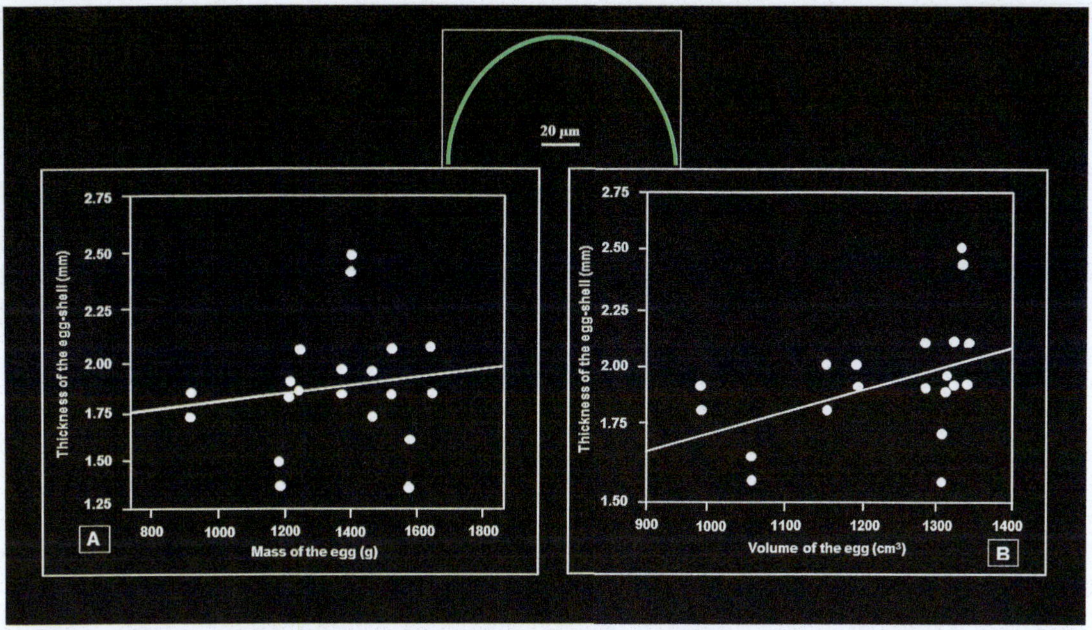

Plate 9.7 Correlation between the thickness of the shell of the egg of the ostrich (*Struthio camelus*) and the mass of the egg (**a**) and that between the thickness of the shell to the volume of the egg (**b**). The insert shows the shell which is notably uniform in thickness

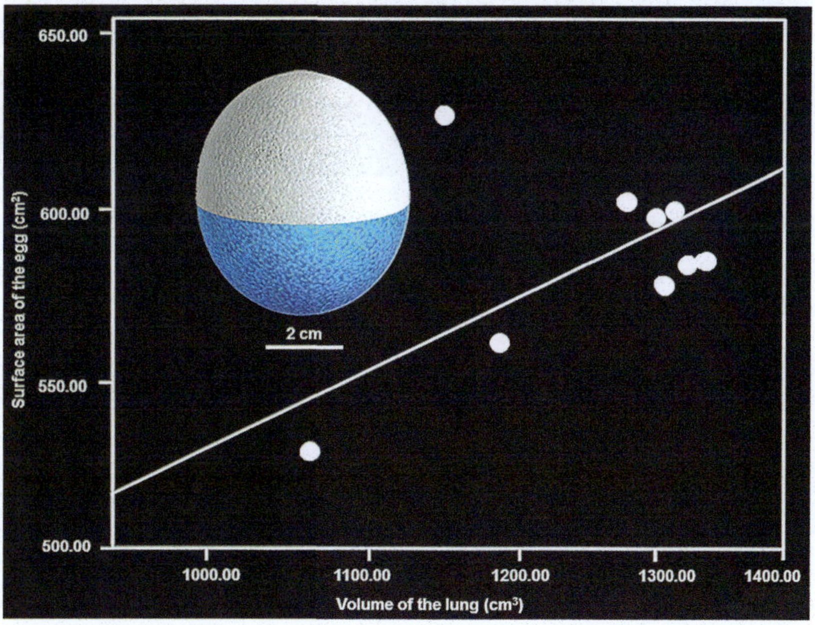

Plate 9.8 Correlation between the surface area of the shell of the ostrich (*Struthio camelus*) egg and the volume of the egg. Insert: Microfocus X-ray computer tomography-generated image of the egg in which the sharp and the blunt halves of the egg are colour-coded

dinosaur, *Megaloolithus siruguei*, has a complicated network of interconnected passages (Rasskin-Gutman et al. 2013) that resemble those in the shells of species of birds such as the ostrich (Tyler and Simkiss 1959; Bronwyn et al. 2016).

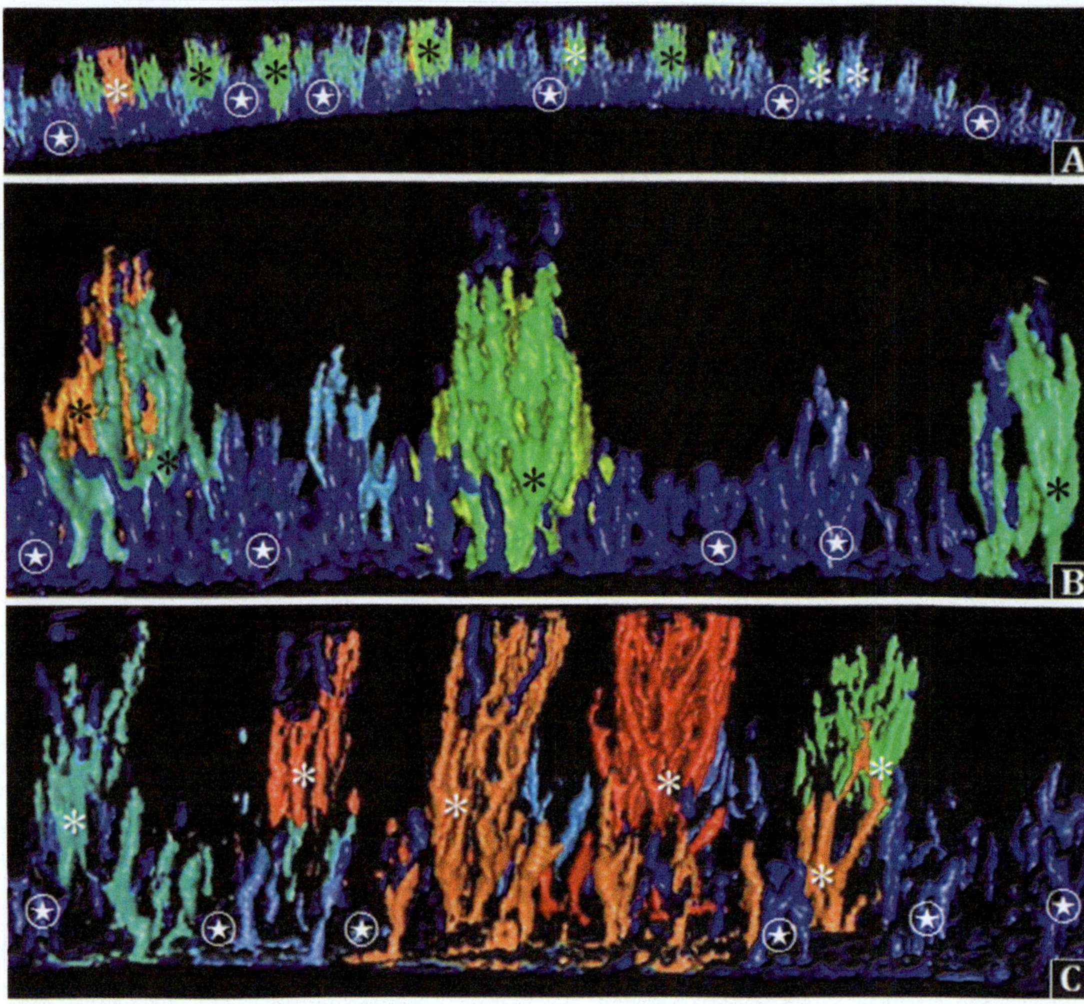

Plate 9.9 All the figures (**a–c**) are from the ostrich (*Struthio camelus*) egg and show the branched (tree-like) morphology of some of the pores in the shell (*asterisks*). Some of the pores perforate the shell wall (*asterisks*), while others do not (*encircled stars*)

Although the thickness of the eggshell provides space for the pores to branch, the complexity of the pores does not considerably correlate with shell thickness (Tyler 1964) and neither is there a strong relationship between the eggshell pore morphology and the phylogeny of birds (Tyler and Simkiss 1959). For example, in the egg of the extinct *Aepyornis*, where the shell is ~3.8 mm thick (Tyler and Simkiss 1959), the pores are much less complicated compared to those of the relatively much thinner shell of the ostrich egg (Tyler and Simkiss 1959; Board et al. 1977; Ar and Rahn 1978; Keffen and Jarvis 1985; Bowsher 1992; Christensen et al. 1996; Brown et al. 1996; Şahan et al. 2003). Interestingly, in the ostrich eggshell, Bronwyn et al. (2016) observed that pores originating from the internal surface of the shell wall did not open to the outer surface of the shell (Plates 9.9 and 9.10a–e). Tyler (1955) supposed that the non-perforate pores of the shell may reduce the weight of the egg, making it manageably light and therefore easy to contain during its development and manipulate during incubation.

Ar and Rahn (1985) determined that the number of pores and the effective pore radius

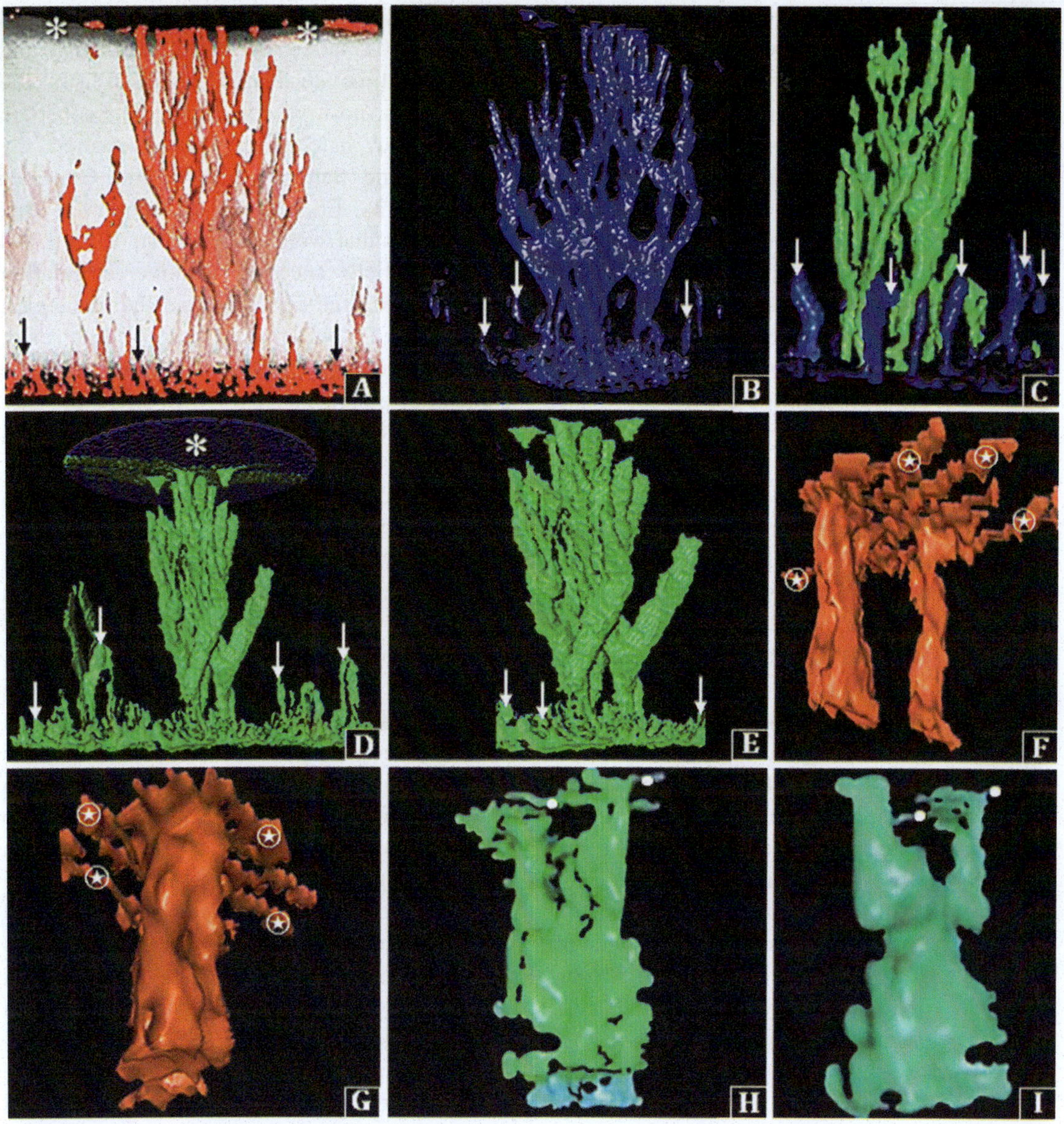

Plate 9.10 All the figures are from the ostrich (*Struthio camelus*) egg and show the different morphologies of the pores. (**a–e**) The multibranched pores; (**f**, **g**) the reticulated pores; (**h**, **i**) the stubby pores. *Asterisk* (**a**, **d**), cuticle; *arrows* (**a–e**), presumptive imperforate pores; *encircled stars* (**f**, **g**), branches that occur close to the outer surface of the shell; *dots* (**h**, **i**), branches that occur close to the outer surface of the shell

increase with egg mass, respectively, from ~300 to 3 μm in 1 g eggs to ~30,000 to 13 μm in 500 g eggs. The total number of pores in the shell of the ostrich egg was reported by Christensen et al. (1996) to be ~11,000, a value close to that of ~10,000 reported in the shell of the egg of the domestic fowl by Romanoff (1964) and Rahn et al. (1979, 1987). Wangensteen and Weibel (1982) estimated that the total number of pores in the shell of the egg of the domestic fowl was ~10,000 and that they were 8 μm wide and 200 μm long. In the ostrich eggshell, Bronwyn et al. (2016) estimated the number of pores at 48,620. Taking into account the differences in the sizes of the eggs where the mean weight of the egg of the domestic fowl is 60 and 1312 g for that of the ostrich, the number of pores reported by Christensen et al. (1996) is possibly a gross underestimate. Quantitation and comparison of the thicknesses of the shells and the numbers and sizes of the pores of the shells of eggs of different species of birds should be made in carefully controlled studies after applying similar techniques. In an Anatidae egg, the average pore density was reported to be 9475 pores per square centimetre of the surface area of the shell by Hoyt et al. (1979). In the ostrich egg, Şahan et al. (2003) reported that shell thickness didn't correlate significantly with the hatchability while pore density did. For the domestic fowl, Liao et al. (2013) reported that hatchability correlated positively with both eggshell and mammillary layer thicknesses. In the ostrich egg, the surface area of pores formed only 28% of that of the shell (Bronwyn et al. 2016). It greatly exceeds the value of only 0.2% that was determined for the egg of the domestic fowl by Rahn et al. (1987). In the eggs of snare penguin (*Eudyptes robustus*), Massaro and Davis (2005) found that the average pore diameter ranged from 21 to 22 μm, a value that is much less than that of 376 μm in the shell of the ostrich egg (Bronwyn et al. 2016).

According to Rodriguez-Navarro (2007), the structure of the eggshell determines its strength. To a great extent, eggshell morphology determines the hatchability of the egg (Bennett 1992; Peebles and McDaniel 2004) since the shell serves as a gas exchanger during the development of the embryo (Wangensteen and Weibel 1982; Papoutsi et al. 2001; Nowak-Sliwinska et al. 2014). Satteneni and Satterlee (1994) and Gonzales et al. (1999) reported a positive correlation between pore density and hatchability of the eggs. The hatchability is reduced mainly because eggs with thicker shells and lower porosity lose less water than normal (Satteneni and Satterlee 1994). In the Magellanic penguin, egg pores and shell thickness did not relate to the incubation time of the eggs (Boersma and Rebstock 2009): synchronous onset of incubation was the more important factor. In species of the Alcidae (a family of marine birds), after controlling for egg mass, Zimmermann and Hipfner (2007) noted that eggshell porosity and incubation time correlated negatively. In the ostrich eggs, positive correlation between pore density and hatchability was reported by Satteneni and Satterlee (1994) and Gonzales et al. (1999). El-Safty (2012) observed that ostrich eggs that weighed less than 1350 g had better hatching success. Positive correlation exists between the egg mass and the incubation time (Worth 1940; Rahn and Ar 1974; Rahn et al. 1977). Rahn and Ar (1974) determined that the incubation period increases with egg size. Egg size is determined by genes, environmental factors and levels of nutrition (Thiele 2012). The large differences between the incubation times of the avian eggs that range from 11 to 90 days have perplexed investigators (Heinroth 1922; Needham 1931; Nice 1954). Bradford and Seymour (1988) determined that oxygen deficiency prolongs incubation time.

Regional differences of the thickness of the eggshell and distribution of the pores in the shell have been reported in various species of birds (Rokitka and Rahn 1987; Seymour and Rahn 1978; Seymour and Visschedijk 1988; Liao et al. 2013; Orłowski et al. 2016; Bronwyn et al. 2016). In contradiction to the assertion made by Hoyt et al. (1979), Booth (1989) and Soliman et al. (1994) that the structural features are fixed at the formation of the eggshell, Orłowski et al. (2016) reported that changes in parameters such as the shell thickness, shell conductance and pore structure occur during the incubation of the bird egg. Rokitka and Rahn (1987) reported that in all avian species, regional eggshell conductance and pore density decreased from the blunt (air cell) end of the eggshell to the pointed (sharp) one. Solomon et al. (1994) and Riley et al. (2014) observed that in the shell of the egg of the domestic fowl, the pores differed in their average size (minimum pore area), with those at the sharp end being significantly narrower than those elsewhere. In the ostrich eggshell, Bronwyn et al. (2016) found that the numbers, volumes and surface areas of the pores did not differ significantly between the different parts of the eggshell, even between the blunt and the sharp ends. For the shells of the Eurasian reed warbler (*Acrocephalus scirpaceus*), Orłowski et al. (2016) observed that at the equator, the shells of eggs with embryos were significantly thinner (on average 8.0% less) compared to the shells of non-embryonated eggs and that the eggshell thickness at the blunt end of the egg was a relatively reliable estimator of the shell thickness, regardless of the stage of embryonic development. It was speculated by Bronwyn et al. (2016) that since the blunt end of the egg contributes more to gas exchange during embryonic development compared to the sharp end (Seymour and Visschedijk 1988), differences in pore number and morphology would exist between the two areas. Tyler and Simkiss (1959) and Board (1982) observed that pore geometry correlates with egg mass, whereas Carter (1971), Tullett (1978a) and van Toledo et al. (1982) stated that porosity can change independently of pore shape.

The conflicting data that have been reported on the shapes, sizes, numbers and distribution of pores in the shells of different bird species may largely arise from factors such as the disparities between the weights and the volumes of the egg, the age of the laying bird, the levels of nutrition and the environmental conditions under which the birds live, the season of laying and the length and condition of storage of the eggs (Tyler and Simkiss 1959; Rahn et al. 1987; Board et al. 1977; Deeming and Ar 1999; Rodriguez-Navarro et al. 2002; Balkan and Biricik 2008; Österström and Lilja 2012; Maurer et al. 2014). High calcium intake increases shell thickness and reduces the pore number (Peebles 1986), and young birds lay eggs with thicker shells and produce smaller chicks, leading to increased embryo mortality after pipping (Pedroso et al. 2005). The lower eggshell conductance of eggs from young layers results in inadequate loss of water vapour during the development of the embryo (Christensen et al. 2005). In the last days of incubation, parent flock age has been reported to influence embryonic metabolism, leading to greater embryonic mortality (Hamidu et al. 2007). In the Pekin duck, Onbaşılar et al. (2014) observed that the hatching weight was affected by the breeder age but hatching results were comparable to those of the breeder age groups.

The thickness of the shell of the ostrich egg has been reported to be 1.94 mm by Şahan et al. (2003), 1.92 mm by Brown et al. (1996), 2.01 mm by Ar and Rahn (1978), 1.90 mm by Christensen et al. (1996), 1.48–2.77 mm by Bowsher (1992), 1.79 to 1.83 mm by Keffen and Jarvis (1985) and 1.92 mm by Bronwyn et al. (2016). In the rhea (*Rhea americana*), the emu, the ostrich, the Magellanic penguin and an Anatidae egg, the eggshell thicknesses were, respectively, 0.917, 1.03, 1.92, 0.510 and 0.45 mm (Tyler and Simkiss 1959; Hoyt et al. 1979; La Scala et al. 2000; Bronwyn et al. 2016). The shell of the egg of the dinosaur, *Megaloolithus siruguei*, was reported by Rasskin-Gutman et al. (2013) to be 2.9 mm thick. In the Austrasian megapodes, *Alectura lathami* and *Leipoa ocellata*, Grellet-Tinner et al. (2016) reported that the average thicknesses of the eggshells respectively ranged from 348 to 359 μm and 258 to 280 μm because of the differences in the heights of the nodes that occur on the surface of the shell. During the incubation of the eggs of the megapodes under the soil, where the partial pressure of oxygen (PO_2) is subatmospheric and that of carbon dioxide (PCO_2) is higher than atmospheric one (Seymour and Ackerman 1980), ~21% of the inner surface of the eggshell erodes to cause a threefold increase in eggshell conductance to O_2 (Booth and Seymour 1987).

9.4 Function of the Eggshell

With the shell being hard and inflexible, the egg cannot allow the developing embryo to provoke ventilatory movements. In a bird, convective gas exchange does not occur until the lungs become functional, i.e. after internal pipping. The egg 'breathes' through the pores that perforate the eggshell (Ar et al. 1974; Paganelli et al. 1978; Carey et al. 1980; Rahn et al. 1979, 1987) (Plates 9.1d–f, 9.2, 9.6, 9.9, and 9.10). Presenting the only means of exchange between the atmospheric air and the developing embryo, the pores provide the only physical communication between the embryo developing in the egg and the external milieu (air) (Rahn et al. 1987; Ar et al. 1974; Richards et al. 2000; Liao et al. 2013). In an average domestic fowl's egg, mean weight 60 g, during the normal incubation time of ~21 days as much as 6 L of O_2 are consumed, 4.5 L of CO_2 are eliminated and 11 L of water vapour are lost through the ~10,000 pores of the shell (Romanoff 1964; Rahn et al. 1979, 1987). In the egg of the same species, Ar and Rahn (1985) determined that up to the time of internal pipping when pulmonary ventilation commences, ~20 L of O_2, CO_2 and water vapour will have passed through the pores of an 80 gm egg.

The presence of pores in the shell of a bird egg was first experimentally demonstrated by John Davy in 1862. After placing an egg in a jar of water and pumping air out of the container, small bubbles of air were seen to form on the surface of the shell as the air escaped through the microscopic perforations (pores) of the shell. According to Denny (1993), in eggs with solid porous shells, embryonic development is not possible in water because of the 10,000-fold lower diffusion of O_2 in water compared to that in air. Theoretically therefore, the largest perforate aquatic egg would have to be no larger than that of the domestic fowl, possess metabolic rates that are very low and the shell would have to be highly perforate. The shapes, sizes and numbers of pores determine the diffusing capacity (the conductance) or its reciprocal, the resistance, of the shell to the respiratory gases and the water vapour (Wangensteen 1972; Rahn et al. 1979; Hoyt et al. 1979).

Movement of respiratory gases and the water vapour occurs across the eggshell by passive diffusion (Ar et al. 1974; Rahn et al. 1979; Seymour and Visschedijk 1988; Mueller et al. 2015). Flux of O_2 can therefore be estimated according to Fick's diffusion equation (Paganelli 1980; Simkiss 1986; Rahn et al. 1987; Wangensteen et al. 1970/1971) where:

$$V = G \cdot \Delta PO_2$$

V is the rate of gas flux, G the conductance of the entire shell and ΔPO_2 is the partial pressure gradient of O_2 across the shell barrier.

The diffusion capacity of the eggshells with simple and narrow pores (the so-called 'trumpet-like' pores) to gases was modelled by Paganelli (1980). It was noted that the boundary layer, i.e. the layer of slowly moving or stagnant air that lies next to the outer surface of the shell, does not play a significant role in gas exchange. In the eggshells that have more elaborate pores, e.g. the multibranched pores, Simkiss (1986) contended that resistance to diffusion is best explained by the Maxwell–Stefan's theory of diffusion in ideal multicomponent systems (Stefan 1871; Maxwell 1965). The fundamental assumption in the theory is that change from equilibrium between molecular friction and thermodynamic interactions leads to diffusional flux where the molecular friction is proportional to the difference in speed and mole fractions (Taylor and Krishna 1993; Cussler 1997; Rehfeldt and Stichlmair 2007; Bird et al. 2007). Compared to the 'classical' Fick's diffusion theory, Maxwell–Stefan's theory is more inclusive as the former does not exclude the possibility of negative diffusion coefficients. Fick's theory states that the flux of a component is proportional to the gradient of the concentration of a particular species and there is no influence from the other components: cross-effects are ignored although they are known to occur. Tøien et al. (1988) mathematically modelled the diffusive resistance that eggshells with branched pores like those of the ostrich (Tyler and Simkiss 1959; Bronwyn et al. 2016) (Plates 9.2, 9.9, and 9.10) offer to O_2. It was determined that the total aperture resistance was less than 6.2% of total pore resistance, the effect of outside aperture was on average only 1.5%, and the pore conductance to water vapour was not dependent of egg mass. Sibly and Simkiss (1987) noted that diffusion in branched pores is largely determined by the inner pore radius and that the thickness of the eggshell and that of the cuticle does have not any role. In thick shells with complex pores, Tyler and Simkiss (1959), Hechenleitner et al. (2016) and Bronwyn et al. (2016) suggested that the branching may provide passageways that supply O_2 to areas such as those contacting the ground or the body of the incubating bird. The actual role of the imperforate pores that have been reported in shells of eggs of many species of birds is unclear (Tyler 1955; Tyler and Simkiss 1959; Riley et al. 2014; Bronwyn et al. 2016).

Structural parameters such as the thickness and the surface area of the shell and particularly the numbers, sizes and the shapes of the pores greatly determine the conductance of the shell to respiratory gases and water vapour (Ar et al. 1974, 1996). The rate of water loss across the eggshell increases with the mass of the egg raised to the power of 0.78 (Ar et al. 1974). As the embryo develops, water is lost and the yolk and albumin decrease in mass (Mueller et al. 2015): increasing the size of the air cell at the blunt end of the egg, the water lost from the egg is replaced by air. For the ostrich egg, the pore density, i.e. the number of pores per unit surface area of the shell in centimetres, and the water vapour conductance are, respectively, 10.02 pore·cm^{-2} and 87.77 $mgH_2O{\cdot}d^{-1}{\cdot}torr^{-1}$ (Şahan et al. 2003). In the cassowary, the domestic fowl, and the ostrich eggs, water loss ranges from 1.7 to 2.3 mg·$day^{-1}{\cdot}cm^{-2}$, while the values for those of the rhea are considerably greater (Tyler and Simkiss 1959). For optimal hatchability, ~12% (15% at high altitude) and as much as 16% of the total egg mass must be lost during incubation (Rahn et al. 1977; Hoyt et al. 1979; Fink et al. 1992; Ar et al. 1996; Nys et al. 1999). In the domestic fowl, with development, the mass of the embryos increases geometrically until growth rate slows down during the last stages of development (Clark et al. 1986; Haque et al. 1996). While the metabolic rate and therefore the VO_2 of the developing embryo increase with growth, i.e. with the incubation time, the porosity and the thickness of the eggshell were reported to change (Soliman et al. 1994). The observation differed from that of Hoyt et al. (1979) who noted that the structural parameters are fixed during the formation of the eggshell. Wangensteen and Rahn (1970) and Wangensteen et al. (1970/1971) also stated that the diffusing capacity of the eggshell is set at oviposition to provide adequate O_2 during the development of the embryo. Maurer et al. (2012) reported that there is normally a small decrease in the eggshell thickness with incubation time due to absorption of calcium from the eggshell mainly for the formation of bones during embryogenesis. During incubation, as the embryo develops, pore geometry becomes the limiting factor in exchange of respiratory gases (Tullett 1978b). Interestingly, in the eggs of the domestic Pekin duck (*Anas platyrhynchos*), Balkan et al. (2006) observed that pore density and shell conductance increased during incubation both in the equatorial part and the blunt end of the eggshell: it was attributed to the increase in pore density and not in thinning of the shell. Tyler (1958), Tyler and Simkiss (1959) and Ar et al. (1974) suggested that increase in the morphological complexity of the pores with increasing egg size may indicate optimization of two conflicting parameters: pore area and pore length. According to Solomon (1991), the risk of pathogenic invasion may be the most important determinant of the largest pore size. In the 7.9 mm thick shell of the ~4,850cm^3 volume of the egg of the Cretaceous Sanagasta neosauropod, Hechenleitner et al. (2016) observed a complex funnel-like pore system with lateral canal network. They suggested that by providing collateral air supply, the intricate canal system may prevent the detrimental effects of pore obstruction during extended incubation. As suggested by Bronwyn et al. (2016), this may also apply to the large ostrich eggs with thick shells.

Eggshell thickness, water-vapour conductance and pore density are important structural and functional properties of eggs which determine healthy embryonic development in poultry (Soliman et al. 1994). In the eggs of the Pekin duck, Balkan et al. (2006) reported that pore density correlated with hatchability. In the eggs of the same species, El-Hanoun and Mossad (2008) reported that during the late stages of incubation, as much as 19.4% of the chicks died because the shells were thick and had few pores. Compared to those of the domestic fowl, Changkang et al. (1999) attributed the relatively

Plate 9.11 A simplified schematic diagram showing the structural components of the chorioallantoic membrane of a bird egg. The outer shell membrane interfaces with eggshell and the subchorionic sinuses, bringing blood in the blood capillaries close to the air in the air cell

low hatchability of the duck eggs to the large size of the eggs, their relatively thick shells and the excessive water loss due to large numbers of pores in the shell. In the ostrich industry, the greatest economic losses emanate from poor hatchability of the eggs (Burger and Bertram 1981; Bowsher 1992; Deeming et al. 1993; Button et al. 1994; Deeming and Ar 1999; Cooper 2001; Richards et al. 2002; Şahan et al. 2003; Brand 2012). Incubation procedures and structural defects of the eggshells are the main factors that reduce hatchability (Burger and Bertram 1981; van Toledo et al. 1982; Bowsher 1992; Deeming et al. 1993; Deeming 1995; Button et al. 1994; Sales et al. 1996; Brown et al. 1996; Deeming and Ar 1999; Richards et al. 2000, 2002; Şahan et al. 2003).

9.5 Chorioallantoic Membrane (CAM)

The avian CAM is a simple densely vascularized extraembryonic membrane that performs various important functions during the development of the embryo. Prior to internal pipping, it is the site of gas exchange, calcium transport from the eggshell, acid–base homeostasis in the embryo and ion and water reabsorption from the allantoic space (Gibert 2003; Maksimov et al. 2006; Chien et al. 2009; Gabrielli and Accili 2010; Mueller et al. 2015). All the functions are performed by the highly specialized chorionic and allantoic epithelia. Being the outermost extraembryonic membrane, the CAM lines the acellular inner eggshell membrane (Plates 9.11 and 9.12). The CAM is formed by the fusion of the splanchnic mesoderm of the allantois and the somatic mesoderm of the chorion during the incubation of the egg (Romanoff 1960; Scott 2003). For the domestic fowl's egg, the fused CAM has totally covered the surface of the inner shell membrane by day 12 of incubation (Romanoff 1960; Leeson and Leeson 1963; Mueller et al. 2015). In the chick embryos, between the external air and the dense capillary plexus of the CAM (Plates 9.13 and 9.14), gas exchange occurs by diffusion. By providing O_2 and nutriments to the developing embryo (Gabrielli and Accili 2010), the CAM is functionally and to a large extent structurally analogous to the eutherian placenta (Plate 9.15). As the embryo develops, demand for O_2 and production of metabolic end products increase (Rahn et al. 1974; Rahn and Ar 1980). The

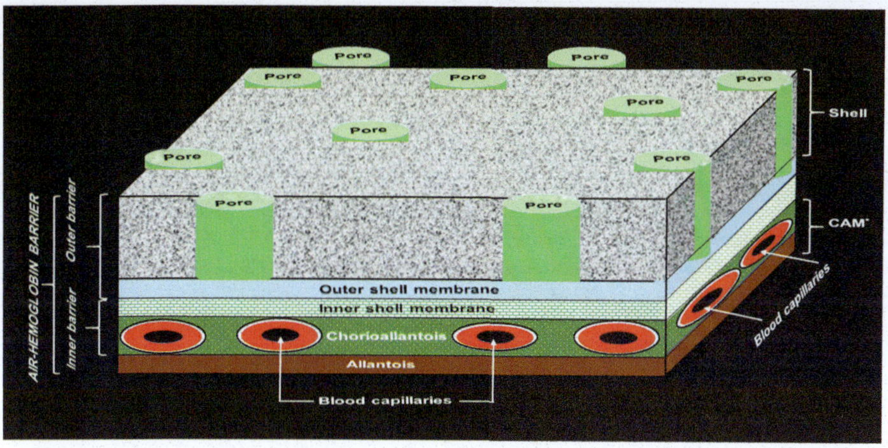

Plate 9.12 A simplified stereogram showing the structural components of the shell of the avian egg. Respiratory gases and water vapour cross the shell wall through the pores. *Asterisk* on CAM, chorioallantoic membrane

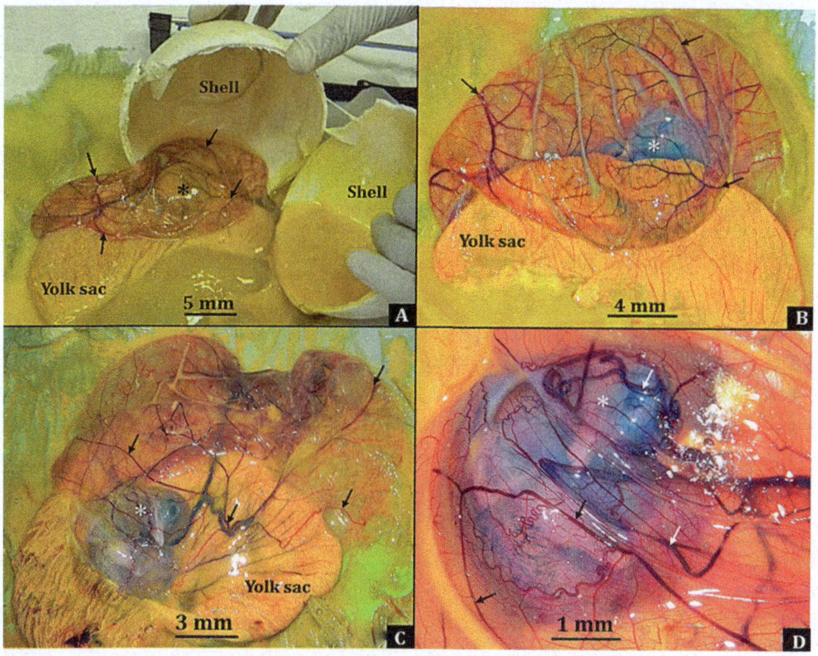

Plate 9.13 All the figures are from the ostrich (*Struthio camelus*) egg. The extremely highly vascularized chorioallantoic membrane is shown on days 8 (**a**), 14 (**b**), 23 (**c**) and 28 (**d**). *Arrows*, blood vessels; *asterisk*, developing embryo

surface area and the intensity of vascularisation of the CAM increase to service the developing embryo (Bronwyn 2015) (Plate 9.16) since it particularly enhances the flux of O_2 and CO_2 (Hollwedel et al. 2008). The prevailing partial pressure of O_2 (PO_2) influences the vascularisation of the CAM, with those of eggs incubated at high PO_2 having lower vascularization intensity.

The CAM supports the extraembryonic respiratory capillaries and also actively transports sodium and chloride from the allantoic sac and calcium from the eggshell to the embryonic circulation (Romanoff 1960; Stewart and Terepka

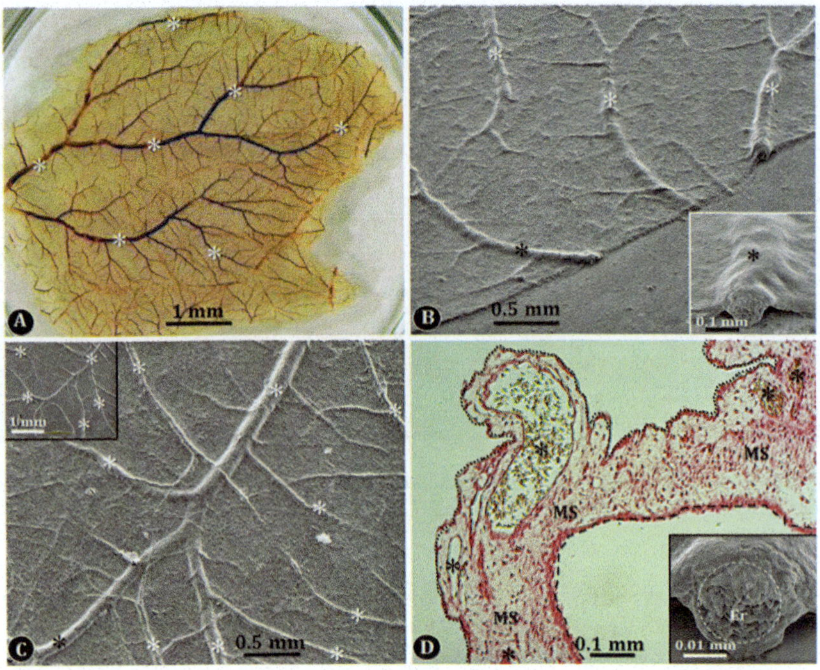

Plate 9.14 All the figures are from the ostrich (*Struthio camelus*) egg. (**a–c**): The chorioallantoic membrane is intensely vascularized. *Asterisks*, blood vessels. Insert to (**b**): Close up of a blood vessel (*asterisk*); insert to (**c**), view of blood vessels on a chorioallantoic membrane (*asterisk*). (**d**) Histological section of the chorioallantoic membrane showing chorionic epithelium (*line with shorter dashes*), mesodermal blood vessels (*asterisks*), mesodermal space (MS) and allantoic epithelium (*line with longer dashes*). Insert: close-up of the large blood vessel (shown in **b**—insert) of a chorioallantoic membrane. *Er* erythrocytes in the vessel lumen (*dashed circle*)

1969; Terepka et al. 1969; Coleman and Terepka 1972). Also, it (CAM) forms part of the wall of the allantoic sac into which excretory products are stored. The structure and function of the CAM has been well-documented (Romanoff 1960; Leeson and Leeson 1963; Ganote et al. 1964; Fáncsi and Fehér 1979; Ribatti et al. 1996; Lusimbo et al. 2000; Bronwyn 2015; Makanya et al. 2016). The following features, properties and processes appertain to the CAM: the chorion is the external layer and lies directly under the inner shell membrane and is composed of epithelial cells that originate from the chorionic ectoderm; the mesoderm comprises highly vascularized mesenchymal tissue; the allantoic epithelium arises from the allantoic endoderm; the chorion consists of two layers of cuboidal epithelial cells between which vascular sinuses form; in the domestic fowl's egg, during the development of the embryo, blood capillaries and sinuses invade the chorionic layer and become lodged between the epithelial cells, and therefore blood comes to lie next to the ambient air, i.e. the air in the air cell; the blood–gas barrier is as thin as 0.2 μm; the mesenchymal cells are stellate in form and; depending on the stage of development, the allantoic epithelium may consist of as many as four layers of fusiform cells with oval-shaped or rod-like nuclei (Makanya et al. 2016).

During the development of the domestic fowl's embryo, the vascular capillaries of the CAM are incorporated into the chorionic epithelium, forming a rich plexus of intra-epithelial vascular spaces, and rapid angiogenesis occurs in the mesoderm (Ganote et al. 1964). In the CAM of the mallard duck, *Anas platyrhynchos*, the intensity of the vascularization in the mesoderm increases from 4.2 vessels per mm on day 12 of incubation to a maximum of 9.4 vessels per mm by day 16 (Lusimbo et al. 2000).

Compared to the gas transport systems that have evolved in the respiratory organs of the

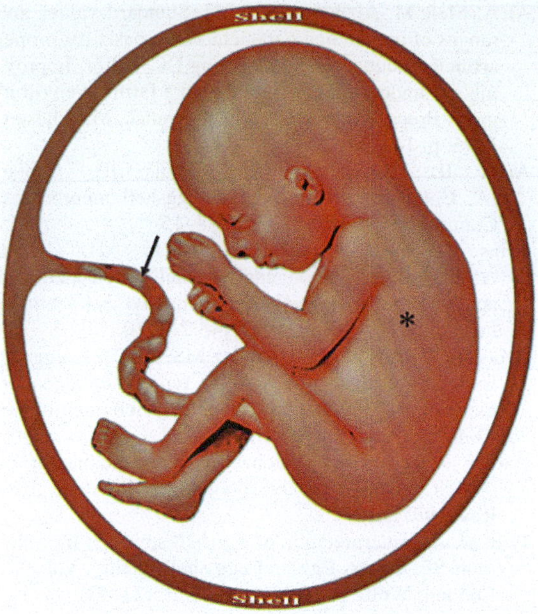

Plate 9.15 A diagram illustrating the correspondence of the placenta (*arrow*) of the eutherian animals and the chorioallantoic membrane of the amniotic eggs. *Asterisk*, embryo

air-breathing vertebrates, including the post-embryonic avian respiratory system (the lung–air sac system) (Chap. 8) in the developing avian embryo, gas exchange is profoundly different. Except for a very small component of it (Paganelli et al. 1978), no convective mechanisms occur between the bulk external environment and the blood capillaries of the CAM (Wangensteen and Weibel 1982): movement of respiratory gases (O_2 and CO_2) and water vapour occurs across the eggshell by passive diffusion along partial pressure gradients. While VO_2 increases with embryonic development and the surface area and the vascularity of the CAM correspondingly increases (Bronwyn 2015) (Plate 9.16), arguably, according to some investigators (e.g. Hoyt et al. 1979), the numbers, sizes and shapes of the pores in the eggshell are determined at the formation of the egg. If that is the case, the structural and functional changes in the CAM are vital to ensure that the embryo is adequately supplied with O_2 throughout its development.

Plate 9.16 Correlation between the surface area of the chorioallantoic membrane (CAM) of the ostrich (*Struthio camelus*) egg (insert) with the incubation time. The surface area and the intensity of the vascularization of the CAM (insert) increases with the development of the embryo

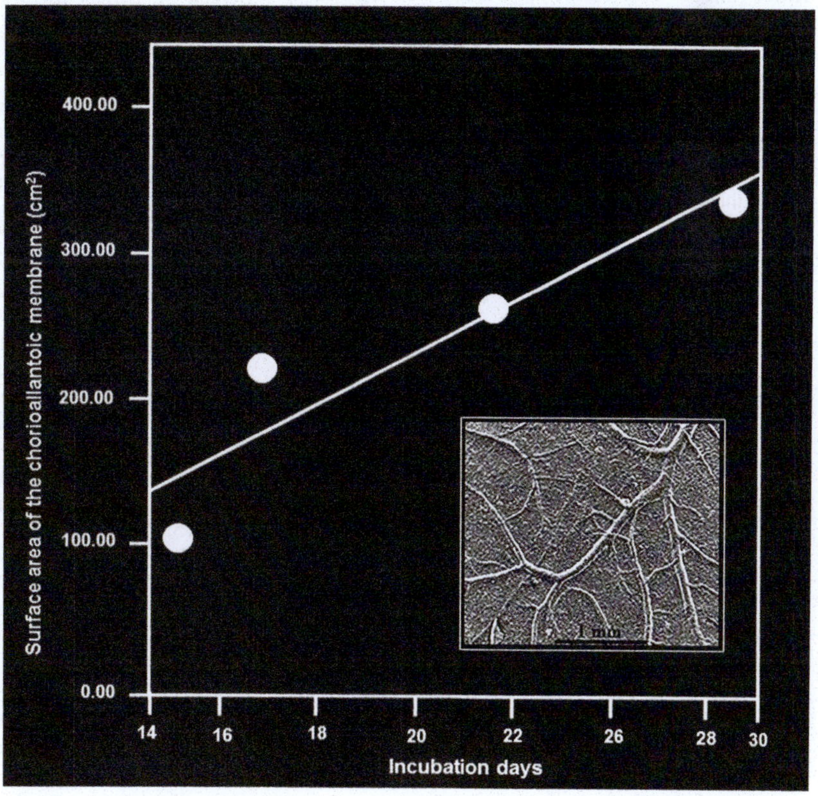

The path that O_2 follows to reach the developing avian embryo comprises of two main resistances that are arranged in series (Wangensteen 1972; Piiper et al. 1980) (Plates 9.11 and 9.12): the outer barrier consist of the porous shell and the outer shell membrane while the inner one is formed by the inner shell membrane and the thin blood–gas barrier of the CAM (Wangensteen and Weibel 1982). Morphometric analysis of the dimensions of the structural components of the gas exchange pathway of the developing domestic fowl's embryo at day 16 of incubation was made by Wangensteen and Weibel (1982). It was found that the O_2 diffusing capacity of the chorioallantoic (DCA) was 6.8 $\mu LO_2\ min^{-1}\ torr^{-1}$ (0.544 $mLO_2\ s^{-1}\ mbar^{-1}$) and the rate limiting factor in chorioallantoic O_2 uptake is the O_2-hemoglobin binding in erythrocytes which is ten times slower than diffusion across the thin (harmonic mean thickness = 0.47 μm) blood gas barrier. At the developmental stage of the chick embryo studied by Wangensteen and Weibel (1982), the inner shell membrane provides a negligible resistance to O_2 conductance.

Acknowledgements The preparation of this work was supported by the National Research Foundation (NRF) of South Africa. The views and opinions expressed here are, however, mine and not those of the NRF. To the many people that have collaborated with me over the years, I am eternally grateful for their sharing of ideas with me, giving me their time unreservedly and particularly for their continued friendship.

References

Ar A, Rahn H. Interdependence of gas conductance, incubation length, and weight of the avian lung. In: Pipper J, editor. Respiratory function in birds, adult, and embryonic. Heidelberg: Springer; 1978. p. 227–36.

Ar A, Rahn H. Pores in avian eggshells: gas conductance, gas exchange and embryonic growth rate. Respir Physiol. 1985;61:1–20.

Ar A, Paganelli CV, Reeves RB, Greene DG, Rahn H. The avian egg: water vapour conductance, shell thickness, and functional pore area. Condor. 1974;76:153–8.

Ar A, Meir M, Aizik N, Campi D. Standard values and ranges of ostrich egg parameters as a basis for proper artificial incubation. In: Deeming DC, editor. Improving our understanding of ratites in a farming environment. Proceedings of Ratite Conference, Manchester. 1996. p. 144–6.

Arias JL, Fernandez MS, Dennis JE, Caplan IA. Collagens of the chicken eggshell membranes. Connect Tissue Res. 1991;26:37–45.

Arias JL, Fink DJ, Xiao S, Heuer AH, Caplan AI. Biomineralization and eggshells: cell-mediated acellular compartments of mineralized extracellular matrix. Int Rev Cytol. 1993;145:217–50.

Arias JL, Nakamura O, Fernandez MS, Wu JJ, Knigge P, Eyre DR, Caplan AI. Role of type X collagen on experimental mineralization of eggshell membrane. Connect Tissue Res. 1997;36:21–33.

Badley AR. Fertility, hatchability and incubation of ostrich (*Struthio camelus*) eggs. Poult Avian Biol Rev. 1997;8:53–76.

Bain M. A reinterpretation of eggshell strength. In: Solomon SE, editor. Egg and egg-shell quality, vol. 9. - London: Wolfe Publishing; 1991. p. 131–41.

Bain MM. Eggshell strength: a relationship between the mechanism of failure and the ultrastructural organization of the mammillary layer. Br Poult Sci. 1992;33:303–19.

Baker JR, Balch DA. A study of the organic material of hen's egg-shell. Biochem J. 1962;82:352–61.

Balkan M, Biricik M. Main eggs characteristics in the Peking Duck, *Anas platyrhynchos* variant *domesticus*. D Ü Ziya Gökalp Eğitim Fakültesi Dergisi. 2008;11:142–50.

Balkan M, Karakas R, Biricik M. Changes in eggshell thickness, shell conductance and pore density during incubation in the Peking duck (*Anas platyrhynchos* f. *dom.*). Ornis Fenn. 2006;83:117–23.

Becking JH. The ultrastructure of the avian eggshell. Ibis. 1975;117:143–51.

Bennett CD. The influence of shell thickness on hatchability in commercial broiler breeder flocks. J Appl Poult Res. 1992;1:16–65.

Bird RB, Stewart WE, Lightfoot EN. Transport phenomena. 2nd ed. New York: Wiley; 2007.

Board RG. Microstructure, water resistance and water repellency of the avian egg-shell. Br Poult Sci. 1974;15:415–9.

Board RG. The microstructure of the cuticle-less shell of the eggs of the domestic hen. Br Poult Sci. 1975;16:89–91.

Board RG. Properties of avian egg shells and their adaptive value. Biol Rev. 1982;57:1–28.

Board RG, Halls NA. The cuticle: a barrier to liquid and particle penetration of the shell of the hen's egg. Br Poult Sci. 1973;14:69–97.

Board RG, Sparks NHC. Shell structure and formation in avian eggs. In: Deeming DC, Ferguson MWJ, editors. Egg incubation: its effects on embryonic development in birds and reptiles. Cambridge: Cambridge University Press; 1991. p. 71–86.

Board RG, Tranter HS. The microbiology of eggs. In: Stadelman WJ, Cotterill OJ, editors. Pages in egg science and technology. Westport: AVI Publishing Company; 1986. p. 209–330.

Board RG, Tullett SG. The pore arrangement in the emu (*Dromaius novaehollandiae*) eggshell as shown by plastic models. J Microsc. 1975;103:281–4.

Board RG, Tullett SG, Perrott HG. An arbitrary classification of pore systems in avian egg shells. J Zool. 1977;182:251–65.

Boersma PD, Rebstock GA. Magellanic penguin eggshell pores: does number matter? Ibis. 2009;151:535–40.

Booth DT. Regional changes in shell thickness, shell conduction and pore structure during incubation in eggs of mute swan. Physiol Zool. 1989;62:607–20.

Booth DT, Seymour RS. Effect of eggshell thinning on water-vapour conductance of Malleefowl eggs. Condor. 1987;89:453–9.

Bowsher MW. Improvement of reproductive efficiency of the ostrich: characterization of late embryo mortality. Ph.D. Thesis, University of Texas (A&M). 1992.

Bradford DF, Seymour RS. Influence of environmental PO_2 on embryonic oxygen consumption, rate of development, and hatching in the frog, *Pseudophryne bibronii*. Physiol Zool. 1988;61:475–82.

Brand Z. Studies on embryonic development and hatchability of the ostrich egg. Ph.D. Thesis, University of Stellenbosch. 2012.

Bronwyn W. Morphological and morphometric study of the ostrich egg-shell with observations on the chorio-allantoic membrane: a μCT-, scanning electron microscope, and histological study. B.Sc. Honours Dissertation, University of Johannesburg. 2015.

Bronwyn W, Lindi S, Lunga B, Olivier AJ, Devey R, Maina JN. Micro-focus x-ray tomography study of the microstructure and the morphometry of the shell of the ostrich, *Struthio camelus*, egg. Anat Rec. 2016;299:1015–26.

Browder LW, Erickson CA, Jeffery WR. Developmental biology. 3rd ed. Philadelphia: Harcourt College Publishing; 1991.

Brown CR, Peinke D, Loveridge A. Mortality in near-term ostrich embryos during artificial incubation. Br Poult Sci. 1996;37:73–85.

Burger AE, Bertram BCR. Ostrich eggs in artificial incubators: could their hatching success be improved? S Afr J Sci. 1981;77:188–9.

Bušs A, Keišs O. Method for identification of avian species by eggshell microstructure: preliminary study. Acta Univ Latv. 2009;753:89–98.

Button K, Moon D, Turner D. Increasing the hatchability of ostrich eggs. J Aust Ostrich Assoc. 1994;14:18–23.

Carey C, Rahn H, Parisi P. Calories, water, lipid and yolk in avian eggs. Condor. 1980;82:335–43.

Carnarius KM, Conrad KM, Mast MG, MacNeil JH. Relationship of eggshell ultrastructure and shell strength to the soundness of shell eggs. Poult Sci. 1996;75:656–63.

Carroll RL. Vertebrate paleontology. New York: WH Freeman Company; 1988.

Carter T. The hen's egg: variation in tensile strength of shell material and its relationship with shearing strength. Br Poult Sci. 1971;12:56–7.

Changkang W, Ang L, Guangying W. Effects of the quantitative characters of hatching eggs on hatchability in Muscovy duck. Proceedings of 1st World Waterfowl Conference, Taichung. 1999. p. 188–92.

Chien YC, Hincke MT, McKee MD. Ultrastructure of avian eggshell during resorption following egg fertilization. J Struct Biol. 2009;168:527–38.

Christensen VL, Davis GS, Lucore LA. Egg shell conductance and other functional qualities of ostrich eggs. Poult Sci. 1996;75:1404–10.

Christensen VL, Wineland MJ, Ort DT, Mann KM. Eggshell conductance and incubator ventilation as factors in embryo survival and poultry quality. Int J Poult Sci. 2005;4:818–26.

Clark EB, Hu N, Dummett JL, Vandekieft GK, Olson C, Tomanek R. Ventricular function and morphology in chick embryo from stages 18 and 29. Am J Physiol. 1986;250:H407–13.

Coleman JR, Terepka AR. Fine structural changes associated with the onset of calcium, sodium and water transport by the chick chorioallantoic membrane. J Membr Biol. 1972;7:111–27.

Cooper RG. Handling, incubation, and hatchability of ostrich (*Struthio camelus* var. *domesticus*) eggs: a review. J Appl Poult Res. 2001;10:262–73.

Creger CR, Phillips H, Scott JT. Formation of an egg shell. Poult Sci. 1976;55:1717–23.

Cusack M, Fraser AC, Stachel T. Magnesium and phosphorus distribution in the avian eggshell. Comp Biochem Physiol B Biochem Mol Biol. 2003;134:63–9.

Cussler EL. Diffusion—mass transfer in fluid systems. 2nd ed. Cambridge: Cambridge University Press; 1997.

Davies SJJF. Elephant birds. In: Hutchins M, editor. Grzimek's animal life encyclopedia: birds. I. Tinamous and ratites to hoatzins. 2nd ed. Farmington Hills: Gale Group; 2003. p. 103–4.

De Reu K, Grijspeerdt K, Messens W, Heyndrickx A, Uyttendaele M. Eggshell factors influencing eggshell penetration and whole egg contamination by different bacteria, including *Salmonella enteritidis*. Int J Food Microbiol. 2006;112:253–60.

Deeming DC. Factors affecting hatchability during commercial incubation of ostrich (*Struthio camelus*) eggs. Br Poult Sci. 1995;36:51–65.

Deeming DC. Ratite egg incubation. Oxford: Ratite Conference; 1997.

Deeming DC, Ar A. Factors affecting the success of commercial incubation. In: Deeming DC, editor. The ostrich: biology, production and health. University of Manchester, CABI Publishing. 1999. p. 159–90.

Deeming DC, Ayres L, Ayres FJ. Observations on the commercial production of ostrich (*Struthio camelus*) eggs in the United Kingdom: incubation. Vet Rec. 1993;132:602–7.

Dejours P. Respiration in water and air: adaptation-regulation-evolution. Amsterdam: Elsevier; 1988.

Dennis JM, Xiao SQ, Agarwal M, Fink DJ, Heuer AH. Microstructure of matrix and mineral components of eggshells from White Leghorn Chickens (*Gallus gallus*). J Morphol. 1996;228:287–306.

Denny MW. Air and water: the biology and physics of life's media. Princeton: Princeton University Press; 1993.

Devine B. God in creation. Chicago: Moody Press; 1982.

Dieckert JW, Dieckert MC, Creger CR. Calcium reserve assembly: a basic structural unit of the calcium reserve system of the hen egg shell. Poult Sci. 1989;68:1569–84.

Dunn IC, Wilson PW, Lu Z, Bain MM, Crossan CL, Talbot RT, Waddington D. New hypothesis on the function of the avian eggshell gland derived from micro array analysis comparing tissue from juvenile and sexually mature hens. Gen Comp Endocrinol. 2009;63:225–32.

El-Hanoun AM, Mossad NA. Hatchability improvement of Peking duck eggs by controlling water evaporation rate from the egg shell. Egypt Poult Sci. 2008;28:767–84.

El-Hanoun AM, Rizk RE, Shahein EHA, Hassan NS, Brake J. Effect of incubation humidity and flock age on hatchability traits and posthatch growth in Pekin ducks. Poult Sci. 2012;91:2390–7.

El-Safty SA. Effect of egg weight grades, porosity and their interaction on some hatching traits of ostrich eggs. Egypt Poult Sci. 2012;32:725–33.

Erben H, Hoefs K, Wedepohl KH. Paleontological and isotopic studies of egg shells from a declining dinosaur species. Paleobiology. 1979;5:380–414.

Fajardo RJ, Hernandez E, O'Connor PM. Postcranial skeletal pneumaticity: a case study in the use of quantitative microCT to assess vertebral structure in birds. J Anat. 2007;211:138–47.

Fáncsi T, Fehér G. Ultrastructural studies of chicken embryo chorioallantoic membrane during incubation. Anat Histol Embryol. 1979;8:151–9.

Fathi MM, El-Dein AZ, El-Safty SA, Radwan LM. Using scanning electron microscopy to detect the ultrastructural variations in eggshell quality for Fayoumi and Dandarawi chicken breeds. Int J Poult Sci. 2007;6:236–41.

Fink DJ, Caplan AI, Heuer AH. Eggshell mineralization: a case study of bioprocessing strategy. MRS Bull. 1992;20:27–31.

Flacourt É. Histoire de la grande isle Madagascar. Paris: Alexandre Lesselin; 1658.

Freeman BM, Vince MA. Development of the avian embryo: a behavioural and physiological study. London: Chapman and Hall; 1974.

Fujii S. Further morphological studies on the formation and structure of hen's eggshell by scanning electron microscopy. J Fac Fish Anim Husb Hiroshima Univ. 1974;13:29–56.

Fuller BR. Tensegrity. Portf Art News Annu. 1961;4:112–27.

Fuller BR, Applewhite EJ. Synergetics. New York: Macmillan; 1975.

Fung YC. Biomechanics: mechanical properties of living tissues. 2nd ed. Berlin: Springer; 1993.

Gabrielli MG, Accili D. The Chick chorioallantoic membrane: a model of molecular, structural, and functional adaptation to transepithelial ion transport and barrier function during embryonic development. J Biomed Biotechnol. 2010;2010:1–12.

Ganote CE, Beaver DL, Moses HL. Ultrastructure of the chick chorioallantoic membrane and its reaction to inoculation trauma. Lab Investig. 1964;13:1575–89.

Gibert SF. Developmental biology. New York: Sinauer Associates; 2003.

Gilbert AB. The formation of the egg in the domestic chicken. In: McLaren A, editor. Advances in reproductive physiology, vol. 1. London; Logos Press 1967; p. 43–87.

Gonzales AM, Satterlee DG, Moherer F, Cadd GG. Factors affecting ostrich egg hatchability. Poult Sci. 1999;78:1257–62.

Göth A, Vogel U. Egg laying and incubation of the Polynesian megapode. Ann Rev. 1997;97:43–54.

Grellet-Tinner G, Lindsay S, Thompson M. The biomechanical, chemical, and physiological adaptations of the eggs of two Australian megapodes to their nesting strategies and their implications for extinct titanosaur dinosaurs. Peer J. 2016. https://doi.org/10.7287/peerj.preprints.2100v1 | CC-BY 4.0 Open Access | rec: 3 Jun 2016, publ: 3 Jun 2016.

Hamidu JA, Fasenko GM, Feddes JJR, O'Dea EE, Ouellette CA, Wineland MJ, Christensen VL. The effect of broiler breeder genetic strain and parent flock age on eggshell conductance and embryonic metabolism. Poult Sci. 2007;86:2420–32.

Hao L, Zhang H, Tand Z, Hu G. Micro-computed tomography for small animal imaging: technological details. Prog Nat Sci. 2008;18:513–21.

Haque MA, Pearson JT, Hou PCL, Tazawa H. Effects of preincubation egg storage on embryonic functions and growth. Respir Physiol. 1996;103:89–98.

Hasiak RJ, Vadehra DV, Baker RC. Lipid composition of the egg exteriors of the chicken *Gallus gallus*. Comp Biochem Physiol. 1970;37:429–35.

Hassan SB, Aigbodion VS. Effects of eggshell on the microstructures and properties of Al-Cu-Mg/eggshell particulate composites. J King Saud Univ Eng Sci. 2013;27:49–56.

Hechenleitner EM, Grellet-Tinner G, Fiorelli LE. What do giant titanosaur dinosaurs and modern Australasian megapodes have in common? Peer J. 2015;3:e1341. doi:10.7717/peerj.1341.

Hechenleitner EM, Grellet-Tinner G, Foley M, Fiorelli LE, Thompson MB. Micro-CT scan reveals an unexpected high-volume and interconnected pore network in a Cretaceous Sanagasta dinosaur eggshell. J R Soc Interface. 2016;13:20160008.

Heinroth O. Die Beziehungen zwischen Vogelgewicht, Eigewicht, Gelegegewicht und Brutdauer. J Ornithol. 1922;70:172–285.

Hincke MT. Ovalbumin is a component of the chicken eggshell matrix. Connect Tissue Res. 1995;31:227–33.

Hincke MT, Nys Y, Gautron J, Mann K, Rodriguez-Navarro AB, McKee MD. The eggshell: structure, composition and mineralization. Front Biosci. 2012;17:1266–80.

Hoffman R, Schultz JA, Schellhorn R, Rybacki E, Keupp H, Gerden SR, Lemanis R, Zachow S. - Non-invasive imaging methods applied to neo- and paleo-ontological cephalopod research. BioGeosiences. 2014;11:2721–39.

Hollwedel D, Walzl M., Kapeller B, Frank J, Schoeffl H, Macfelda K, Losert U. The hen's egg testing on the chorioallantoic membrane—a model for multiple purposes. Poster Presentation. 2008. http://www.altex.ch/resources/Hollwedel.pdf. Accessed 28 May 2014.

Hoyt DF, Board RG, Rahn H, Paganelli CV. The eggs of Anatidae: conductance, pore structure and metabolism. Physiol Zool. 1979;52:438–50.

Hunton P. Understanding the architecture of the egg shell. Worlds Poult Sci J. 1995;51:141–7.

Johnston PM, Comar CL. Distribution and contribution of calcium from the albumen, yolk and shell to the developing chick embryo. Am J Physiol. 1955;183:365–70.

Jonchère V, Réhault-Godbert S, Hennequet-Antier C, Cabau C, Sibut V, Cogburn LA, Nys Y, Gautron J. Gene expression profiling to identify eggshell proteins involved in physical defense of the chicken egg. BMC Genomics. 2010;11:57. doi:10.1186/1471-2164-11-57.

Jones DN. Construction and maintenances of the incubation mounds of the Australian brush-Turkey *Alectura lathami*. Emu. 1988;88:210–8.

Jones DN, Dekker RWRJ, Roselaar CS. The megapodes. Oxford: Oxford University Press; 1995.

Kaplan S, Siegesmund KA. The structure of the chicken egg shell and shell membranes as studied with the scanning electron microscope and energy dispersive X-Ray microanalysis. Poult Sci. 1973;52:1798–801. doi:10.3382/ps.0521798.

Karlson O, Lilja C. Eggshell structure, mode of development and growth rate in birds. Zoology (Jena). 2008;111:494–502.

Keffen RH, Jarvis MJF. Some physical requirements for ostrich egg incubation. Ostrich. 1985;56:42–51.

Khatkar MS, Sandhu JS, Brah GS, Chaudhary ML. Estimation of egg shell breaking strength from egg characteristics in layer chickens. Indian J Poult Sci. 1997;32:111–3.

King AS, McLelland J. Birds: their structure and function. London: Baillière Tindall; 1984.

Kutchai H, Steen JB. Permeability of the shell and shell membranes of hens' eggs during development. Respir Physiol. 1971;11:265–78.

La Scala N, Boleli IC, Ribeiro LT, Freitas D, Macari M. Pore size distribution in chicken eggs as determined by mercury porosimetry. Rev Bras Cienc Avic. 2000;2:177–81.

Lammie D, Bain MM, Wess T. Microfocus X-ray scattering investigations of egg shell nanostructure. J Synchrotron Radiat. 2006;12:721–6.

Leeson TS, Leeson CR. The chorio-allantois of the chick: light and electron microscopic observations at various times of incubation. J Anat. 1963;97:585–95.

Li H, Zhang H, Tang Z, Guangshu H. Micro-computed tomography for small animal imaging: technological details. Prog Nat Sci. 2008;18:513–21.

Liao B, Qiao HG, Zhao XY, Bao M, Liu L, Zheng CW, Li CF, Ning ZH. Influence of egg-shell ultrastructural organization on hatchability. Poult Sci. 2013;92:2236–9.

Little C. The colonisation of land: origins and adaptations of terrestrial animals. Cambridge: Cambridge University Press; 1983.

Little C. Comparative physiology as a tool for investigating the evolutionary routes of animals on to land. Trans R Soc Edinb Earth Environ Sci. 1989;80:201–18.

Little C. The terrestrial invasion: an ecophysiological approach to the origins of land animals. Cambridge: Cambridge University Press; 1990.

Luckett WP. Ontogeny of amniote fetal membranes and their application to phylogeny. In: Hecht MK, Goody PC, Hecht BM, editors. Major patterns in vertebrate evolution. London: Plenum; 1976. p. 439–516.

Lunam CA, Ruiz J. Ultrastructural analysis of the eggshell: contribution of the individual calcified layers and the cuticle to hatchability and egg viability breeders. Br Poult Sci. 2000;41:584–92.

Lusimbo WS, Leighton FA, Wobeser GA. Histology and ultrastructure of the chorioallantoic membrane of the mallard duck (*Anas platyrhynchos*). Anat Rec. 2000;259:25–34.

Makanya AN, Dimova I, Koller T, Styp-Rekowska B, Djonov V. Dynamics of the developing chick chorioallantoic membrane assessed by stereology, allometry, immunohistochemistry and molecular analysis. PLoS One. 2016;11:e0152821. doi:10.1371/journal. pone. 0152821.

Maksimov VF, Korostyshevskaya IM, Kurganov SA. Functional morphology of chorioallantoic vascular network in chicken. Bull Exp Biol Med. 2006;142:367–71.

Mann K, Olsen JV, Maček B, Gnad F, Mann M. Identification of new chicken egg proteins by mass spectrometry-based proteomic analysis. Worlds Poult Sci J. 2008;64:209–18.

Mao KM, Murakami A, Iwasawa A, Yoshizaki N. The asymmetry of avian egg-shape: an adaptation for reproduction on dry land. J Anat. 2007;210:741–8.

Massaro M, Davis LS. Differences in egg size, shell thickness, pore density, pore diameter and water vapour conductance between first and second eggs of Snares penguins, *Eudyptes robustus*, and their influence on hatching asynchrony. Ibis. 2005;147:251–8.

Maurer G, Portugal S, Cassey P. A comparison of indices and measured values of eggshell thickness of different shell regions using museum eggs of 230 European bird species. Ibis. 2012;154:714–24.

Maurer G, Portugal SJ, Hauber ME, Mikšík I, Russel DGD, Cassey P. First light for avian embryos: eggshell thickness and pigmentation mediate variation in development and UV exposure in wild bird eggs. Funct Ecol. 2014;29(2):209–18. doi:10.1111/1365-2435.12314.

Maxwell JC. On the dynamical theory of gases. The Scientific Papers of JC Maxwell. 1965;2:26–78.

Mazzuco H, Bertechini AG. Critical points on egg production: causes, importance and incidence of eggshell breakage and defects. Ciênc Agrotecnol. 2014;38:7–14.

McDaniel GR, Roland DA, Coleman MA. The effect of egg shell quality on hatchability and embryonic mortality. Poult Sci. 1979;58:10–3.

Meiri LN, Pines M. New insight in eggshell formation. Poult Sci. 2000;79:1014–7.

Messens W, Grijspeerdt K, Herman L. Eggshell penetration by Salmonella: a review. Worlds Poult Sci J. 2005;61:71–85.

Mikhailov KE. Classification of fossil egg shells of amniotic vertebrates. Acta Paleontol Pol. 1991;36:193–238.

Mîndrilă I, Pârvănescu H, Pirici D, Niculescu M, Buteică SA, Taisescu O, Tarniţă DN. The chick chorioallantoic membrane: a model of short-term study of Dupuytren's disease. Rom J Morphol Embryol. 2014;55:377–82.

Mizutani R, Suzuki Y. X-ray microtomography in biology. Micron. 2012;43:104–15.

Mooney SJ, Pridmore TP, Helliwell J, Bennett MJ. Developing X-ray computed tomography to non-invasively image 3D root systems architecture in soil. Plant Soil. 2012;352:1–22.

Mortola JP, Cooney E. Cost of growth and maintenance in chicken embryos during normoxic or hypoxic conditions. Respir Physiol Neurobiol. 2008;162:223–9.

Mueller CA, Burggren WW, Tazawa H. The physiology of the avian embryo. In: Scanes CG, editor. Sturkie's avian physiology. 6th ed. New York: Elsevier; 2015.

Nascimento VP, Cranstoun S, Solomon SE. Relationship between shell structure and movement of *Salmonella enteritidis* across the eggshell wall. Br Poult Sci. 1992;33:37–48.

Nathusius W. Uber die Hullen, welche den Dotter des Vogeleies umgeben. Z Wiss Zool. 1868;18:225–70.

Needham J. Chemical embryology. Annu Rev Biochem. 1931;1:507–26.

Nice MM. Incubation periods throughout the ages. Centaurus. 1954;3:311–59.

Nowak-Sliwinska P, Segura T, Iruela-Arispe ML. The chicken chorioallantoic membrane model in biology, medicine and bioengineering. Angiogenesis. 2014;17:779–804.

Nys Y, Hincke MT, Arias JL, Garcia-Luis JM, Solomon S. Avian eggshell mineralization. Poult Sci Avian Biol Rev. 1999;10:143–66.

Nys Y, Gautron J, Garcia-Luis JM, Hincke MT. Avian egg mineralization: biochemical and functional characterization of matrix proteins. Gen Paleontol. 2004;3:549–62.

Onbaşılar EE, Erdem E, Hacan Ö, Yalçın S. Effects of breeder age on mineral contents and weight of yolk sac, embryo development, and hatchability in Pekin ducks. Poult Sci. 2014;93:473–8.

Ono T, Wakasugi N. Mineral content of quail embryos cultured in mineral-rich and mineral-free conditions. Poult Sci. 1984;63:159–66.

Orłowski G, Hałupka L, Klimczuk E, Sztwiertnia H. Shell thinning due to embryo development in eggs of a small passerine bird. J Ornithol. 2016;157:565–72.

Österström O, Lilja C. Evolution of avian eggshell structure. J Morphol. 2012;273:241–7.

Packard MJ, Seymour RS. Evolution of the amniote egg. In: Sumida SS, KLM M, editors. Amniotic origins. London: Academic; 1997.

Paganelli CV. The physics of gas exchange across the avian egg-shell. Am Zool. 1980;20:329–38.

Paganelli CV, Olszowka A, Ar A. The avian egg: surface area, volume, and density. Condor. 1974;76:319–25.

Paganelli CV, Ackerman RA, Rahn H. The avian egg: *in vivo* conductances to oxygen, carbon dioxide, and water vapour in late development. In: Piiper J, editor. Respiratory function in birds, adult and embryonic. Berlin: Springer; 1978. p. 212–8.

Panheleux M, Bain M, Fernandez MS, Morales I, Gautron J, Arias JL, Solomon JE, Hinke M, Nys Y. Organic matrix composition and ultrastructure of eggshell: a comparative study. Br Poult Sci. 1999;40:240–52.

Papoutsi M, Tomarev SI, Eichmann A, Prols F, Christ B, Wilting J. Endogenous origin of the lymphatics in the avian chorioallantoic membrane. Dev Dyn. 2001;222:238–51.

Pedroso AA, Andrade MA, Cafe MB, Leandro NS, Menten JFM, Stringhini JH. Fertility and hatchability of eggs laid in the pullet-to-breeder transition period and in the initial production period. Anim Reprod Sci. 2005;90:355–64.

Peebles ED. Factors affecting broiler hatching egg production. Ph.D. Dissertation, North Carolina State University (Raleigh). 1986.

Peebles ED, Brake J. Eggshell quality and hatchability in broiler breeder eggs. Poult Sci. 1987;6:596–604.

Peebles ED, McDaniel CD. A practical manual for understanding the shell structure of broiler hatching eggs and measurements of their quality. Mississippi State University, Mississippi Agricultural and Forestry Experimental Station, Bulletin 1139. 2004. p. 1–16.

Piiper J, Tazawa H, Ar A, Rahn H. Analysis of chorioallantoic gas exchange in the chick embryo. Respir Physiol. 1980;39:273–84.

Pough FH, Heisler JB, McFarland WN. Vertebrate life. 3rd ed. New York: Macmillan; 1989.

Rahn H, Ar A. The avian egg: incubation time and water loss. Condor. 1974;76:147–52.

Rahn H, Ar A. Gas exchange of the avian lung: time, structure and function. Am Zool. 1980;20:477–84.

Rahn H, Paganelli CV. Gas exchange in avian eggs: publications in gas exchange, physical properties and dimensions of bird eggs. Buffalo: Department of Physiology, State University of New York at Buffalo; 1981.

Rahn H, Paganelli CV, Ar A. The avian egg: air cell gas tension, metabolism and incubation time. Respir Physiol. 1974;22:297–309.

Rahn H, Carey C, Balmas K, Bhatia B, Paganelli CV, Ar A. Reduction of pore area of the avian egg shell as an adaptation to altitude. Proc Natl Acad Sci USA. 1977;74:3095–8.

Rahn H, Ar A, Paganelli CV. How bird eggs breathe. Sci Am. 1979;240:46–55.

Rahn H, Paganelli CV, Ar A. Pores and gas exchange of avian eggs: a review. J Exp Zool. 1987;1:165–72.

Randall DJ, Burggren WW, Farrell A, Haswell MS. The evolution of air breathing in vertebrates. Cambridge: Cambridge University Press; 1981.

Rasskin-Gutman D, Elez J, Esteve-Altava B, López-Martinez N. Reconstruction of the internal structure of the pore system of a complex dinosaur egg shell (*Megaloothinus siruguei*). Span J Paleontol. 2013;28:61–8.

Rees A. Dictionary of zoo biology and animal management. New York: Wiley-Blackwell; 2013.

Rehfeldt S, Stichlmair J. Measurement and calculation of multicomponent diffusion coefficients in liquids. Fluid Phase Equilib. 2007;256:99–104.

Ribatti D, Vacca A, Roncali L, Dammmacco F. The chick embryo chorioallantoic membrane as a model for in vivo research on angiogenesis. Int J Dev Biol. 1996;40:1189–97.

Richards PD, Richards PA, Lee ME. Ultrastructural characteristics of ostrich egg shell: outer shell membrane and the calcified layers. J S Afr Vet Assoc. 2000;71:97–102.

Richards PD, Botha A, Richards PA. Morphological and histochemical observations of the organic components of ostrich egg shell. J S Afr Vet Assoc. 2002;73:13–22.

Riley A, Sturrock CJ, Mooney SJ, Luck MR. Quantitation of egg shell microstructure using X-ray micro computed tomography. Br Poult Sci. 2014;3:311–20.

Ritman EL. Micro-computed tomography: current status and developments. Annu Rev Biomed Eng. 2004;6:185–208.

Robb AR. X-ray computed tomography: from basic principles to applicants. Annu Rev Biophys Bioeng. 1982;11:177–201.

Rodriguez-Navarro A. Rapid quantification of avian eggshell microstructure and crystallographic-texture using two-dimensional X-ray diffraction. Br Poult Sci. 2007;48:133–44.

Rodriguez-Navarro A, Kalin O, Nys Y, Garcia-Ruiz JM. Influence of the microstructure on the shell strength of eggs laid by hens of different ages. Br Poult Sci. 2002;43:395–403.

Rodríguez-Navarro AB, Domínguez-Gasca N, Muñoz A, Ortega-Huertas M. Change in the chicken eggshell cuticle with hen age and egg freshness. Poult Sci. 2013;92:3026–35.

Rokitka MA, Rahn H. Regional differences in shell conductance and pore density of avian eggs. Respir Physiol. 1987;68:371–6.

Romanoff AL. The extraembryonic membranes. In: Romanoff AL, editor. The avian embryo: structural and functional development. New York: The Macmillan Company; 1960. p. 1039–140.

Romanoff AL. Biochemistry of the avian embryo. New York: Wiley; 1964.

Romanoff AL, Romanoff AJ. The avian egg. New York: Wiley; 1949.

Rose-Martel M, Du J, Hincke MT. Proteomic analysis provides new insight into the chicken eggshell cuticle. J Proteomics. 2012;75:2697–706.

Rossi M, Nys Y, Anton M, Bain M, De Ketelaere B, De Reu K, Dunn I, Gautron J, Hammershoj M, Hidalgo A, Meluzzi A, Mertens K, Nau F, Sirri F. Developments in understanding and assessment of egg and egg product quality over the last century. Worlds Poult Sci J. 2013;69:414–29.

Ruiz J, Groves P, Glatz P, Meldrum J, Lunam C. Ultrastructural analysis of eggshell laminae. Proceedings of Australian Poultry Science Symposium, Sydney. 1998. p. 200.

Şahan Ü, Altan Ö, Ipek A, Yilman B. Effects of some characteristics on the mass loss and hatchability of ostrich (*Struthio camelus*) eggs. Br Poult Sci. 2003;44:380–5.

Sahni A, Kumar G, Bajpais S, Srinivasan S. Ultrastructure and taxonomy of ostrich eggshells from upper Palaeolithic sites of India. J Palaeontol Soc India. 1989;34:91–8.

Sales J, Poggenpoel DG, Cilliers SC. Comparative physical and nutritive characteristics of ostrich eggs. Worlds Poult Sci J. 1996;52:45–52.

Saleuddin ASM, Kyriakides CPM, Peacock A, Simkiss A. Physiological and ultrastructural aspects of ion movements across the chorioallantois. Comp Biochem Physiol. 1976;54:7–12.

Sato T, Chieda O, Yamakashi Y. X-ray tomography for microstructural objects. Appl Opt. 1981;20:3880–3.

Satteneni G, Satterlee DG. Factors affecting hatchability of ostrich eggs. Poult Sci. 1994;93:38.
Sauer EGF. Ratite eggshells and phylogenetic questions. Bonn Zool Beitr. 1972;23:3–48.
Sauter EA, Petersen CF. The effect of eggshell quality on penetration by various salmonellae. Poult Sci. 1974;53:2159–62.
Schmalhausen II. The origin of terrestrial vertebrates. London: Academic; 1968.
Scott GF. Developmental biology. Sunderland: Sinauer Associates; 2003.
Seymour RS, Ackerman RA. Adaptations to underground nesting in birds and reptiles. Am Zool. 1980;20:437–47.
Seymour RS, Rahn H. Gas conductance in the eggshell of the mould-building Turkey. In: Piiper J, editor. Respiratory function in birds, adult and embryonic. Berlin: Springer; 1978. p. 243–6.
Seymour RS, Visschedijk AHJ. Effects of variation in total and regional shell conductance on air cell gas tensions and regional gas exchange in chicken eggs. J Comp Physiol B. 1988;158:229–36.
Shanawany MM, Dingle J. Ostrich production systems. FAO Animal Production and Health Paper 144. 1999. ISSN 0254-6019.
Sibly RM, Simkiss K. Gas diffusion through non-tubular pores. J Exp Biol. 1987;1:87–91.
Simkiss K. Calcium metabolism and avian reproduction. Biol Rev. 1961;36:321–67.
Simkiss K. Eggshell conductance—Fick's or Stefan's law? Respir Physiol. 1986;65:213–22.
Sochava AV. Dinosaur eggs from the Upper Cretaceous of the Gobi Desert. Paleontol J. 1969;4:517–27.
Soliman FNK, Rizk RE, Brake J. Relationship between shell porosity, shell thickness, egg weight loss and embryonic development. Poult Sci. 1994;73:1607–11.
Solomon SE. Egg and egg shell quality. London: Wolfe; 1991.
Solomon SE. The eggshell: strength, structure and function. Br Poult Sci. 2010;51:52–9.
Solomon SE, Bain MM, Cranstoun S, Nascimento V. Hen's egg structure and function. In: Board RG, Fuller R, editors. Microbiology of the avian egg. London: Chapman and Hall; 1994. p. 1–24.
Sparks N. Shell structure and hatchability. Int Hatch Pract. 1995;10:21.
Sparks NHC, Deeming DC. Ostrich egg shell ultrastructure—a study using electron microscopy and X-ray diffraction. In: Deeming DC, editor. Improving our understanding of ratites in a farming environment. Proceedings of Ratite Conference, Oxford. 1996. p. 164–5.
Stefan J. Über das Gleichgewicht und Bewegung, insbesondere die Diffusion von Gemischen. Sitzungsberichte Kaiserlichen Akad Wissenschaften, Wien 2te Abteilung a 1871;63:63–124.
Stemberger BH, Mueller WJ, Leach RM. Microscopic study of the initial stages of egg shell calcification. Poult Sci. 1977;56:537–43.
Stewart ME, Terepka AR. Transport functions of the chick chorio-allantoic membrane. I. Normal histology and evidence for active electrolyte transport from the allantoic fluid, *in vivo*. Exp Cell Res. 1969;58:93–106.
Szarski H. The origin of vertebrate foetal membranes. Evolution. 1968;22:211–3.
Szarski H. Sarcopterygii and the origin of the tetrapods. In: Hecht MK, Goody PC, Hecht BM, editors. Major patterns in vertebrate evolution. London: Plenum; 1976. p. 517–40.
Talbot CJ, Tyler C. A study of the progressive deposition of shell in the shell gland of the domestic hen. Br Poult Sci. 1974;15:217–24.
Taylor TG. How an egg-shell is made. Sci Am. 1970;222:88–95.
Taylor R, Krishna R. Multicomponent mass transfer. New York: Wiley; 1993.
Terepka AR, Stewart ME, Merkel N. Transport functions of the chick chorioallantoic membrane. II. Active transport, *in vitro*. Exp Cell Res. 1969;58:107–17.
Thiele HH. Management tools to influence egg weight in commercial layers. Lohmann Inf. 2012;47:21–31.
Tøien O, Paganelli CV, Rahn H, Johnson RR. Diffusive resistance of avian egg-shell pores. Respir Physiol. 1988;74:345–54.
van Toledo B, Parsons AH, Combs GF. Role of ultrastructure in determining eggshell strength. Poult Sci. 1982;61:569–72.
Tullett SG. Regulation of eggshell porosity. J Zool. 1978a;177:339–48.
Tullett SG. Pore size versus pore number in avian egg shells. In: Piiper J, editor. Respiratory function in birds, adult and embryonic. Berlin: Springer; 1978b. p. 219–26.
Tumova E, Gous RM. Interaction of hen production type, age, and temperature on laying pattern and egg quality. Poult Sci. 2012;91:1269–75.
Tyler C. Studies on eggshells VI. The distribution of pores in eggshells. J Sci Food Agric. 1955;6:170.
Tyler C. Studies on egg-shells. VII. Some aspects of structure of the pores in the egg shell of the hen's egg. J Sci Food Agric. 1956;7:483–96.
Tyler C. Studies on eggshells: a method for determining variations in characteristics in different parts of the same shell. J Sci Food Agric. 1958;9:584.
Tyler C. A study of the eggshells of Anatidae. Proc Zool Soc Lond. 1964;142:547–83.
Tyler C. Avian eggs: their structure and characteristics. Int Rev Gen Exp Zool. 1969;4:82–127.
Tyler C, Simkiss K. A study of the egg shells of ratite birds. J Zool (Lond). 1959;133:201–43.
Uni Z, Yadgary L, Yair R. Nutritional limitations during poultry embryonic development. J Appl Poult Res. 2012;21:175–84.
Vleck CM, Vleck D, Hoyt DF. Patterns of metabolism and growth in avian embryos. Am Zool. 1980;20:405–16.
Wang G, Vannier M. Micro-CT scanners for biomedical applications: an overview. Adv Imaging. 2001;16:18–27.

Wangensteen OD. Gas exchange by a bird's embryo. Respir Physiol. 1972;14:64–74.

Wangensteen OD, Rahn H. Gas exchange in the avian embryo. Fed Proc. 1970;29:376.

Wangensteen D, Weibel ER. Morphometric evaluation of chorioallantoic oxygen transport in the chick embryo. Respir Physiol. 1982;47:1–20.

Wangensteen OD, Wilson D, Rahn H. Diffusion of gases across the shell of the hen's egg. Respir Physiol. 1970;11:16–30.

Wedral EM, Vadehra DV, Baker RC. Chemical composition of the cuticle, and the inner and outer shell membranes from eggs of *Gallus gallus*. Comp Biochem Physiol B. 1974;47:631–40.

Wellman-Labadie O, Picman J, Hincke MT. Antimicrobial activity of the Anseriform outer eggshell and cuticle. Comp Biochem Physiol B Biochem Mol Biol. 2008;149:640–9.

Wellman-Labadie O, Lemaire S, Mann K, Picman J, Hincke MT. Antimicrobial activity of lipophilic avian eggshell surface extracts. J Agric Food Chem. 2010;58:10156–61.

Westneat MW, Socha JJ, Lee WK. Advances in biological structure, function, and physiology using synchrotron X-ray imaging. Annu Rev Physiol. 2008;70:119–42.

Wilson HR. Interrelationships of egg size, chick size, posthatching growth and hatchability. Worlds Poult Sci J. 1991;47:5–20.

Worth CB. Egg volumes and incubation periods. Auk. 1940;57:44–60.

Yoshizaki N, Saito H. Changes in shell membranes during development of quail embryos. Poult Sci. 2002;81:246–51.

Yuan J-E, Xu K, Wu W, Luo Q, Yu J-L. Application of the chick embryo chorioallantoic membrane in neurosurgery disease. Int J Med Sci. 2014;11:1275–81.

Zimmermann K, Hipfner JM. Egg size, eggshell porosity, and incubation period in the marine bird family Alcidae. Auk. 2007;124:307–15.

Index

J.N. Maina (ed.), *The Biology of the Avian Respiratory System*, DOI 10.1007/978-3-319-44153-5

The manufacturer's authorised representative in the EU is Springer Nature Customer Service Centre GmbH, Europaplatz 3, 69115 Heidelberg, Germany. If you have any concerns regarding our products, please contact ProductSafety@springernature.com

Printed and bound by CPI Group (UK) Ltd, Croydon, CR0 4YY
30/09/2026
02225454-0019